Climatology and Meteorology: Advanced Researches

Climatology and Meteorology: Advanced Researches

Editor: Mary D'souza

R CALLISTO REFERENCE

www.callistoreference.com

Callisto Reference,
118-35 Queens Blvd., Suite 400,
Forest Hills, NY 11375, USA

Visit us on the World Wide Web at:
www.callistoreference.com

ISBN: 978-1-63239-929-8 (Hardback)

Cataloging-in-Publication Data

Climatology and meteorology : advanced researches / edited by Mary D'Souza.
        p. cm.
Includes bibliographical references and index.
ISBN 978-1-63239-929-8
1. Climatology. 2. Meteorology. 3. Climatic changes. I. D'Souza, Mary.
QC861.3 .C55 2018
551.6--dc23

Table of Contents

Preface

In my initial years as a student, I used to run to the library at every possible instance to grab a book and learn something new. Books were my primary source of knowledge and I would not have come such a long way without all that I learnt from them. Thus, when I was approached to edit this book; I became understandably nostalgic. It was an absolute honor to be considered worthy of guiding the current generation as well as those to come. I put all my knowledge and hard work into making this book most beneficial for its readers.

Climatology studies the climate patterns of a defined area over a period of time while meteorology studies changes in atmospheric pressure and precipitation in order to predict and forecast the weather. This book on climatology and meteorology discusses topics related to climatic shifts and patterns in relation to global warming, remote sensing techniques for mapping atmospheric phenomena, etc. The book studies, analyses and upholds the pillars of climatology and meteorology and their utmost significance in modern times. The various studies that are constantly contributing towards advancing technologies and evolution of these fields are examined in detail. It will be of great help to students and researchers in the fields of Earth sciences, climatology and physical geography.

I wish to thank my publisher for supporting me at every step. I would also like to thank all the authors who have contributed their researches in this book. I hope this book will be a valuable contribution to the progress of the field.

Editor

An overview of the phenological observation network and the phenological database of Germany's national meteorological service (Deutscher Wetterdienst)

F. Kaspar, K. Zimmermann, and C. Polte-Rudolf

Deutscher Wetterdienst, National Climate Monitoring, Frankfurter Str. 135, 63067 Offenbach, Germany

Correspondence to: F. Kaspar (frank.kaspar@dwd.de)

Abstract. First phenological observations have been performed in Germany already in the 18th century. The onset dates of characteristic phases of plant development (phenological phases) are observed and recorded. Today, Germany's national meteorological service (Deutscher Wetterdienst, DWD) maintains a dense phenological observation network and a database with phenological observations. The data are used in many applications, esp. for advisory activities to agriculture or pollen dispersion information.

1 Introduction and history of phenological observations of plants in Germany

Phenology is the study of periodically recurring patterns of growth and development of plants and animal behaviour during the year (Lieth, 1974). These are closely related to weather and their long-term trends are therefore important indicators of the impact of climate change on the biosphere (Menzel, 2002; Settele et al., 2014). Phenological data are therefore a valuable source of information for agricultural advisory activities as well as for climate research (Richardson et al., 2013). Systematic collection of phenological data started in several European countries in the 18th century (see all contributions to Nekovář et al., 2008): in the UK Robert Marsham started to recorded 27 "indications of spring" in 1736. These included first flowering, first leafing and migrant bird records. Successive generations of the family continued with these recordings until 1958 (Sparks and Carey, 1995; Sparks and Collinson, 2008). In the middle of the 18th century the Swedish naturalist Carl von Linné initiated the first known phenological network with 18 stations in Sweden (Dahl and Langvall, 2008). The Economic Society of Bern founded the first phenological network of Switzerland in 1759 (Defila, 2008). About 30 years later the Societas Meteorologica Palatina carried out systematic phenological observations at several sites in Europe, with focus on Germany, between 1781 and 1792 (e.g. Ephemerides Societatis

Meteorologicae Palatinae. observationes anni 1782; Wege, 2002). The first modern phenological network in the Austro-Hungarian Monarchy which included also parts of today's Czech Republic, Slovakia, Romania and Croatia was established by Carl Fritsch in the first half of the 19th century. Furthermore he developed und published several guidelines for phenological observations (Reiss, 1959; Koch et al., 2008).

In Germany the real breakthrough in phenology was achieved in 1882, when university professor Hermann Hoffmann and his colleagues prepared and published guidelines for the observation of phenological phases. These guidelines provided the basis for standardized observations in Germany and beyond. It gave rise to a Europe-wide monitoring network, as well as a number of regional networks in Germany.

The first long-term nationwide phenological observation network in Germany was set up by the Biological Institute for Agriculture and Forestry of the Reich in 1922. The network was run by this institute until 1936, when it was taken over by the Meteorological Service of the Reich and further developed by Fritz Schnelle (Schnelle, 1955).

Today, the phenological observation network is operated by Germany's national meteorological service (Deutscher Wetterdienst, DWD). An overview on the current status in European countries can be found in Nekovář et al. (2008).

Table 1. Plants and their phenological phases in the current observation programme of Deutscher Wetterdienst. Abbreviations are explained in Table 2. German names are provided for users accessing the data via DWD's website.

Latin	English	German	observed phases (abbreviation)
wild growing plants, forest and ornamental woody plants			
Acer platanoides	norway maple	Spitz-Ahorn	B
Aesculus hippocastanum	horse chestnut	Rosskastanie	A, BO, B, F, BV, BF
Alnus glutinosa	european alder	Schwarz-Erle	B, BO
Alopecurus pratensis	meadow foxtail	Wiesen-Fuchsschwanz	B, AB
Anemone nemorosa	wood anemone	Busch-Windröschen	B
Artemisia vulgaris	mugwort	Beifuß	B
Betula pendula, B. Verucosa, B. alba	european white birch	Hänge-Birke	A, BO, B, BV, BF
Calluna vulgaris	heather	Heidekraut	B
Colchicum autumnale	autumn crocus	Herbstzeitlose	B
Cornus mas	cornelian cherry	Kornelkirsche	B, F
Corylus avellana	hazel	Hasel	B
Crataegus laevigata, C. oxyacantha	hawthorn	Zweigriffeliger Weißdorn	B, F
Dactylis glomerata	orchad grass	Wiesen-Knäuelgras	AB
Fagus sylvatica	european beech	Rotbuche	BO, BV, BF
Forsythia suspensa	forsythia	Forsythie	B
Fraxinus excelsior	ash	Esche	B, BO
Galanthus nivalis	snowdrop	Schneeglöckchen	B
Larix decidua, L. europaea	european larch	Europäische Lärche	BO, BV, BF
Picea abies, P. excelsa	norway spruce	Fichte	M
Pinus sylvestris	scots pine	Kiefer	M, B
Prunus spinosa	blackthorne	Schlehe	B
Quercus robur, Q. pendunculata	pedunculate oak	Stiel-Eiche	BO, F, BV, BF
Robinia pseudoacacia	black locust	Robinie	B
Rosa canina	dog rose	Hunds-Rose	B, F
Salix caprea	goat willow	Sal-Weide	B
Sambucus nigra	black elder	Schwarzer Holunder	B, F
Sorbus aucuparia	rowan	Eberesche	A, BO, B, F, BF
Syringa vulgaris	lilac	Flieder	B
Taraxacum officinale, T. vulgare	dandelion	Löwenzahn	B
Tilia platyphyllos, T. grandifolia	large leaved lime	Sommer-Linde	B
Tussilago farfara	coltsfoot	Huflattich	B
agricultural plants			
Avena sativa	summer oats	Sommerhafer	BST, AU, SCH, AE, MR, GR, E
Beta vulgaris	sugarbeet or mangold	Zucker- oder Futterrübe	BST, AU, BG, E
Brassica napus, var.napus	winter rape	Winterraps	SCH, KNO, B, VR, E, BST, AU, RO
Hordeum vulgare	summer barley	Sommergerste	BST, AU, SCH, AE, GR, E
Hordeum vulgare	winter barley	Wintergerste	SCH, AE, GR, E, BST, AU
Secale cereale	winter rye	Winterroggen	SCH, AE, B, AB, GR, E, BST, AU
Triticum aestivum	winter wheat	Winterweizen	SCH, AE, MR, GR, E, BST, AU
Zea mays	corn	Mais	FAO, BST, AU, SCH, AE, B, MR, TR, GR, E
–	permanent grassland	Dauergrünland	ERG, E
fruit and vine			
Malus domestica	apple	Apfel	SKZ, A, B, AB, EB, F, BF
Prunus avium	cherry	Süßkirsche	SKZ, B, AB, EB, F, BV
Prunus cerasus	sour cherry	Sauerkirsche	SKZ, B, AB, EB, F
Pyrus communis	pear	Birne	SKZ, B, AB, EB, F
Ribes rubrum, R.sylvestre	redcurrant	Rote Johannisbeere	SKZ, B, F
Ribes uva-crispa, Ribes grossularia	gooseberry	Stachelbeere	SKZ, A, BO, B, F
Vitis vinifera	vine	Wein	SKZ, BL, A, BO, B, AB, EB, F, L, BV, BF

2 The current status of the German phenological observation network

Currently, about 1200 observers contribute to this network, the large majority of them on a voluntary basis. At approx. 60 sites the observations are performed by weather observers of DWD. Approx. 160 phenological phases of wild plants, agricultural crops, fruit trees and bushes, and grape vines are observed and archived in the phenological database of DWD. Tables 1 and 2 show the list of currently observed phases.

Two types of reporting strategies are applied: "Yearly reporters" transmit their data once a year in a report sheet and approx. 400 observers submit their reports by phone to the next weather station immediately after occurrence of the phenological phase.

The observation programme of the yearly reporters contains widely distributed wild plants, forest trees and ornamental shrubs, the most important agricultural crops, as well as frequently grown fruit plants and grape vines. The phenomena to be observed expand over the whole vegetation period, e.g. unfolding of leaves, flowering or leaf fall. These phases are clearly visible events in the development cycle of the plants and represent a change in their physiological condition.

The main purpose of the immediate reporting network is to provide up-to-date information on the current stage of plant development. This is especially important for advice to farmers and pollen information services. The network is still sufficiently dense for these tasks, but measures are taken to avoid further degradation, e.g. advertising efforts in cooperation with other government agencies.

A detailed instruction manual serves as a plant identification book for the plants contained in the observation programme (Deutscher Wetterdienst, 1991). It contains a picture of the habit of each plant and a precise photographic presentation of each phenological phase.

Typically, the observer defines the size of the observation area himself. An area with a radius of 1.5 to 2 km is typically sufficient and the maximum distance to the origin should not exceed 5 km. The altitude of the sites of the plants should not differ by more than ±50 m from the base altitude. The area should preferably be open and plain, i.e. valleys, etc. should be avoided. Observations should be performed two to three times a week (Deutscher Wetterdienst, 1991).

3 The phenological database and quality control

The phenological reports have been archived since 1951. They are a component of the national climate database of DWD.

Until 1991, observations in Western and Eastern Germany have been carried out according to programmes that agreed in large parts. But in East Germany not all phases were integrated into the archives. However, activities have been started to also include these data, too. After the German reunifica-

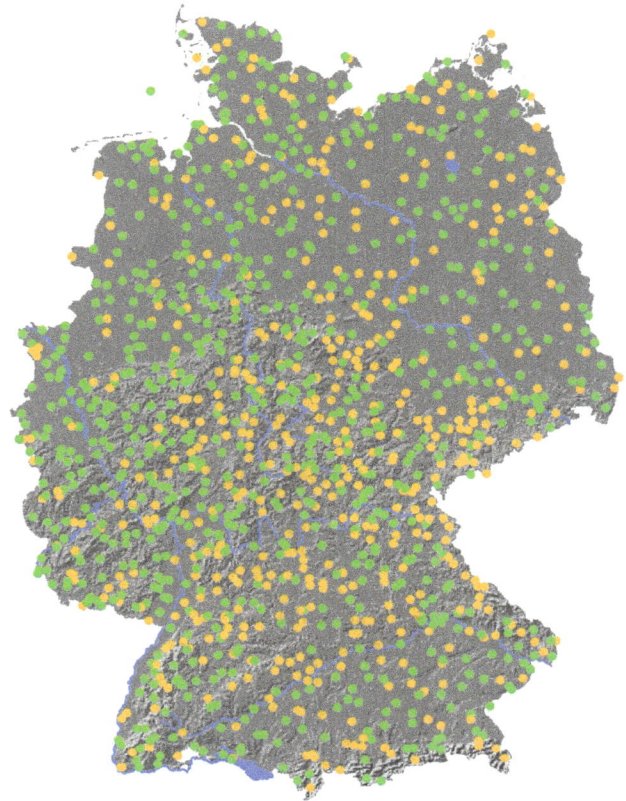

Figure 1. Locations of phenological observation sites and length of their time series. Green: at least 1 year of observations. Orange: 50 or more years of observations.

tion, a new observation programme with detailed observation instructions was defined in 1992 and has been used since then.

474 of the stations in the database cover a period of 50 years (or more; see Figs. 1 and 2), 164 stations cover a period of 60 years (within the period 1951–2010). However, not all phases have been observed over that period. Based on historical documents the database was expanded to the time before 1951. For the location Geisenheim these observations are available back to 1896 (Fig. 3).

An automatic quality control procedure is applied to the data (Zimmermann and Polte-Rudolf, 2013). In the first step, a test with predefined thresholds is applied, the temporal order of the phases is checked and it is tested if the temporal differences between subsequent phases are in a plausible range. In the next step, a spatial test is applied. Within a specific spatial range, the observed date of the phase at an individual site is compared with a statistically derived value for that area. The final result of the tests is provided as a quality indicator (e.g. "rejected" or "wrong") together with the data. But even if a value is flagged as "rejected", it might still be correct and useable. Frequently errors have causes that can

Table 2. Observed phases and their abbreviations.

Code	Name of the phase (English)	Name of the phase (German)
A	sprouting of leaves	Beginn des Austriebs
AE	panide emergence	Beginn des Ährenschiebens
	beginning of ear	Beginn des Rispenschiebens
AB	full blossom; general blossom	Vollblüte
AU	beginning of emergence	Beginn des Auflaufens
B	beginning of flowering; beginning of blossom	Beginn der Blüte
BB	beginning of formation of leaves	Beginn der Blattbildung
BF	leaf fall	Blattfall
	needle fall	Nadelfall
BG	closed stand	Bestand geschlossen
BL	first bleeding of the vines	Erstes Bluten
BO	unfolding of leaves	Beginn der Blattentfaltung
		Beginn der Nadelentfaltung
BST	tilling, sowing, drilling	Beginn der Bestellung
BV	colouring of leaves	Blattverfärbung
	colouing of needles	Nadelverfärbung
E	harvest	Ernte
	crop	Schnitt
EB	end of flowering	Ende der Blüte
ERG	beginning of turning green	Beginn des Ergrünens
F	first ripe fruits (forest and ornamental woody plants)	erste reife Früchte
F	fruit ripe for picking (fruit)	Beginn der Pflückreife
F	beginning of ripening (grape vines)	Beginn der Reife
GR	beginning of yellow ripeness	Beginn der Gelbreife
KNO	beginning of bud formation	Beginn der Knospenbildung
L	grape harvest	Lese
M	beginning of may sprouting	Maitrieb
MR	beginning of milk ripeness	Beginn der Milchreife
RO	beginning of rosette formation	Beginn des Rosettenbildung
SCH	beginning of growth in height	Beginn des Längenwachstums
	beginning of shooting	Beginn des Schossens
TR	beginning of wax-ripe stage	Beginn der Teigreife
VR	beginning of full ripeness	Beginn der Vollreife
SKZ	index-number of variety	Sortenkennziffer

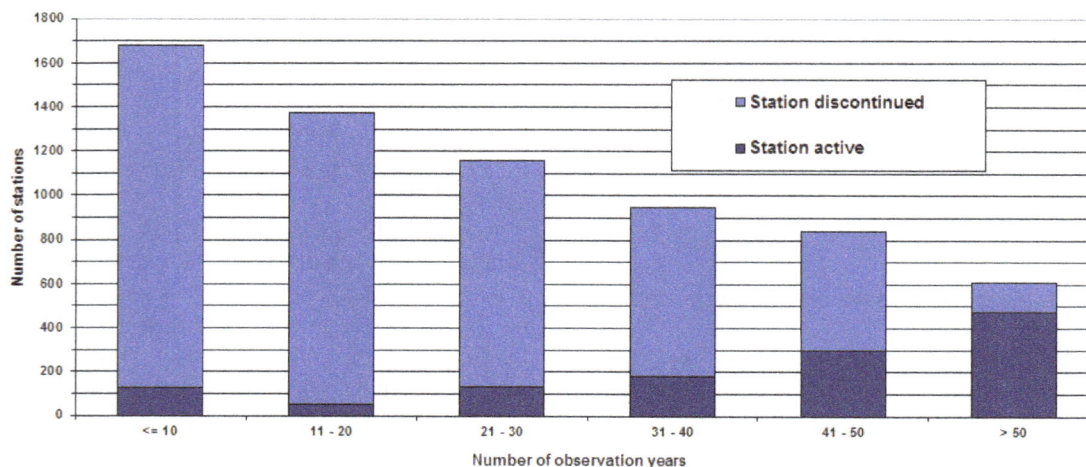

Figure 2. Number of stations in the database and length of their time series (within the period 1951 to 2010).

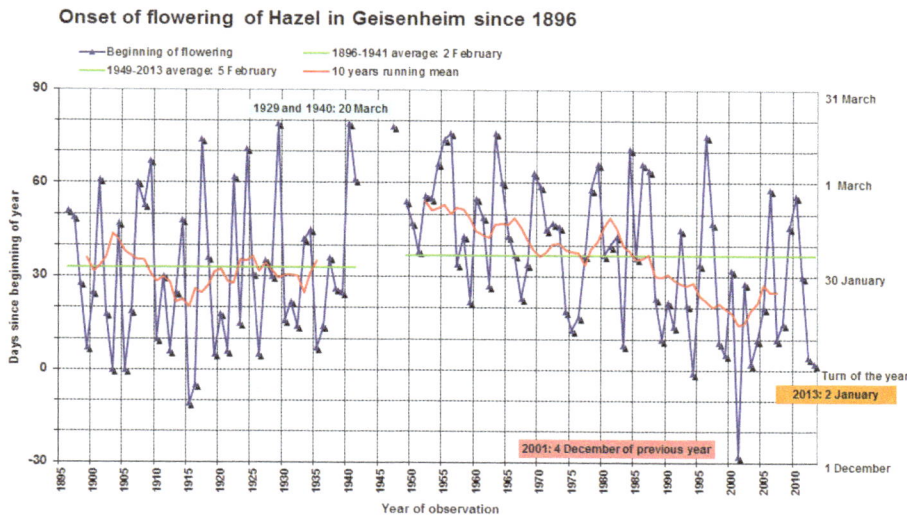

Onset of flowering of Hazel in Geisenheim since 1896

Figure 3. Some phenological records held by the research station at Geisenheim go back to 1896. These show a marked shift towards earlier flowering times over the past 60 years.

be corrected afterwards (e.g. caused during technical conversion).

Starting backwards from 2012, also the data of previous years are controlled. For 2012, 1.8 % of the data have been flagged as "wrong". For some stations, the error rate is significantly above the average rate. These cases are carefully analysed and where appropriate discussed with the observers.

4   Applications, phenological seasons and "clocks"

The data of the intermediate observers are used by DWD for pollen information service and agrometorological advice. In combination with the weather forecasts they provide a basis for pest management and field irrigation as well as for a decision on favourable harvest dates or to allow the estimation of pollen dispersion. The data are also provided to universities, research institutes, public authorities, the media and many other customers.

The data are also used by DWD to create different types of charts that provide easy to understand visualisations of the temporal course of vegetation development. Such diagrams can be used to illustrate the year-to-year development but also long term shifts.

One simple example is the definition of a phenological "growing season". The preferred choice for this purpose are wild growing plants that occur in all regions of Germany. DWD defines the growing season as the period between beginning of flowering of Forsythia and leaf fall of penduncu-late oak. Figure 4 shows an example.

A visualization with more detailed information are the "phenological clocks" that are based on the definition of the "seasons" of a "phenological year". For that purpose ten different "phenological seasons" are defined based on specific

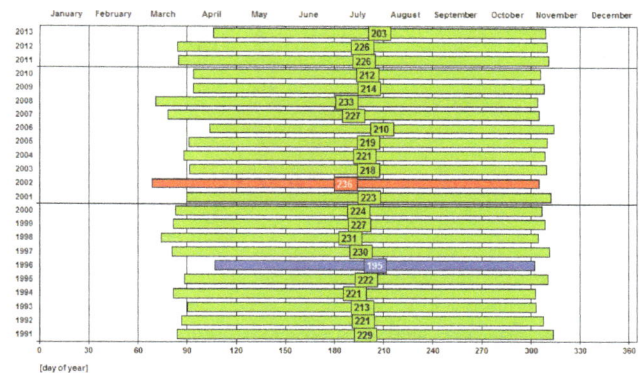

Figure 4. Growing seasons in Germany since 1991: mean beginning, duration (days) and ending defined by the period between "Forsythia (Forsythia suspensa) – beginning of flowering" and "Pedunculate Oak (Quercus robur) leaf fall".

growth stages of selected plants that are used as "indicator phases" (Table 3). This is typically done for a wider region but can also be done for individual sites. Each season begins with an indicator phase and ends with the beginning of the next season. When visualized as "double clocks" (concentric; see Fig. 5 for an example), these diagrams allow to easily compare a single year with the long-term mean or two climatological reference periods.

Table 3. Indicator phases used to define the phenological seasons. Each season begins with an indicator phase and ends with the beginning of the next season. The alternative phase can be used if the indicator phase is not available.

phenological seasons	indicator plants/phases	alternative plants/phases
prespring	hazel (flowering)	snowdrop (flowering)
first spring	forsythia (flowering)	gooseberry (leaf unfolding)
full spring	apple, mainly early ripeness (flowering)	pedunculate oak (leaf unfolding)
early summer	black elder (flowering)	black locust (flowering)
midsummer	large leaved lime (flowering)	red currant (fruit)
late summer	apple, early ripeness (fruit)	rowan (fruit)
early autumn	black elder (fruit)	cornelian cherry (fruit)
full autumn	pedunculate oak (fruit)	horse chestnut (fruit)
late autumn	pedunculate oak (leaf colouring)	rowan (leaf fall)
winter	pedunculate oak (leaf fall)	1. apple, late ripeness (leaf fall)
		2. european larch (needle fall)

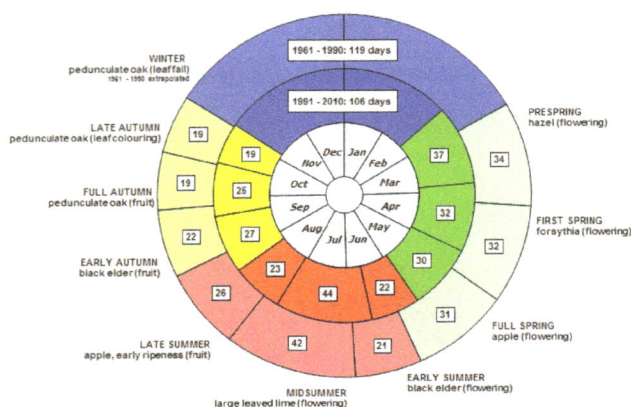

Figure 5. Phenological clock for Germany: indicator phases, mean beginning and duration of the phenological seasons. In this clock, the periods 1961–1990 and 1991–2010 are compared.

5   Data access and co-operation in Europe

The phenological data are freely available via the website of DWD; e.g.

- ftp://ftp-cdc.dwd.de/pub/CDC/observations_germany/ phenology/ or

- http://www.dwd.de/phaenologie.

DWD contributes to the Pan European Phenology database (PEP725; successor of COST-725 "Establishing a European Phenological Data Platform for Climatological Applications", see Sparks et al., 2009), which was set up by the Austrian Central Institute for Meteorology and Geodynamics (ZAMG), the Austrian Federal Ministry of Science and Research and the Economic Interest Grouping of European National Meteorological Services (EUMETNET). The main aim of PEP725 is to promote and support phenological research by making available an annually updated pan-European database, providing open access to phenological data for science, research and training.

Together with additional data sources, the phenological observations of DWD are integrated into the Plant-Phenological Online Database PPODB (http://www.ppodb. de/). The merged database comprises plant-phenological observations collected in Central Europe between 1880 and 2009 with an emphasis on Germany (Dierenbach et al., 2013).

6   Conclusions

Germany's national meteorological service maintains a dense phenological observation network. Observations are currently performed at about 1200 locations, mainly by voluntary observers. Approx. 400 observers submit their reports immediately after occurrence of the phenological phase. The information is used for several tasks of DWD, e.g. advice to farmers and pollen information services. The data are quality-controlled and stored in a database. The database contains times series covering several decades and is therefore a valuable source for research activities. The data are publicly available via the Climate Data Center of DWD and also integrated into international collections.

Acknowledgements. The authors are especially grateful to the many voluntary observers. Two reviewers helped to improve the manuscript.

References

Dahl, Å. and Langvall, O.: Observations on phenology in Sweden: past and presen, edited by: Nekovář, J., Koch, E., Kubin, E., Nejedki, P., Sparks, T., and Wielgolaski, F. E., COST Action 725: The history and current status of plant phenology in Europe, COST office, 2008.

Defila, C.: Plant phenological observations in Switzerland, edited by: Nekovář, J., Koch, E., Kubin, E., Nejedki, P., Sparks, T., and

Wielgolaski, F. E., COST Action 725: The history and current status of plant phenology in Europe, COST office, 2008.

Deutscher Wetterdienst: Anleitung für die phänologischen Beobachter des Deutschen Wetterdienstes, 3rd Edn., Deutscher Wetterdienst, 1991.

Dierenbach, J., Badeck, F. W., and Schaber, J.: The plant phenological online database (PPODB): an online database for long-term phenological data, Int. J. Biometeorol., 57, 805–812, doi:10.1007/s00484-013-0650-2, 2013.

Koch, E., Lipa, W., Neumcke, R., and Zach, S.: The history and current status of the Austrian phenology network, edited by: Nekovář, J., Koch, E., Kubin, E., Nejedki, P., Sparks, T., and Wielgolaski, F. E., COST Action 725: The history and current status of plant phenology in Europe, COST office, 2008.

Lieth, H. (Ed.): Phenology and Seasonality Modeling. Springer Verlag, Berlin, Heidelberg, 444 pp., doi:10.1007/978-3-642-51863-8, 1974.

Menzel, A.: Phenology, its importance to the Global Change Community, Clim. Change, 54, 379–385, 2002.

Nekovář, J., Koch, E., Kubin, E., Nejedki, P., Sparks, T., and Wielgolaski, F. E.: COST Action 725: The history and current status of plant phenology in Europe, COST office, 2008.

Reiss, M.: Die Phänologie in Österreich seit 1826 und ihre Beziehungen zur Klimakunde, Wetter und Leben, 11, Österreichische Gesellschaft für Meteorologie, 1959.

Richardson, A. D., Keenan, T. F., Migliavacca, M., Ryu, Y., Sonnentag, O., and Toomey, M.: Climate change, phenology, and phenological control of vegetation feedbacks to the climate system, Agr. Forest Meteorol., 169, 156–173, doi:10.1016/j.agrformet.2012.09.012, 2013.

Schnelle, F.: Pflanzen-Phänologie, Akademische Verlagsgesellschaft Geest und Portig, Leipzig, 1955.

Settele, J., Scholes, R., Betts, R., Bunn, S., Leadley, P., Nepstad, D., Overpeck, J. T., and Taboada, M. A.: Terrestrial and inland water systems, in: Climate Change 2014: Impacts, Adaptation, and Vulnerability. Part A: Global and Sectoral Aspects. Contribution of Working Group II to the Fifth Assessment Report of the Intergovernmental Panel on Climate Change. Cambridge University Press, Cambridge, United Kingdom and New York, NY, USA, 271–359, 2014.

Sparks, T. H. and Carey, P. D.: The responses of species to climate over two centuries: an analysis of the Marsham phenological record, 1736–1947, J. Ecol., 83, 321–329, doi:10.2307/2261570, 1995.

Sparks, T. H. and Collinson, N.: The history and current status of phenological recording in the UK, edited by: Nekovář, J., Koch, E., Kubin, E., Nejedki, P., Sparks, T., and Wielgolaski, F. E., COST Action 725: The history and current status of plant phenology in Europe, COST office, 2008.

Sparks, T. H., Menzel, A., and Stenseth, N. C.: European cooperation in plant phenology, Clim. Res., 39, 175–177, 2009.

Wege, K.: Die Entwicklung der meteorologischen Dienste in Deutschland. Geschichte der Meteorologie in Deutschland – Band 5. Selbstverlag des Deutschen Wetterdienstes, Offenbach am Main, 2002.

Zimmermann, K. and Polte-Rudolf, C.: Prüfung und Korrektur phänologischer Daten, Phänologie-Journal, 41, Deutscher Wetterdienst, 2013.

2

An overview of the use of Twitter in National Weather Services

S. Gaztelumendi[1,2], **M. Martija**[1,2], **O. Principe**[1,2], **and V. Palacio**[1,2]

[1]TECNALIA, Energy and Environment Division, Meteorology Area, Miñano, Basque Country, Spain
[2]Basque Meteorology Agency (EUSKALMET), Miñano, Basque Country, Spain

Correspondence to: S. Gaztelumendi (santiago.gaztelumendi@tecnalia.com)

Abstract. Twitter is a service that enables users to post messages ("tweets") of up to 140 characters supporting a variety of communicative practices. In this paper we analyze different aspects related to the use of Twitter in different National Meteorological Services (NMS) worldwide. Firstly, we will review the general position of NMS worldwide regarding the use of Twitter technology. Secondly, we will focus on different practices of some selected meteorological services. Thirdly, we will deal specifically with the Basque Meteorology Agency (Euskalmet) case. Finally some conclusions are presented.

1 Introduction

National Meteorological Services use different "classical" tools for meteorological information dissemination, including television, radio, newspaper, phone, e-mail and public/private web (WMO, 2001, 2007, 2010). In recent years, new technologies, and in particular the rapid expansion of Twitter, have caused that the transmission of information can be virtually instantaneous and accessible to large segments of population (Orbe, 2012; Weller et al., 2013).

We can consider Twitter as the perfect place for quick and efficient communication with audience (Boyd et al., 2010; Rodríguez, 2011). As a consequence many NMS have developed new communication strategies and incorporated this tool for different purposes. Some NMS do not only provide forecast and other remarkable information routinely, but gives real-time observed data, forecast and relevant information continuously before and during severe-weather episodes, as in the Basque Meteorology Service (Euskalmet) case (Gaztelumendi et al., 2013a).

A simple methodology is used during this study. Firstly the general position regarding the use of Twitter technology of National Meteorological Services worldwide (WMO members) and some local/regional non-members, are analyzed checking general available accounts information. On the other hand, usual practices are examined during six month, from February to August 2014, analyzing the daily Twitter

activity of every NMS in their publicly available tweets. During this period the accounts evolution is followed and information from their public timelines is collected. A Twitter analytics tool (twitonomy.com) is used to extract information, including followers, tweets, retweets, replies, mentions, hashtags, etc. Excel is used to compare all the information collected and to analyze the evolution of specific features from accounts of different NMS.

2 Twitter and meteorological services

In the last years, most of the meteorological services have joined social networks as YouTube, Facebook, Twitter, etc. The increasing availability of mobile phones with access to these networks is an opportunity not only to spread messages, but to real time interaction. More than 40 % of global NMS have an account on Twitter, although this percentage is reduced in Africa, Asia and Oceania, as shown in Fig. 1, due to the low diffusion of mobile technologies and the limited access to internet in these areas, among other factors (Palacio et al., 2014).

The first centre to join Twitter (in an active way) was the New Zealand Meteorological National Service, on 28 December 2008. The following years, the number of meteorological centres who joined Twitter progressively increased

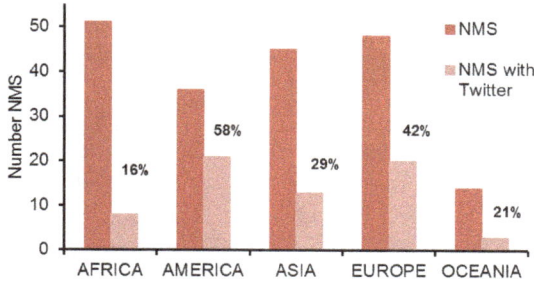

Figure 1. Percentages of NMS who have accounts on Twitter by continents (January 2015).

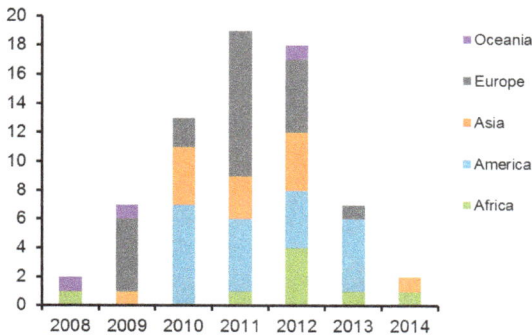

Figure 2. Yearly distribution of NMS incorporation on Twitter.

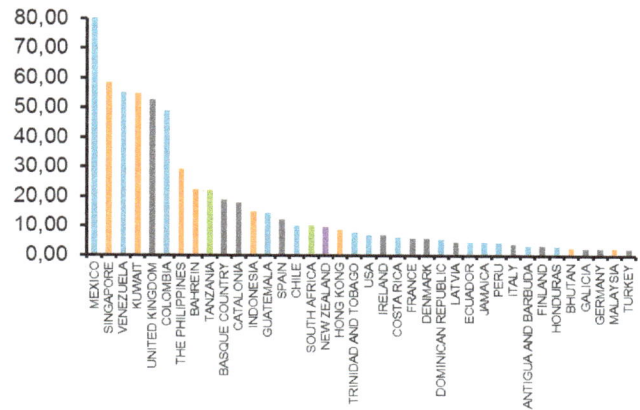

Figure 3. Average number of tweets posted per day by NMS (from beginning to 30 August 2014).

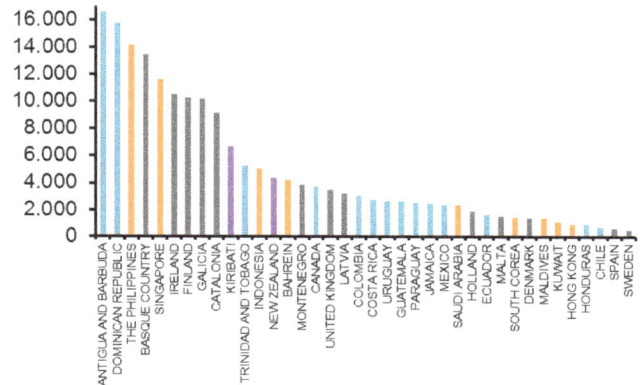

Figure 4. Followers by million inhabitants of NMS's Twitter accounts (from beginning to 30 August 2014).

until the end of 2013, when it almost stopped, as can be seen in Fig. 2.

In Africa, there are five active accounts. It is worth noting South Africa case because of its number of followers, the number of tweets sent and the degree of influence. In Oceania, the most active account is the one of New Zealand. In Asia, there are thirteen active accounts. Indonesia and Philippines are remarkable because of its high activity and the number of followers. In the American continent, it is relevant the NOAA's activity, with 150 active accounts (the main one, the National Weather Service of USA, is the most influential one). On the other hand, the Mexican service has almost 300 000 followers (2.4 per thousand inhabitants), whereas the Venezuelan centre has the largest number of tweets. In Europe, most of relevant centres have accounts on Twitter. MetOffice (UK) stands out for the large number of followers, its great activity and interaction with followers. In Fig. 3, we show the number of tweets sent per day for different meteorological services. In Fig. 4, we represent the number of followers considering the potential influence (millions of inhabitants for the concerned country).

3 Usual practices

In this section, we analyze different Twitter practices, focusing on a selection of NMS, which are representative of general situation worldwide (see Table 1) (Twitonomy, 2015). In

general, NMS use Twitter for fast and efficient communication with users, including different kind of information, as routine weather forecast and observations, general weather, climate and scientific information, events communications and others. In some cases, this platform is used for real time severe weather data dissemination and warnings. At the end, NMS look for bidirectional communication and users fidelization, in some cases with a relatively high interaction with users (not usually).

Regarding sent tweets and followers, it is worth to mention the account of Mexico, with 91 tweets daily and 291 000 followers (2.4 per thousand inhabitants), well above the next one which is UK, with 55 tweets per day and 221 000 followers (3.5 per thousand inhabitants). Nonetheless, the Mexican account barely interacts with users, while the UK service is the one that more does it. The 85 % of sent tweets mention another user (they usually answer questions or doubts of the users by direct mentions to them). They receive a mention of some user for almost every tweet (90 %). Instead, most of tweets sent by the account of Mexico do not have any specific

Table 1. Relevant data about Twitter usage in different NMS (considering, at 30 August 2014, last 3200 tweets).

Date	USA	Canada	Mexico	Germany	Spain	UK	France	Hong Kong	Basque Country	Catalonia
Followers	139 000	131 000	291 000	4625	30 000	221 000	6062	6768	29 500	69 400
Tweets/day	7.2	1.1	91.4	0.8	8.4	55.2	2.3	6.1	18.8	17.9
Mention/tweet	0.1	0.2	0.0	0.2	0.1	0.9	0.5	0.0	0.1	0.1
Link/tweet	0.2	0.6	0.3	0.8	0.4	0.6	0.6	0.9	0.1	0.5
Retweets	76 %	16 %	2 %	3 %	26 %	0 %	8 %	0 %	15 %	1 %
Replies	2 %	5 %	0 %	17 %	5 %	85 %	8 %	0 %	0 %	3 %
Hashtag/tweet	0.2	1.6	0.6	2.6	0.4	0.1	1.1	0.0	0.3	1.1
Tweets retweeted	24 %	75 %	96 %	66 %	72 %	20 %	67 %	55 %	68 %	81 %
Tweets favorited	23 %	52 %	93 %	51 %	48 %	25 %	45 %	8 %	48 %	64 %

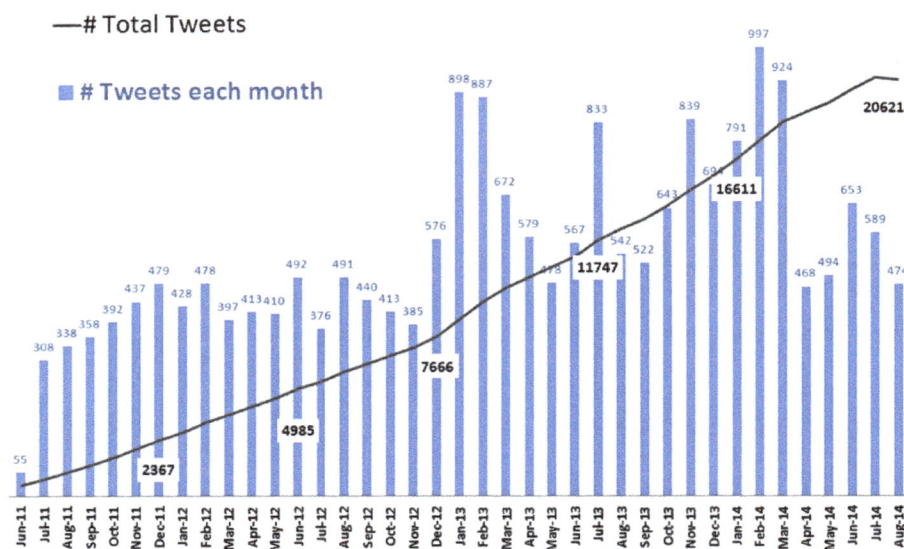

Figure 5. Evolution of tweets posted by Euskalmet.

mention to any of its followers, they rarely receive mentions or mention other users.

Concerning the ratio of NWS information vs non-own content of posted tweets, users' tweets are not usually retweeted. It is important to mention the NWS of USA, because it is the one that retweets more (75 % of sent tweets are retweeted of other users); however, in this case they do it of their own regional accounts (they have more than 140).

Talking about the hashtags (they ease the organization and the access to the information), most of the centres make a limited use of them, apart from German DWD (2.6 hashtags per tweet).

In relation to the use of links, most of the meteorological services use them. In the case of Hong Kong, most of the tweets (90 %) have a link to their own webpage, where there is more detailed information.

In relation to the awaken interest among users by retweeted and favorited tweets (we suppose that the higher number of them, the more relevance of the tweet), it is note-

worthy the Mexican account with a 96 % of retweeted tweets and a 93 % of favorited tweets. On the other hand, MetOffice has a 20 and 25 %, respectively.

4 Euskalmet case

In the case of Basque Meteorology Agency (Euskalmet), one of the most powerful reason to use Twitter is the fast and efficient communication in high impact weather scenarios (Gaztelumendi et al., 2012; Martija et al., 2014), and especially in those associated with rapid development processes that occur in less than an hour such as storms, coastal trapped disturbances or flash floods case (Gaztelumendi et al., 2013b).

The first tweet was sent by Euskalmet on 24 June 2011 dealing with a yellow warning due to a heat wave, since that time, more than 22 200 tweets have been sent (see Fig. 5). The number of followers is over 30 800 for a total Basque Population of 2.2 million (see Fig. 6). During months with

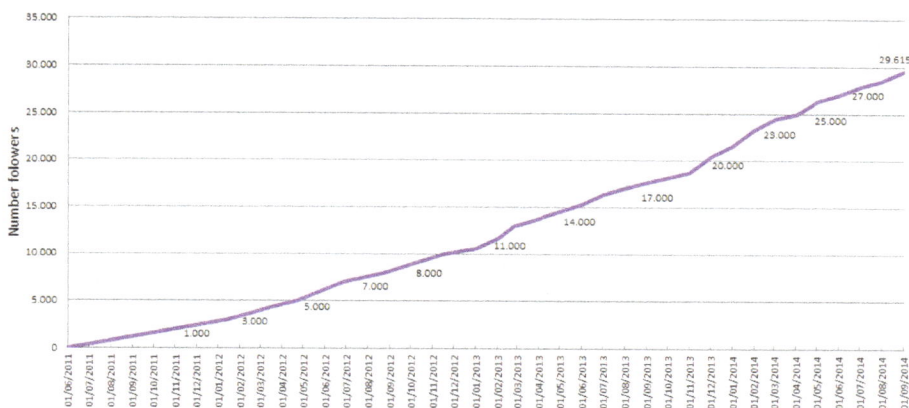

Figure 6. Time evolution of total followers of Euskalmet's Twitter account.

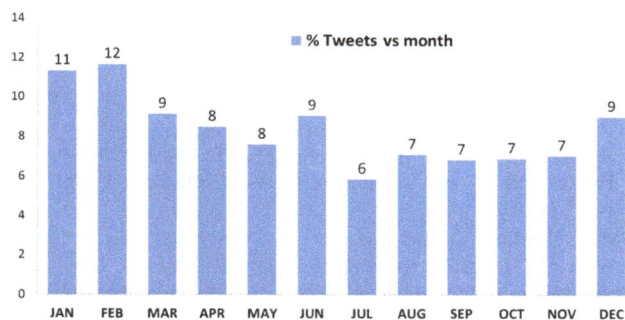

Figure 7. Monthly distribution of tweets posted by Euskalmet (from July 2011 to August 2014).

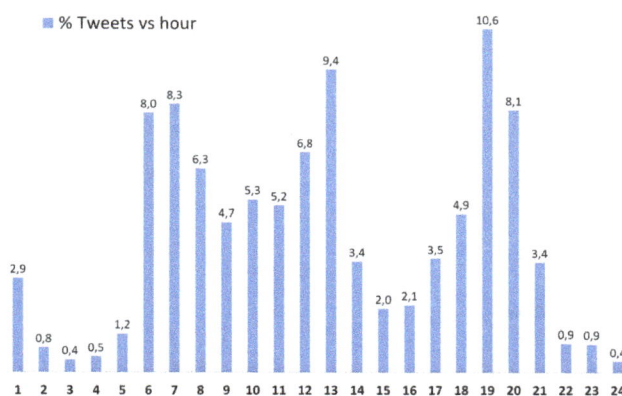

Figure 8. Hourly distribution of tweets posted by Euskalmet (from July 2011 to August 2014).

more severe weather events an increase of tweets are produced (see Fig. 7). The effect of routinely information emission at fixed times logically affect the hourly distribution (see Fig. 8).

In Euskalmet case, the number of tweets per day is around 20. Retweets (15 %) are used only for relevant and reliable information, especially if it is accompanied with photos or videos. One third of messages contain "attached" extra information, using links (graphs, maps, radar images, pdf documents, photos, web path, etc). The 24 % of tweets are accompanied with photos or videos. The 63 % of tweets deal with forecast aspects, the 31 % with observations and the 5 % with other subjects. We have done the 43 % of communications related to high impact weather and the 52 % related to routine weather. Direct messages suppose less than 2 %, this private messages to followers are restricted to technical clarifications and to thank followers for providing useful information (more details in Gaztelumendi et al., 2013a).

5 Conclusions

New technologies and social networks have increased the available meteorological information and the opportunities to share it. In particular, Twitter is one of the most used social network and communication tool. For that reason, most of the leading meteorological centres have an active account.

The most influential Twitter accounts of NWS are in Europe and North America, but the most followed in Asia. However, Twitter is not very widespread in Africa, the still low mobiles presence in this continent could be one of the main reasons.

The use of this tool is not homogenous in the different NWS. The operational routines are very different, with a variety of information content and different relationships with followers.

Attracting followers and maintaining them is a very complex task, considering that there is no direct relation between the number of sent tweets (effort) and the number of followers of an account (award).

This communication tool has a personal and direct character; therefore, locality character has a great importance, as we can see by the fact that big centres use regional accounts and the high number of followers that small centres with more local character have (Euskalmet, Meteocat, etc.).

Twitter is one of the most promising tool for information transmission in situations of severe weather both for its rapid expansion and for its collaborative nature (Hughes and Palen, 2009). Euskalmet is a good example at regional level, due to its high transmission capacity in severe weather situations, among other factors, because of the dense instrumentation present in Basque Country and its operational integration in the emergency department of the Basque Government.

Acknowledgements. The authors would like to thank the Emergencies and Meteorology Directorate – Security Department – Basque Government for public provision of data and operational service financial support. We also would like to thank all our colleagues from EUSKALMET for their daily effort in promoting valuable services for the Basque community, and particularly for those of then that participate in twitter operational tasks.

References

Boyd, D., Golder, S., and Lotan, G.: Tweet, Tweet, Retweet: Conversational aspects of retweeting on Twitter, Proceedings of HICSS-43, IEEE, Kauai, HI, 2010.

Gaztelumendi, S., Egaña, J., Otxoa-de-Alda, K., Hernandez, R., Aranda, J., and Anitua, P.: An overview of a regional meteorology warning system, Adv. Sci. Res., 8, 157–166, doi:10.5194/asr-8-157-2012, 2012.

Gaztelumendi, S., Orbe, I., Lopez, A., Aranda, J. A., and Anitua, P.: Social media and high impact weather communication in Basque Meteorology Agency, 13th EMS, 11th ECAM, 9–13 September 2013, Reading, UK, 2013a.

Gaztelumendi, S., Egaña, J., Pierna, D., Aranda, J. A., and Anitua, P.: The Basque Country Severe Weather Warning System in perspective, 13th EMS, 11th ECAM 9–13 September 2013, Reading, UK, 2013b.

Hughes, A. L. and Palen, L.: Twitter adoption and use in mass convergence and emergency events, Proceedings of the 6th ISCRAM Conference, Gothenburg, Sweden, 2009.

Martija, M., Palacio, V., Príncipe, O., and Gaztelumendi, S.: Meteo adversa y su comunicación vía radio y Twitter, XXXIII Jornadas científicas de la AME, Oviedo, Spain, 2014.

Orbe, I.: Emergencias y medios de comunicación, Academia Vasca de Policía y Emergencias, Vitoria-Gasteiz, Spain, 2012.

Palacio, V., Martija, M., Príncipe, O., and Gaztelumendi, S.: Servicios meteorológicos y Twitter, XXXIII Jornadas de la AME, Oviedo, Spain, 2014.

Rodríguez, O.: Twitter, Aplicaciones profesionales y de empresa, Gurús Press, Anaya, 2011.

Twitonomy: Twitter #analytics and much more …, http://www.twitonomy.com (last access: 30 August 2014), 2015.

Weller, K., Bruns, A., Burgess, J., Mahrt, M., and Puschmann, C.: Twitter and Society, Digital Formations, 89. Peter Lang, New York, 2013.

WMO: Weather on the internet and other new technologies, WMO/TD No. 1084, Geneva, Switzerland, 2001.

WMO: Examples of best practice in communicating weather information, WMO/TD No. 1409, Geneva, Switzerland, 2007.

WMO: Guidelines on early warning systems and application of nowcasting and warning operations, WMO/TD No. 1559, Geneva, Switzerland, 2010.

3

Precipitation climate maps of Belgium

M. Journée, C. Delvaux, and C. Bertrand

Royal Meteorological Institute of Belgium, Brussels, Belgium

Correspondence to: M. Journée (michel.journee@meteo.be)

Abstract. Investigations are conducted to best estimate precipitation climate maps over Belgium from daily observations available for the period 1981–2010. Several mapping approaches are compared in a cross-validation exercise. These approaches differ by several aspects and in particular by the order in which the temporal aggregation (i.e. computation of climate mean values from daily data) and spatial interpolation steps are performed, and by the integration of ancillary information in the spatial interpolation method. The selected approach is used to derive a large panel of climate maps. In particular, the main spatio-temporal features of the annual cycle of rainfall in Belgium are extracted by principal component analysis (PCA).

1 Introduction

The Royal Meteorological Institute of Belgium (RMI) has recently updated the climate maps of Belgium for various meteorological parameters. These maps represent the spatial distribution over Belgium of 30-years climate mean values based on the reference period 1981–2010, in accordance with the recommendation of the World Meteorological Organization (WMO, 2014). Such climate maps have already been derived for several regions of the world, e.g. The Netherlands (Sluiter, 2012), the Alps (Frei and Schär, 1998), the USA (Daly et al., 2008) and New Zealand (Tait and Zheng , 2007) to name a few examples.

The precipitation climate maps of Belgium rely on daily rain gauges observations from the Belgian network of climatological stations. In the past, such maps were derived manually on the basis of climate mean values derived from continuous time series of observations (Dupriez and Sneyers, 1979). Nowadays, climate maps are processed with reproducible methods without any subjective manual intervention (Daly, 2006).

This paper summarizes the investigations conducted to update the precipitation climate maps of Belgium. These investigations have two objectives: first, to determine the best method to generate climate maps from daily observations and second, to highlight the main characteristics of precipitation regimes along the year in Belgium. To reach the first objective, various methods have been considered and evaluated by cross-validation. For the second objective, a spatio-temporal analysis of rainfall by principal component analysis (PCA) was performed.

2 Materials and methods

2.1 Precipitation data

Monitoring of the Belgian climate is a key responsability of RMI, which implies the regular observation of several climate variables. RMI operates since the end of the 19th century a network of climatological stations providing observations of the daily precipitation quantities and the daily extreme air temperatures. Regarding precipitation, the number of these climatological stations has varied during the period 1981–2010 between 333 sites in 1981 and 247 sites in 1997 (see Fig. 1, left panel). The locations of the climatological stations (with at least 50 % of daily data available for the period 1981–2010) are illustrated in Fig. 2.

Although the climate maps are mainly derived from data acquired by the network of climatological stations, it is worth to mention that, since the begin of the 2000s, additional networks of automatic rain gauges have been deployed by the regional Belgian hydrological services. In particular, a network of 90 well-maintained and regularly calibrated tipping bucket rain gauges is operated by the Walloon hydrological services (WHS) since 2005 (see Fig. 2 for the location of these rain gauges). This data may be valuable ancillary information to improve the spatial representation of rainfall in

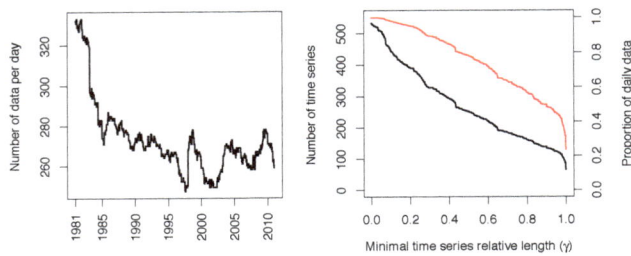

Figure 1. Evolution of the number of observations from the climatological stations available per day from 1981 to 2010 (left panel) and number of time series with a relative length greater or equal to a level γ (right panel, blue line and left axis). The corresponding fraction of daily observations is given by the red line (right axis).

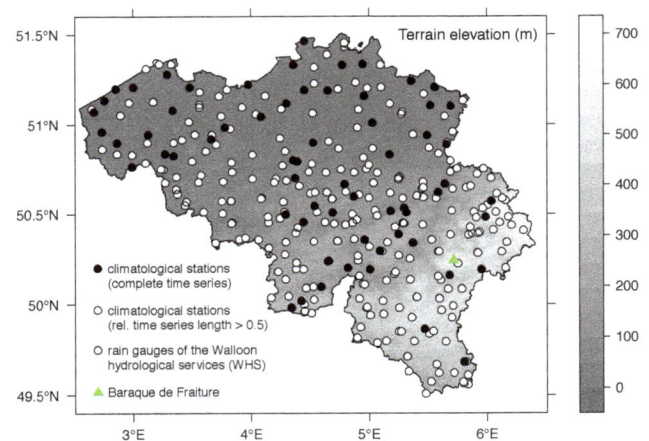

Figure 2. Elevation map of Belgium and spatial distribution of the climatological stations with a relative observations time series length larger or equal to 50 % (empty red circles) and 100 % (filled red circles) for 1981–2010 and the WHS rain gauges (blue circles). The green triangle indicates the location of "Baraque de Fraiture".

the region covered by the WHS network. This point will be discussed in the following.

All measurements made within the networks operated by RMI and WHS are routinely quality controlled by RMI.

2.2 Spatial interpolation methods

Climate maps typically result from the spatial interpolation of irregularly distributed station data to a regular high-resolution grid (Daly, 2006). The commonly used methods for the spatial analysis of climate data include inverse-distance weighting (IDW), thin plate smoothing splines, local regression models and geostatistical interpolation techniques (Sluiter, 2012). These methods can be divided into exact or approximate interpolation, depending on whether the interpolate equals or not the observation at the stations' locations. In this study, only exact interpolation methods are considered namely IDW and two geostatistical methods.

To derive climate maps from daily observations, two approaches are possible. First, the observations can be spatially interpolated to generate a set of daily precipitation grids, which are then temporally aggregated to provide the climate mean at each grid point (i.e. "interpolation-aggregation" approach). Second, climate mean values estimated from the observations time series can be spatially interpolated (i.e. "aggregation-interpolation" approach). These two approaches distinguish themselves by several aspects.

First, the "aggregation-interpolation" approach (denoted AI in the following) requires observations time series that are continuous on the entire 30-years period to allow a proper computation of the climate mean values, while the "interpolation-aggregation" approach (denoted IA in the following) can include all observations time series, even the short ones. As illustrated in Fig. 1 (right panel), only 64 time series are complete from 1981 to 2010, which represents 24 % of all daily data available for the 1981–2010 period. This drawback of the AI approach can be mitigated if one allows gaps in the time series used to compute the climate mean values. These gaps can be filled by estimations (ob-

tained, for instance, by spatial interpolation of neighboring daily data) before temporal aggregation. This enables to significantly increase the number of involved time series, as illustrated in Fig. 1 (right panel). For instance, 138 time series are complete at 90 % and they represent half of the database of daily precipitation data.

Second, the spatial interpolation of climate mean values is generally less prone to uncertainties when compared to daily observations, as the spatial variability of precipitation fields reduces when the temporal aggregation horizon increases. Furthermore, since precipitation quantities are, on climate average, affected by physiographic features of the Earth's surface such as the terrain elevation (Goovaerts, 2000; Daly, 2006), the spatial interpolation of the climate mean values can be improved with the knowledge of densely sampled covariate data. As an example, Fig. 3 (left panel) illustrates the strong correlation between mean annual precipitation amount and terrain elevation.

Finally, the daily precipitation data acquired within the ancillary network of tipping bucket rain gauges operated by WHS can contribute to improve the representation of the climate patterns of rainfall in a region where they are the most complex. In the IA approach, these data can improve the estimation of the daily precipitation fields from 2005 to 2010, i.e. for only 6 years of the 30-years reference period. In case of the AI approach, a covariate of the 1981–2010 mean climate values can be derived if one assumes that the climate patterns of rainfall in Belgium can already be identified from a few years long time series. Such a covariate is defined by the spatial interpolation at high resolution of the 2005–2013 (9 years) mean values derived from the WHS data. This interpolation integrates terrain elevation data in a kriging with external drift approach (KED, Wackernagel, 1995). Figure 3

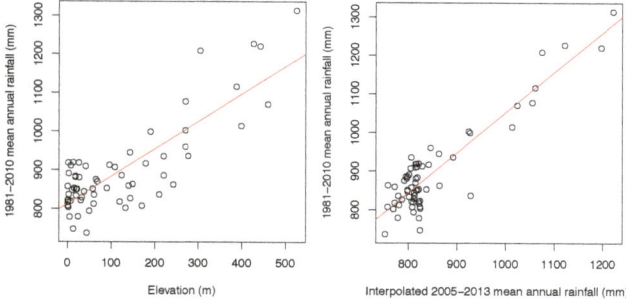

Figure 3. Scatter plots of the 1981–2010 mean annual rainfall computed for the 64 complete time series versus the terrain elevation (left panel) and the interpolated 2005–2013 mean annual rainfall based on the WHS rain gauges (right panel). The regression line is represented in red. The corresponding correlation coefficients are 0.825 and 0.915, respectively.

(right panel) illustrates the correlation between the 1981–2010 mean annual rainfall and the covariate based on the WHS network of rain gauges.

To sum up, the following methods to derive precipitation climate maps for Belgium are considered and evaluated in this study:

– Estimation of daily precipitation fields based on the maximum number of observations available each day, followed by temporal aggregation at each grid point (method "IA"). The spatial interpolation at the daily time scale is performed by inverse distance weighting (IDW) based on a maximum number of 20 neighbors located within a radius of 50 km.

– Completion of the gaps by IDW in all time series exhibiting a relative length equal or larger than a level $\gamma \in [0, 1]$, computation of the corresponding climate mean values and finally spatial interpolation. This spatial interpolation of climate mean values can be done by IDW, ordinary kriging (OK, Wackernagel, 1995), KED with the terrain elevation as drift or KED with the covariate based on the WHS network of rain gauges as drift. These methods will be respectively denoted by AI_{IDW}, AI_{OK}, AI_{KED1} and AI_{KED2}. For the geostatistical methods, a variogram model is systematically fitted to the data. A zero nugget is imposed to estimate a precipitation field without any discontinuity at the stations' locations.

2.3 Validation methodology

The considered climate mapping methods are evaluated and compared by cross-validation. The validation relies on the $n = 64$ time series that are complete on the entire 1981–2010 period (filled red circles in Fig. 2). Because some of these stations are very close from each other (i.e. less than 5 km), the classical leave-one-out cross-validation (i.e. one station

is systematically used to validate the estimation derived from the remaining stations' values) may lead to overly optimistic results. Therefore, a larger validation data set of size $k = 20$ was considered. The set of all daily observations is thus repeatedly splitted in separate training and validation data sets, with the validation data set being a random selection of k out of the n complete time series. The training data contains the remaining complete and incomplete time series. For the AI approaches, a condition on the minimal time series length (i.e. γ) is furthermore imposed. All methods are then applied on this training data set to estimate the mean annual and monthly precipitation quantities at the k validation locations. This process is restarted several times until each of the n complete time series has been used 10 times for validation. For each validation location $i \in (1, \ldots, n)$, the average estimation over the various reinitializations, $\hat{Y}_i$, is compared against the actual climate mean value, Y_i. The performance of the various methods is then summarized by the following indices:

– the mean bias error $MBE = \frac{1}{n} \sum_{i=1}^{n} (\hat{Y}_i - Y_i)$,

– the mean absolute error $MAE = \frac{1}{n} \sum_{i=1}^{n} |\hat{Y}_i - Y_i|$,

– the root mean square error $RMSE = \sqrt{\frac{1}{n} \sum_{i=1}^{n} (\hat{Y}_i - Y_i)^2}$ and

– the maximum absolute bias $\max_{i=1,\ldots,n} |\hat{Y}_i - Y_i|$.

MBE provides information about the average bias of the methods, while both MAE and RMSE characterize the average error magnitude. The maximum absolute bias indicates the largest deviation that is locally reached. These indices are assumed to provide sufficient evidence to select the best method.

2.4 Spatio-temporal analysis method

Principal component analysis (PCA) has been successfully used in several studies in order to improve the understanding of precipitation regimes at the scale of a country (Sneyers et al., 1989; Baeriswyl and Rebetez, 1997; Benzi et al., 1997). PCA is in general directly applied on stations's values, e.g. daily values (Benzi et al., 1997), monthly values (Baeriswyl and Rebetez, 1997) or monthly climate mean values (Sneyers et al., 1989). In contrast, in this study, PCA is used after spatial interpolation in order to more specifically highlight spatial patterns related to the precipitation regimes in Belgium. To reach that objective, we compiled a set of 365 maps of the mean rainfall per periods of 60-days in a sliding window approach by steps of one day. These maps were derived by the selected spatial interpolation method. This data set characterizes the spatial variability of the mean

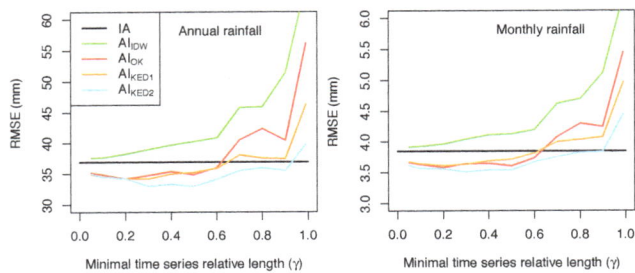

Figure 4. Cross-validation root mean square error (RMSE) for the considered mapping methods in the case of mean annual rainfall (left panel) and mean monthly rainfall (right panel). In case of the monthly rainfall, the results are averaged over the 12 months of the year. For the AI methods, the RMSE is illustrated as a function of the minimal time series length considered to define the training set (γ).

seasonal cycle of rainfall in Belgium. A PCA analysis of this data set was performed to extract its main spatio-temporal features. The PCA decomposition was evaluated after centering (i.e. substraction of the mean value for each map) in order to remove the spatial average of the annual cycle and highlight relative differences between different areas. The dominant components can highlight specific spatial patterns, while the corresponding loadings represent the seasonal variability of these patterns.

3 Results

3.1 Cross-validation analysis

The cross-validation exercise described in Sect. 2.3 enabled to evaluate the ability of the considered mapping methods to estimate the mean annual rainfall and the 12 mean monthly rainfall. Figure 4 provides the RMSE of the various methods for the mean annual rainfall and on average for the mean monthly rainfall (i.e. average over the 12 months of the year). Several observations can be made from these results.

First, the AI methods, and especially AI_{IDW} and AI_{OK}, are very sensitive to the minimal time series length γ. Globally, the performance of these methods increases when γ decreases, which highlights the importance of filling gaps in incomplete time series. The high sensitivity to γ between 0.9 and 1 is related to the large number of time series that are actually almost complete (see Fig. 1, right panel), i.e. 64 time series are complete (24 % of all daily data) while 138 time series are complete at 90 % (50 % of all daily data). Although the RMSE of AI_{IDW} continuously decreases towards the one of IA when γ goes to zero, for the other AI methods, there is an optimal γ below which the performance does improve anymore and even slightly degrades. In the case of AI_{KED1} and AI_{KED2}, this means that the ancillary data contribute more than short time series to the estimation of the precipitation fields.

Table 1. Cross-validation scores (in mm) for the considered mapping methods in case of the mean annual rainfall and the mean monthly rainfall (average over the 12 months of the year). The AI approaches rely on time series with at least 30 % of data available over the period 1981–2010 (i.e. $\gamma = 0.3$).

Mean annual rainfall			
	MBE	MAE	RMSE	max. abs. bias
IA	−2.30	31.20	36.90	100.00
AI_{IDW}	−3.00	33.20	39.00	98.10
AI_{OK}	−2.00	29.30	34.70	82.40
AI_{KED1}	−1.40	29.00	34.20	74.60
AI_{KED2}	−0.00	27.70	33.00	70.90

Mean monthly rainfall			
	MBE	MAE	RMSE	max. abs. bias
IA	−0.20	3.10	3.80	13.00
AI_{IDW}	−0.30	3.30	4.00	12.80
AI_{OK}	−0.20	2.90	3.60	13.60
AI_{KED1}	−0.10	2.90	3.60	12.70
AI_{KED2}	−0.00	2.80	3.50	11.70

Second, the geostatistical methods (AI_{OK}, AI_{KED1} and AI_{KED2}) outperform the simple IDW method (AI_{IDW}). There is furthermore a clear benefit in using ancillary data (topography and WHS data). Overall, the best performance is obtained with AI_{KED2} and γ between 0.3 and 0.5. This observation is confirmed by the other scores in Table 1. In particular, the maximum absolute bias, which provides an idea of the maximum error that can be locally reached, is significantly reduced by AI_{KED2} in the case of the mean annual rainfall. The sensitivity with respect to the method is however higher in the annual case than in the monthly one.

3.2 Qualitative validation

In parallel to this cross-validation analysis, a qualitative evaluation was performed by comparing the spatial distribution of the 1981–2010 average spatial distribution of rainfall estimated by the various methods against the climate maps related to the period 1833–1975 previously published by Dupriez and Sneyers (1979). As expected, the climate patterns are globally similar between the new and the former climate maps. However, several small-scale differences can be highlighted, which are related to the evolution of the climatological stations network. As an example, the 1833–1975 climate map indicates a local increase of the mean annual rainfall above 1200 mm per year in the area of the location called "Baraque de Fraiture" (indicated by the green triangle in Fig. 2). This local pattern is however not reproduced by the methods IA, AI_{IDW} and AI_{OK} because no station has been in operation in this area during a significant period between 1981 and 2010. Nevertheless, it can be estimated by

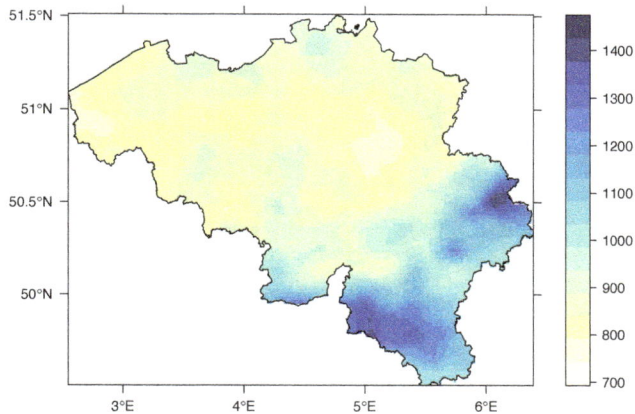

Figure 5. Map of the mean annual rainfall in Belgium for the period 1981–2010 (in mm).

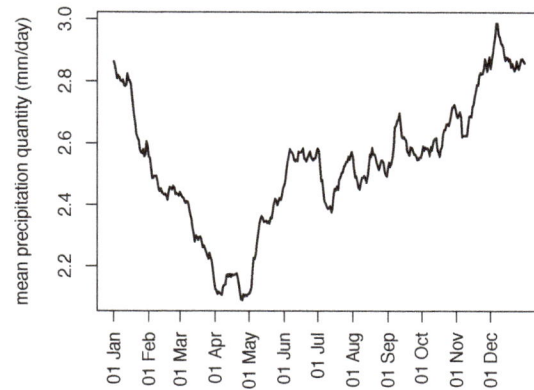

Figure 6. Annual cycle of the Belgian spatial average of the 60-days mean rainfall.

the method AI_{KED1} thanks to the correlation with the topography (Baraque de Fraiture is located at 652 m a.s.l. (above sea level) and is the third highest point of Belgium). The same is true for AI_{KED2}, which integrates topographic data as well as measurements from a rain gauge operated by WHS since 2005 in Baraque de Fraiture. This comparative analysis against the 1833–1975 climate maps favors thus the methods that integrate ancillary information to estimate the spatial distribution of the mean annual rainfall.

4 Discussion

The results presented in the previous section indicates that the best mapping performance for both the annual and the monthly average precipitation quantities is obtained with the method AI_{KED2} based on all time series with a minimal relative length γ between 0.3 and 0.5. This method with $\gamma = 0.3$ is therefore selected to derive a large panel of 1981–2010 climate maps of Belgium related to precipitation: maps of the mean rainfall per year (Fig. 5), season and month, as well as maps of related climate indices such as the average number of days per year, season or month with daily precipitation accumulation exceeding a given threshold (e.g. 1 and 10 mm).

As illustrated in Fig. 5, the mean annual rainfall varies in Belgium from 700 to 1400 mm. The largest values are observed in the east and south parts of Belgium. As expected, a local maximum is present in the area of Baraque de Fraiture. Figure 6, which represents the Belgian spatial average computed for each of the 60-days mean rainfall map derived in a sliding window approach, provides an idea of the average seasonal cycle of rainfall in Belgium. The mean rainfall reaches its minimum in April and its maximum in December. The amplitude of this annual cycle is of the order of 14 % with respect to its mean value.

Finally, the three dominant components resulting from the PCA decomposition of the set of 60-days mean rainfall maps are illustrated in Fig. 7 with the corresponding loadings (nor-

malized to unit norm vectors) and fractions of explained variance. These results have to be considered in a relative manner only, both spatially and temporally, e.g. locations and periods of time can be compared against each other but the absolute level can not be directly interpreted. The first principal component indicates that the southern part of Belgium receives more precipitation in winter and less in summer when compared to the rest of the country. Similarly, the North-West (coastal area) is subject to lower precipitation quantities in spring and larger ones in autumn, as highlighted by the second principal component. Since the first two components account for more than 90 % of the variability, the interpretation of the other components is less significant.

5 Conclusions

In order to update the precipitation climate maps of Belgium with respect to the reference period 1981–2010, a study was conducted to evaluate several mapping approaches. The "interpolation–aggregation" method is able to integrate the entire database of daily observations but requires a spatial interpolation at the daily time scale. On the other hand, the "aggregation-interpolation" approaches perform a spatial interpolation of climate mean values, which is less prone to uncertainties and can furthermore integrate ancillary information, such as terrain elevation data. The benefit of these ancillary topographic information has been highlighted in the study. The observations made within the ancillary network of rain gauges operated by WHS furthermore improves the mapping performance, although this network is operational since 2005 only. However, the integration of incomplete time series has the strongest impact on the performance of the "aggregation-interpolation" methods. The approach selected to derive the precipitation climate maps of Belgium is thus a geostatistical "aggregation-interpolation" method involving all time series complete at 30 % with ancillary terrain elevation and WHS data. In addition to the classical maps about

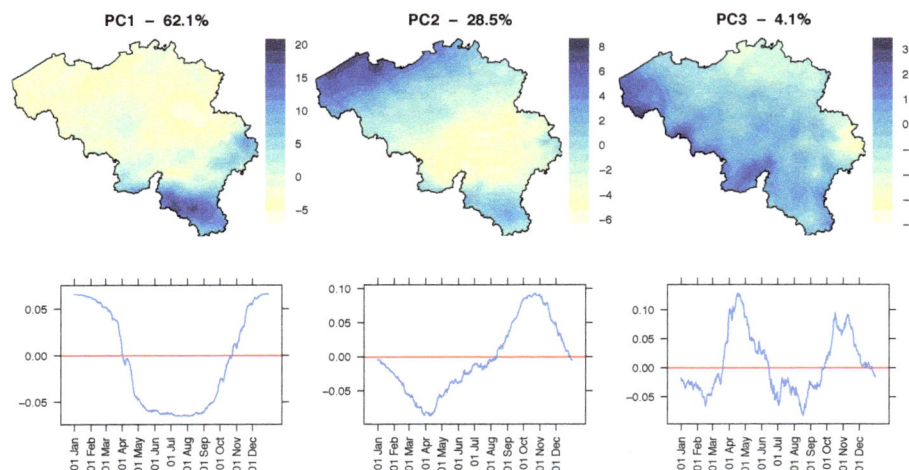

Figure 7. Illustation of the three dominant principal components of the set of 60-days mean rainfall maps with the corresponding fractions of explained variance and loadings normalized to unit norm vectors.

mean rainfall and average number of precipitation days, a set of maps charaterizing the spatial variability of the mean seasonal cycle of rainfall in Belgium was compiled and analyzed by PCA to highlight the main spatio-temporal patterns of the precipitation regimes in Belgium.

Acknowledgements. The present study was conducted in R (3.1.2) and the gstat package was used for geostatistical interpolation (Pebesma, 2004).

References

Baeriswyl, P. A. and Rebetez, M.: Regionalization of precipitation in Switzerland by means of principal component analysis, Theor. Appl. Climatol., 58, 31–41, 1997.

Benzi, R., Deidda, R., and Marrocu, M.: Characterization of temperature and precipitation fields over Sardinia with principal component analysis and singular spectrum analysis, Int. J. Climatol., 17, 1231–1262, 1997.

Daly, C.: Guidelines for assessing the suitability of spatial climate data sets, Int. J. Climatol., 26, 707–721, 2006.

Daly, C., Halbleib, M., Smith, J., Gibson, W., Doggett, M., Taylor, G., Curtis, J., and Pasteris, P.: Physiographically sensitive mapping of climatological temperature and precipitation across the conterminous United States, Int. J. Climatol., 28, 2031–2064, 2008.

Dupriez, G. L. and Sneyers, R.: Les nouvelles cartes pluviométriques de la Belgique, Publications de l'Institut Royal Météorologique de Belgique, Brussels, Belgium, 1979.

Frei, C. and Schär, C.: A precipitation climatology of the Alps from high-resolution rain-gauge observations, Int. J. Climatol., 18, 873–900, 1998.

Goovaerts, P.: Geostatistical approaches for incorporating elevation into the spatial interpolation of rainfall, J. Hydrol., 228, 113–129, 2000.

Pebesma, E.J.: Multivariable geostatistics in S: the gstat package, Comput. Geosci., 30, 683–691, 2004.

Sluiter, R.: Interpolation Methods for the Climate Atlas, KNMI technical rapport TR–335, Royal Netherlands Meteorological Institute, De Bilt, 1–71, 2012.

Sneyers, R., Vandiepenbeeck, M., and Vanlierde, R.: Principal component analysis of Belgian rainfall, Theor. Appl. Climatol., 39, 199–204, 1989.

Tait, A. and Zheng, X.: Analysis of the Spatial Interpolation Error associated with Maps of Median Annual Climate Variables, National Institute of Water & Atmospheric Research (NIWA), Wellington, New Zealand, 1–21, 2007.

Wackernagel, H.: Multivariate geostatistics: an introduction with applications, Springer-Verlag, Berlin, 1995.

WMO – World Meteorological Organization: CCI-16 Session report, WMO No. 1137, Geneva, Switzerland, 2014.

4

Comparison of HOMER and ACMANT homogenization methods using a central Pyrenees temperature dataset

N. Pérez-Zanón, J. Sigró, P. Domonkos, and L. Ashcroft

Center for Climate Change (C3), Campus Terres de l'Ebre, Universitat Rovira i Virgili, Tortosa, Spain

Correspondence to: N. Pérez-Zanón (nuria.perez@urv.cat)

Abstract. The aim of this research is to compare the results of two modern multiple break point homogenization methods, namely ACMANT and HOMER, over a Pyrenees temperature dataset in order to detect differences between their outputs which can affect future studies. Both methods are applied to a dataset of 44 monthly maximum and minimum temperature series placed around central Pyrenees and covering the 1910–2013 period. The results indicate that the automatic method ACMANT produces credible results. While HOMER detects more breaks supported by metadata, this method is also more dependent on the user skill and thus sensitive to subjective errors.

1 Introduction

The latest report of the Intergovernmental Panel on Climate Change indicates that the Mediterranean region is one of the most vulnerable areas of the Earth to global warming (Barros et al., 2014).

The Pyrenees in southwestern Europe is a particularly valuable mountain range because of its biodiversity and water resources which allow population development through agriculture, hydropower energy and tourism (López-Moreno et al., 2008). Future climate scenarios indicate that a decrease in snow cover in this region is likely during the next century (López-Moreno et al., 2009). Therefore it is essentially important to learn and understand more precisely the climate and climate change of this important area.

For any type of climate analysis, including paleoclimatology studies, which use climate series for calibration of the proxies (Bradley, 1999), the use of a high-quality observed dataset is essential. There are several reasons why inhomogeneities (changes in the meteorological records due to non-climatic factors) occur in observational series, such as stations relocation, changes in the environment around the station, changes in the observing time (Aguilar et al., 2003; Brunet et al., 2008).

To detect and correct inhomogeneities, a large number of methods have been developed (Venema et al., 2012). In the present study, two modern multiple break point ho-

mogenization methods developed during the Action COST-ES0601 (HOME), namely HOMER (HOMogenization software in R, Mestre et al., 2013) and ACMANT (Adapted Caussinus-Mestre Algorithm for Networks of Temperature series, Domonkos, 2011b), are used. We considered it important to select methods that treat multiple occurences of inhomogeneities with adequate statistical tools, as observed temperature series usually contain 5 or more inhomogeneities per 100 years on average (Domonkos, 2011a; Venema et al., 2012; Willett et al., 2014). While ACMANT is fully automatic, and thus convenient to use for large datasets, the use of HOMER can include the consideration of metadata (document information about the geographical and technical evolution of the observations) because HOMER is an interactive method.

Previous studies in the Pyrenees have developed homogenized series for different spatial and temporal coverage: Bücher and Dessens (1991) examined one station using a bivariate test to detect systematic changes in the mean, while Esteban et al. (2012) used HOMER to homogenize three stations in Andorra. Cuadrat et al. (2013) have recently produced a homogenized dataset for the whole Pyrenees but only for the 1950–2010 period.

For this study, 123 series of either automatic or manual observations were gathered around the Pyrenees covering 1910–2013. Section 2 describes the data in more detail, while

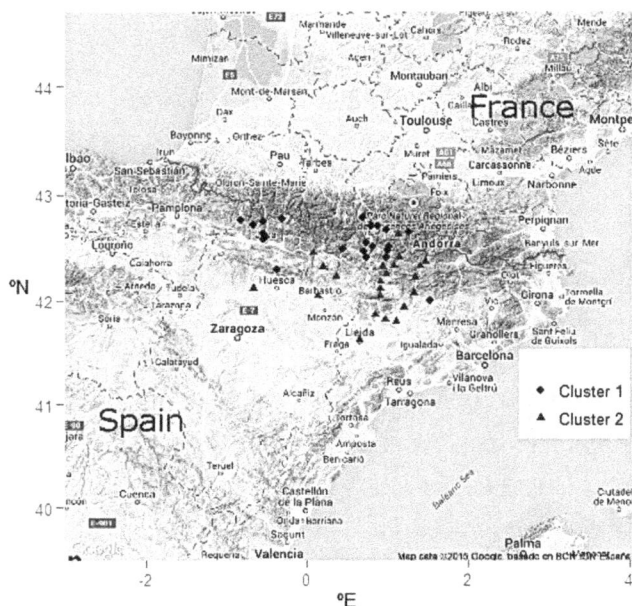

Figure 1. Location of the study area and stations in cluster 1 (black circles) and cluster 2 (grey triangles) using RGoogleMaps package (Kilibarda and Bajat, 2012).

Sect. 3 presents the homogenization methods employed. In Sect. 4, we show the comparison between the two homogenization techniques before drawing conclusions in Sect. 5.

2 Data

The Pyrenees is a mountainous region located in the southwest of Europe and influenced by Atlantic and Mediterranean climate features (e.g. López-Moreno et al., 2008). For this region, 123 series were obtained from the Spanish National Meteorology Agency (AEMET) and the Catalonia Meteorological Service (SMC). The series are of various lengths and cover the period 1910–2013. To develop long-term series, short periods of neighbouring observations were merged and the date of the combination was stored as metadata. The maximum distance accepted in combinations was 7.7 km (in Boí), while the mean difference of altitude was 147 m due to the complexity of the terrain.

Internal consistencies and temporal coherency quality control tests were run on the daily data, following Brunet et al. (2008). More than 834 000 values for each variable (maximum and minimum daily temperature) were examined, with 61 values corrected and 573 set to missing.

Monthly averages of maximum (T_{max}), minimum (T_{min}) and mean temperature (T_{mean}), calculated as the average of T_{max} and T_{min}, were calculated with the missing data tolerance that no more than 7 non-consecutive or 5 consecutive missing daily values for a month were allowed.

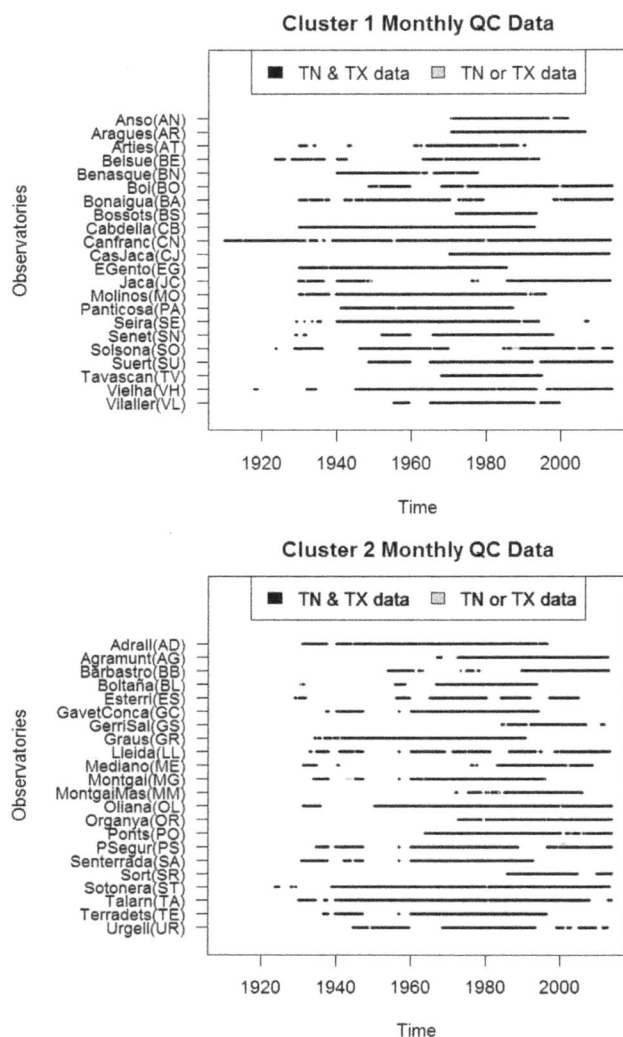

Figure 2. The temporal coverage of monthly maximum (T_X) and minimum (T_N) temperature data for each climatological cluster.

Monthly data must satisfy minimum requirements in terms of the length and completeness to run both HOMER and AC-MANT. Finally, 44 T_{min} and 44 T_{max} series are included in the dataset prepared for the homogenization.

Next, the time series were sorted into two climatological clusters, applying a cluster analysis on monthly T_{mean} (see Sect. 3.1). Each cluster contained 22 series. For each cluster, monthly quality control was applied to detect additional outliers using the Fast QC routine included in HOMER. A total of 13 monthly values were removed from Cluster 1 and 34 from Cluster 2. The obtained series will be referred as QC data and the place and period of data coverage for each station is shown in Figs. 1 and 2.

To perform the comparison between the homogenization methods, only the periods homogenized by ACMANT (varies according to series) are taken into account (see

Cluster Dendrogram

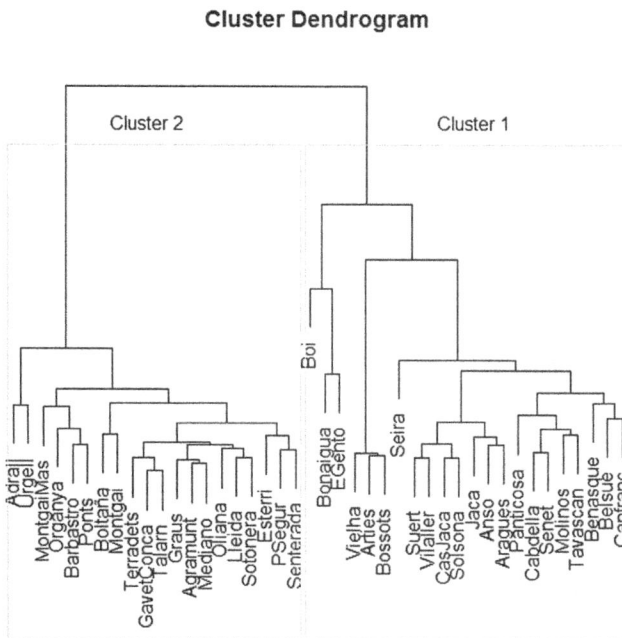

Figure 3. Result of the hierarchical cluster analysis on the first step of computation (see Sect. 3.1).

Sect. 3.3). For these periods, the percentages of monthly missing values ranges between 0.5 and 44 % and 31 series have less than 20 % of missing values.

3 Methodology

3.1 Cluster analysis

As HOMER is interactive software, in order to make the homogenization process more easily manageable, the series were split into two groups. Cluster analysis was applied on T_{mean} data, and not T_{max} or T_{min}, to reduce the likelihood of simultaneous breaks being clustered together. Euclidean distance was calculated between T_{mean} series after normalization (mean $= 0$; standard deviation $= 1$). A hierarchical cluster analysis was applied using Ward agglomeration method, where the analysis starts with as many clusters as the number of time series and the clusters are built by adding T_{mean} series of the least distance in each step (Wilks, 2011).

The Pyrenees series cover different sections of the period 1910–2013, and two pairs of series (EGento and Benasque, with Sort and GerriSal) do not have overlapping sections (see Fig. 2) making cluster analysis impossible. For this reason, the cluster analysis had to be performed twice: first, without the two non-overlapping series to define the two clusters, and second, changing the pairs of series, to include the series that were excluded in the first step to determine to which cluster they belong. The result of cluster analysis for the first step is shown in Fig. 3, and the excluded series were assigned to cluster 2.

3.2 Homogenization methods

Two modern homogenization methods, HOMER (Mestre et al., 2013) and ACMANT (Domonkos, 2011b, 2014) were applied. Both methods were developed during the HOME Cost Action (Venema et al., 2012). HOMER was designed to include the best segments and features of some other state-of-the-art methods: PRODIGE (Caussinus and Mestre, 2004), ACMANT and Joint Detection (Picard et al., 2011).

ACMANT can be applied on both daily and monthly data, but as HOMER works only with monthly data, we ran both methods on monthly data. The two methods have several similarities: both methods are based on the optimal step function fitting (Hawkins, 1972) with the Caussinus Lyazrhi criterion (Caussinus and Lyazrhi, 1997) for optimizing the number of steps (also referred to as "breaks"). Both methods also include the bivariate detection for shifts in the annual means and the summer–winter differences (Domonkos, 2011b), and the minimization of the residual variance (ANOVA, Caussinus and Mestre, 2004) in finding the optimal adjustment terms.

On the other hand, the two methods differ in several other aspects: while HOMER implements a pairwise comparison and a network-wide harmonization in the break detection, ACMANT uses weighted reference time series. A new feature of the most recent ACMANT version (Domonkos, 2014) is that it can also detect relatively short-term inhomogeneities, which are known to be important for long-term data quality (Domonkos, 2011a).

As ACMANT is fully automatic, it can easily be applied to large datasets, while the interactive HOMER allows human intervention to the homogenization procedure and it is possible to decide about the significance of indicated breaks, based on metadata or research experience (Mestre et al., 2013).

HOMER was run comparing all the stations from each cluster separately with annual and seasonal detection, while using ACMANT, again all the time series within clusters were used together and the outlier filtering "off" option was selected, as the input dataset had been quality controlled earlier. HOMER is a user dependent method, and the main way of running the program can be summarised in three steps. First, big break points are identified and corrected. The second step is to repeat the detection in order to evaluate which break points are identified by metadata, and detect those breaks which have smaller amplitude than previously corrected. Finally, annual series are compared by plotting QC data with the homogenized series output. These plots (not shown) allow the user to understand and review the corrections applied. For detection, the three available methods (pairwise detection, joint-segmentation method and ACMANT detection) were considered.

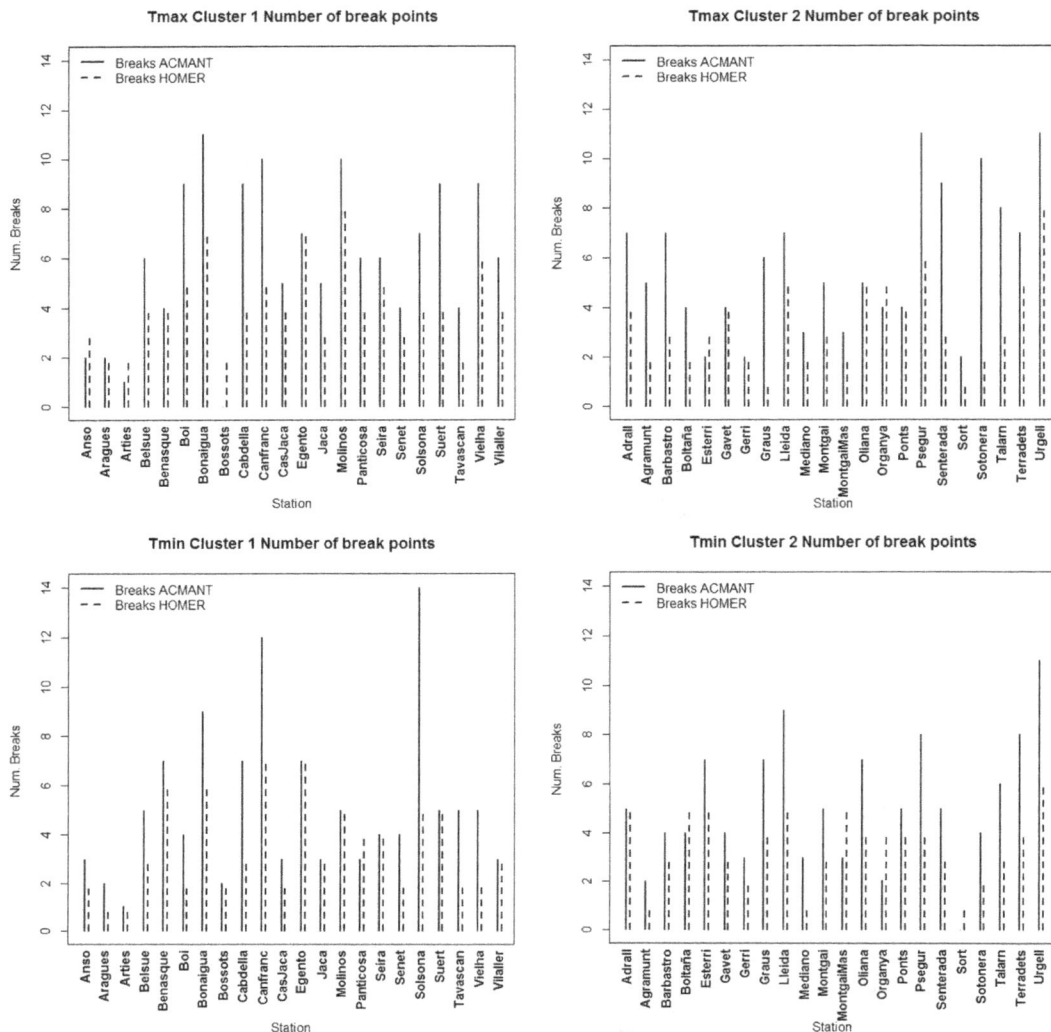

Figure 4. Number of break points detected by ACMANT (continuous line) and HOMER (dashed line) for each station for T_{max} (top panels) and T_{min} (bottom panels) in cluster 1 (left panels) and cluster 2 (right panels).

3.3 Comparison methods

Only periods with data in the QC dataset are considered to compare the output of the two homogenization methods. The period of examination was determined by ACMANT, because ACMANT needs at least 4 spatially comparable series for each section of the homogenization period and this minimum condition is stricter than that of HOMER.

The number of breaks detected by HOMER and AC-MANT, spatial connections of homogenized data, as well as trend slopes of homogenized series were analysed. Spatial connections were examined using Spearman Correlation Coefficients (SCC) calculated between all pairs of monthly series for QC data, data homogenized by HOMER and data homogenized by ACMANT in each cluster. The trend analysis was calculated for the period 1961–1990 in all those stations with more than 80 % of the monthly data available during

this period. 12 series met with this condition in each cluster. This analysis was performed by linear regression on the annual series including years in which all monthly values were available. Significance of trends was evaluated using the Student's t test ($p < 0.05$) (Wilks, 2011).

4 Results

4.1 Break point analysis

HOMER detected at least 1 break in all series, while AC-MANT did not detect any break in two of the 88. However, the maximum number of breaks detected with AC-MANT (14) was much higher than that with HOMER (8) as shown in Fig. 4. In 90% of the series ACMANT detected more breaks than HOMER. The average difference was 2 more breaks per series using ACMANT, which de-

Table 1. Slope trends in °C decade^{-1} for the period 1961–1990 for each station with more than 80 % of monthly data in this period. Significant trends (evaluated using the Student's t test with a significance level of $p < 0.05$) are shown in bold.

Cluster	Station	QC-T_{max} (°C dec^{-1})	HO-T_{max} (°C dec^{-1})	AC-T_{max} (°C dec^{-1})	QC-T_{min} (°C dec^{-1})	HO-T_{min} (°C dec^{-1})	AC-T_{min} (°C dec^{-1})
1	Arties	−0.20	0.01	0.27	−0.25	−0.22	0.00
1	Belsue	**0.94**	0.21	**0.68**	**0.64**	0.00	0.16
1	Cabdella	−0.17	0.06	**0.52**	0.00	−0.09	0.09
1	Canfranc	**0.82**	0.04	**0.59**	**0.88**	−0.03	0.15
1	Egento	**0.68**	−0.24	0.18	**1.75**	−0.27	0.03
1	Molinos	0.31	0.05	0.30	−0.10	−0.01	0.20
1	Panticosa	−0.16	0.05	**0.57**	−0.41	−0.13	0.15
1	Seira	0.17	−0.27	**0.44**	0.12	**−0.25**	0.00
1	Senet	0.24	0.15	**0.61**	−0.25	0.04	**0.45**
1	Suert	**0.70**	0.19	**0.75**	−0.21	0.06	0.20
1	Vielha	0.17	0.40	**0.63**	**0.26**	**0.25**	**0.26**
1	Vilaller	**1.21**	**0.39**	**0.77**	−0.01	0.06	0.16
2	Adrall	**1.60**	0.18	**0.32**	**1.25**	−0.01	−0.08
2	Boltaña	**1.04**	0.21	0.19	−0.27	0.24	−0.10
2	Gavet	**0.65**	0.23	**0.32**	−0.27	0.08	−0.12
2	Graus	−0.05	−0.05	0.28	0.29	0.12	−0.15
2	Montgai	**−1.06**	0.21	**0.32**	0.20	0.20	−0.01
2	Oliana	**0.59**	0.15	0.28	**1.15**	0.12	−0.11
2	Ponts	**1.32**	0.27	**0.47**	**−0.73**	**0.38**	0.21
2	Psegur	0.31	−0.01	0.21	**−0.83**	−0.21	−0.22
2	Senterada	**1.54**	0.25	**0.33**	−0.13	0.00	−0.06
2	Sotonera	−0.12	0.11	**0.37**	**0.26**	0.16	−0.05
2	Talarn	−0.20	0.12	**0.37**	−0.01	0.00	−0.01
2	Terradets	0.09	0.09	**0.35**	0.07	0.02	−0.09

tected 5.5 break points per series on average (10 break points per century), than with HOMER, which detected 3.6 (7 break points per century).

To evaluate the similarity between the results obtained by these homogenization methods, break points per year for each station were compared, considering that the timing of a break point can differ by up to 1 year between both methods due to the difference in the rest of break points. For T_{max} in cluster 1 (2), ACMANT detected 143 (126) break points of which 59 (51) were also detected by HOMER. For T_{min} in cluster 1 (2), ACMANT detected 113 (117) break points of which 49 (42) were also detected by HOMER.

As HOMER is an interactive method that allows the user to introduce known break points, all dates identified in the metadata were included. However, some were removed during the homogenization procedure because the magnitude of the break was less than 0.05 °C and their presence didn't show an improvement in the correction of the series compared with the QC data. From the 8 (15) metadata-supported breaks stored for cluster 1 (2), ACMANT detected 2 (4) break points for T_{min} and 4 (6) T_{max} (all of them detected also by HOMER), while HOMER detected 7 (12) for T_{min} and 6 (13) for T_{max}.

4.2 SCC comparison

SCC values are useful indicators to visualize the temporal linear relationship between time series before and after homogenization (Freitas et al., 2013), and in indicating the presence of large inhomogeneities when they exist. In general, the variance of the SCC for the series in cluster 1 was greater than that for cluster 2. The minimum correlation value for the first cluster was 0.65 while for the second cluster the minimum was 0.90, as shown in Figs. 5 and 6. One reason for this may be a large error in the Vielha observatory data (cluster 1) that was detected after the homogenization process: from 2004 to 2007 the seasonal cycle of temperature seems to be inverted or lagged by a few months for both T_{max} and T_{min} (Fig. 7), although the origin of this error is unknown. This error was detected and corrected adequately only for T_{max} with ACMANT. With HOMER, we failed to be recognized due to the smoothness of the annual values. This type of oversight could be avoided using the CLIMATOL QC check that is also included in HOMER. In ACMANT homogenization of T_{min} the seasonal cycle error remained untouched, while using HOMER, it was even propagated to earlier sections of the series.

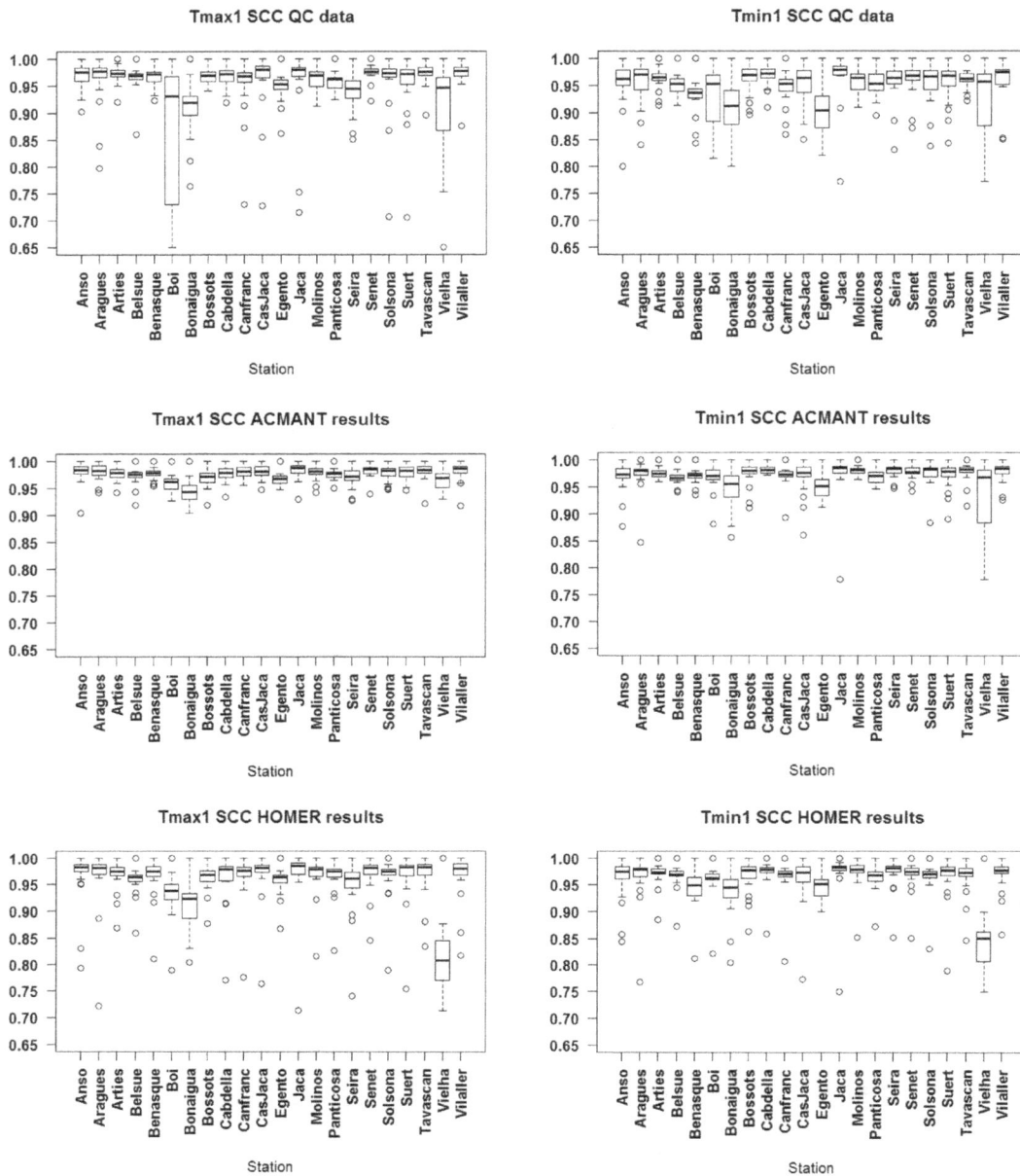

Figure 5. Boxplots of Spearman Correlation Coefficients of QC data (top panels), and ACMANT (middle panels) and HOMER (bottom panels) homogenized data for maximum (left panels) and minimum (right panels) temperature for stations in cluster 1.

4.3 Trend analysis

After homogenization, spatial gradients of trend slopes became smaller, and the number of significant positive trends was reduced as shown in Table 1.

For T_{max}, all 5 significant trends of the 12 series in cluster 1 were positive in the QC data. After homogenization with HOMER, the number of significant and positive trends decreased to 1, while with ACMANT it increased to 9. For cluster 2, 6 of the 12 series had positive and significant trends in the QC data. HOMER didn't return any significant trends

for this cluster, while with ACMANT 8 significant positive trends were obtained. None of the homogenization methods returns negative and significant trends for T_{max}. For T_{min} of cluster 1, only 4 positive significant trends occurred in the QC data. HOMER returned 1 positive and 1 negative significant trends, while the ACMANT homogenized series produced 2 positive and zero negative trends. In cluster 2, the QC data presented 3 positive and 2 negative significant trends. After homogenizing, HOMER kept 1 positive but zero negative significant trends, while with ACMANT all trends were not significant.

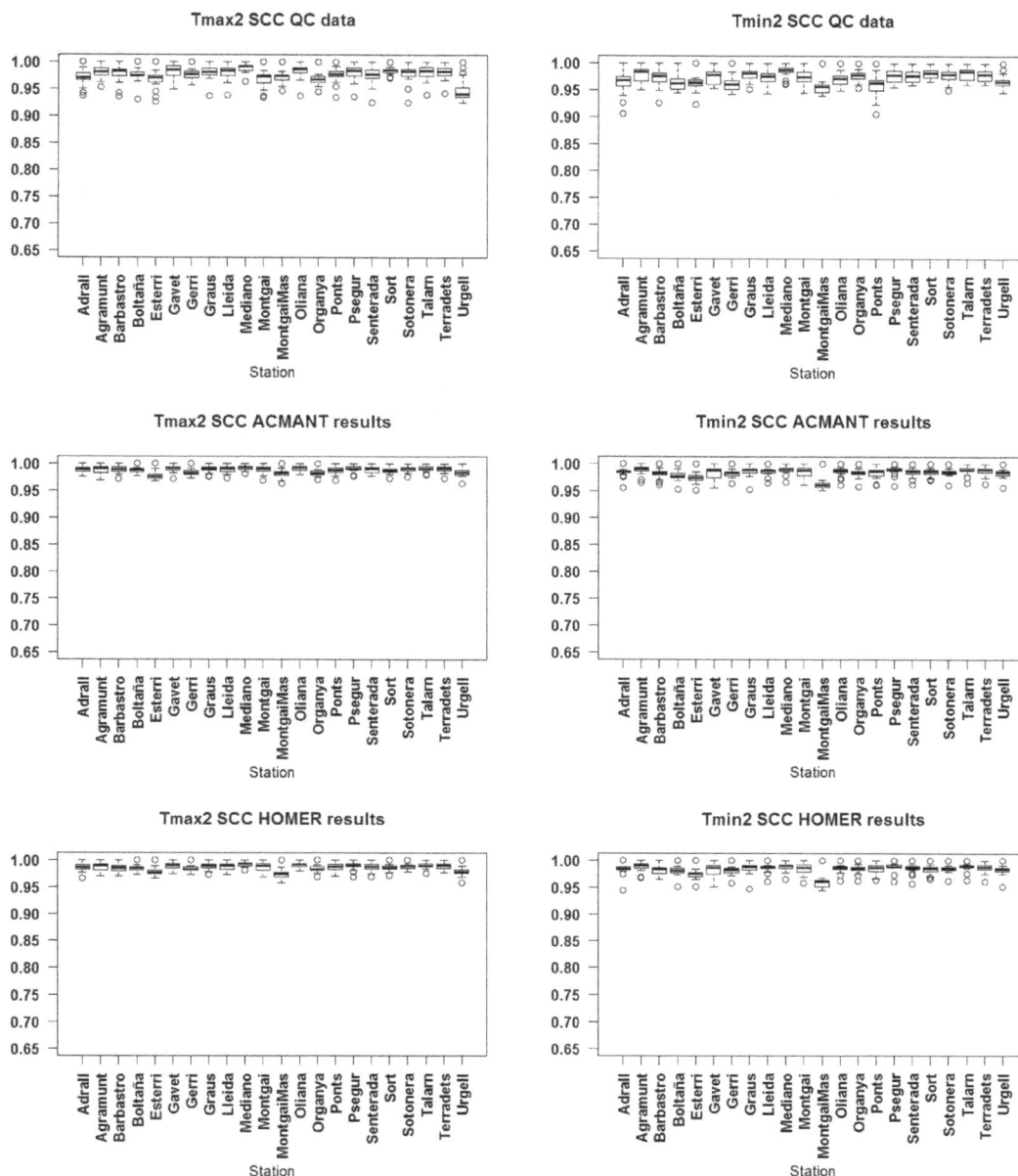

Figure 6. Boxplots of Spearman Correlation Coefficients of QC data (top panels), ACMANT (middle panels) and HOMER (bottom panels) homogenized data for maximum (left panels) and minimum (right panels) temperature for stations in cluster 2.

The relatively large differences between the HOMER homogenization results and ACMANT homogenization results in the mean $T_{\max}$ trends and the number of significant positive trends in $T_{\max}$ series are unexpected results and their origin requires further analysis.

5 Discussion and conclusions

ACMANT and HOMER are two modern, partly similar, multiple break point homogenization methods, but they have distinct strengths and weaknesses. While automatic methods such as ACMANT are easy to use for large datasets, human intervention and the consideration of metadata is possible only with interactive methods like HOMER.

In this case study of Pyrenees temperatures, ACMANT detected and corrected more breaks than HOMER, which is in agreement with the developed sensitivity of ACMANT to detect short-term biases. Concerning breaks justified by metadata, HOMER detected a larger number than ACMANT, showing the advantage of using interactive homogenization methods. Note however, that one cannot conclude on the accuracy of methods from the number of detected breaks, since

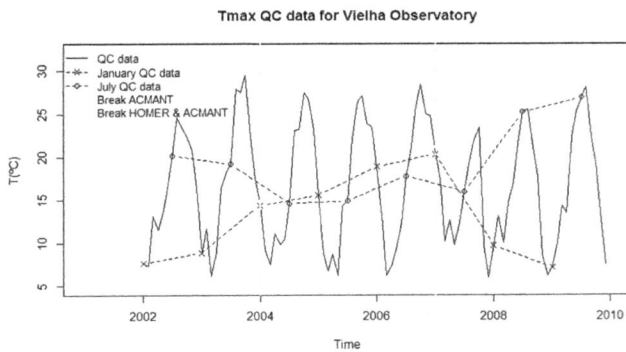

Figure 7. Vielha QC series for T_{max} from 2002 to 2009 continuous line), January data (dashed line and asterisks) and July data (dashed line and circles). Detected breaks of ACMANT (HOMER and AC-MANT) are indicated by gray continuous (dashed) vertical lines.

we do not know the number and exact position and size of breaks in the observed dataset. Detailed evaluation of efficiencies requires the use of artificially developed benchmark datasets.

We have identified a serious error in Vielha T_{min} and T_{max} series, which spectacularly affected the SCC values in cluster 1. This error was corrected well with ACMANT for T_{max}, but not for T_{min}, since the control of seasonal changes is included only in the homogenization of T_{max} with AC-MANT. Concerning the homogenization with HOMER, the program outputs indicated the error, but the indications were left out of consideration, due to the smoothness of the annual means. This rare error in the data handling of Vielha time series points to the necessity of applying a thorough, multi-functional data quality control, since ideally, homogenization procedures should be applied on datasets that are free from such large errors.

In this case study, average trends for all stations in the period 1961–1990 for T_{max} using HOMER (AC-MANT) are $0.12\,°C\,decade^{-1}$ ($0.43\,°C\,decade^{-1}$) and, for T_{min}, using HOMER (ACMANT) the average trend slope is $0.03\,°C\,decade^{-1}$ ($0.05\,°C\,decade^{-1}$). Comparing these trends with other studies of homogenized Pyrenees temperature data reveals mixed results. A single-station study of annual T_{max} and T_{min} for 1882–1970 (Bücher and Dessens, 1991) described the opposite to what is found here, with a negative trend identified for T_{max} and a positive for T_{min}. Previous homogenization of three stations using HOMER (Esteban et al., 2012) showed significant and positive trends on annual T_{max} for all stations, although no significant trend was identified for annual T_{min} in the period 1935–2008; during the shorter 1950–2008 period however, T_{max} and T_{min} were found to be positive and significant in all of the three stations (Esteban et al., 2012). Finally, a study of annual T_{mean} for the Pyrenees over 1950–2010 showed an increase of $0.2\,°C\,decade^{-1}$ (Cuadrat et al., 2013). Three factors can explain these differences in the detected trends. First, the

different time periods in focus; second, the homogenization methods applied, and third, the differences in the number and geographical distribution of stations.

In conclusion, the high SCC results achieved indicate that the homogenization was generally successful with both HOMER and ACMANT, although the difference in T_{max} trend slopes and particularly the handling of Vielha error points to the need of further methodological analysis.

Acknowledgements. This work was supported by the FP7-SPACE-2013-1 project (Uncertainties in Ensembles of Regional Reanalyses, UERRA) and the MONTCORTES project (CGL2012-33665). We grateful to three anonymous reviewers for comments that helped improve this manuscript.

References

Aguilar, E., Auer, I., Brunet, M., Peterson, T. C., and Wieringa, J.: Guidelines on climate metadata and homogenization, WCDMP Report No. 53, WMO-TD 1186, World Meteorological Organization, Geneva, 2003.

Barros, V. R., Field, C. B., Dokken, D. J., Mastrandrea, M. D., Mach, K. J., Bilir, T. E., Chatterjee, M., Ebi, K. L., Estrada, Y. O., Genova, R. C., Girma, B., Kissel, E. S., Levy, A. N., MacCracken, S., Mastrandrea, P. R., and White, L. L. (Eds.): IPCC 2014: Climate Change 2014: Impacts, Adaptation, and Vulnerability. Part B: Regional Aspects, in: Contribution of Working Group II to the Fifth Assessment Report of the Intergovernmental Panel on Climate Change Cambridge University Press, Cambridge, UK and New York, NY, USA, 2014.

Bradley, R. S.: Paleoclimatology: reconstructing climates of the Quaternary, Academic Press, University of Massachusetts, Amherst, USA, 1999.

Brunet, M., Saladié, O., Jones, P., Sigró, J., Aguilar, E., Moberg, A., Lister, D., Whalter, A., and Almarza, C.: A case-study/guidance on the development of long-term daily adjusted temperature datasets, World Meteorological Organization WCDMP Report No. 66, WMO-TD-1425, World Meteorological Organization, Geneva, 2008.

Bücher, A. and Dessens, J.: Secular Trend of Surface Temperature at an Elevated Observatory in the Pyrenees, J. Climate, 4, 859–868, 1991.

Caussinus, H. and Lyazrhi, F.: Choosing a linear model with a random number of change-points and outliers, Ann. Inst. Statist. Math., 49, 761–775, 1997.

Caussinus, H. and Mestre, O.: Detection and correction of artificial shifts in climate series, J. Roy. Stat. Soc. C, 53, 405–425, 2004.

Cuadrat, J. M., Serrano, R., Saz, M.Á., Tejedor, E., Prohom, M., Cunillera, J., Esteban, P., Soubeyroux, J. M., and Deaux, N.: Creación de una base de datos homogeneizada de temperaturas para los Pirineos (1950–2010), Geographicalia, 63–64, 63–74, 2013.

Domonkos, P.: Efficiency evaluation for detecting inhomogeneities by objective homogenisation methods, Theor. Appl. Climatol., 105, 455–467, 2011a.

Domonkos, P.: Adapted Caussinus-Mestre algorithm for networks of temperature series (ACMANT), Int. J. Geosci., 2, 293–309, 2011b.

Domonkos, P.: The ACMANT2 software package, in: Eigth Seminar for Homogenization and Quality Control in Climatological Databases, World Climate Monitoring Program (WCDMP), www.c3.urv.cat/publications/publications.html (last access: 1 March 2015), 2014.

Esteban Vea, P., Prohom Duran, M., and Aguilar E.: Tendencias recientes e índices de cambio climático de la temperatura y la precipitación en Andorra, Pirineos (1935–2008), Pirineos, 167, 87–106, 2012.

Freitas, L., Gonzalez Pereira, M., Caramelo, L., Mendes, M., and Filipe Nunes, L.: Homogeneity of monthly air temperature in Portugal with HOMER and MASH, Idojaras, 117, 69–90, 2013.

Hawkins, D.M.: On the choice of segments in piecewise approximation, J. Inst. Math. Appl., 9, 250–256, 1972.

Kilibarda, M. and Bajat, B.: plotGoogleMaps: the R-based web-mapping tool for thematic spatial data, Geomatica, 66, 37–49, 2012.

López-Moreno, J. I., Goyette, S., and Beniston, M.: Climate change prediction over complex areas: spatial variability of uncertainties and predictions over the Pyrenees from a set of regional climate models, Int. J. Climatol., 28, 1535–1550, 2008.

López-Moreno, J. I., Goyette, S., and Beniston, M.: Impact of climate change on snowpack in the Pyrenees: Horizontal spatial variability and vertical gradients, J. Hydrol., 374, 384–396, 2009.

Mestre, O., Domonkos, P., Picard, F., Auer, I., Robin, S., Lebarbier, E., Böhm, R., Aguilar, E., Guijarro, J., Vertachnik, G., Klancar, M., Dubuisson, B., and Stepanek, P.: HOMER: a homogenization software – methods and applications, Idojaras, 117, 47–67, 2013.

Picard, F., Lebarbier, E., Hoebeke, M., Rigaill, G., Thiam, B., and Robin, S.: Joint segmentation, calling, and normalization of multiple CGH profiles, Biostatistics, 12, 413–428, 2011.

Venema, V. K. C., Mestre, O., Aguilar, E., Auer, I., Guijarro, J. A., Domonkos, P., Vertacnik, G., Szentimrey, T., Stepanek, P., Zahradnicek, P., Viarre, J., Müller-Westermeier, G., Lakatos, M., Williams, C. N., Menne, M. J., Fratianni, S., Cheval, S., Klancar, M., Brunetti, M., Gruber, C., Prohom, M., Lisko, T., Esteban, P., and Brandsma, T.: Benchmarking homogenization algorithms for monthly data, Clim. Past, 8, 89–115, 2012, http://www.clim-past.net/8/89/2012/.

Wilks, D. S.: Statistical methods in the atmospheric sciences, Academic Press, Cornell University, Ithaca, New Yourk, USA, 2011.

Willett, K., Williams, C., Jolliffe, I. T., Lund, R., Alexander, L. V., Brönnimann, S., Vincent, L. A., Easterbrook, S., Venema, V. K. C., Berry, D., Warren, R. E., Lopardo, G., Auchmann, R., Aguilar, E., Menne, M. J., Gallagher, C., Hausfather, Z., Thorarinsdottir, T., and Thorne, P. W.: A framework for benchmarking of homogenisation algorithm performance on the global scale, Geosci. Instrum. Method. Data Syst., 3, 187–200, doi:10.5194/gi-3-187-2014, 2014.

5

Large-eddy simulation of turbulent winds during the Fukushima Daiichi Nuclear Power Plant accident by coupling with a meso-scale meteorological simulation model

H. Nakayama[1], **T. Takemi**[2], **and H. Nagai**[1]

[1]Japan Atomic Energy Agency, Ibaraki, Japan
[2]Disaster Prevention Research Institute, Kyoto University, Kyoto, Japan

Correspondence to: H. Nakayama (nakayama.hiromasa@jaea.go.jp)

Abstract. A significant amount of radioactive material was accidentally discharged into the atmosphere from the Fukushima Dai-ichi Nuclear Power Plant from 12 March 2011, which produced high contaminated areas over a wide region in Japan. In conducting regional-scale atmospheric dispersion simulations, the computer-based nuclear emergency response system WSPEEDI-II developed by Japan Atomic Energy Agency was used. Because this system is driven by a meso-scale meteorological (MM) model, it is difficult to reproduce small-scale wind fluctuations due to the effects of local terrain variability and buildings within a nuclear facility that are not explicitly represented in MM models. In this study, we propose a computational approach to couple an LES-based CFD model with a MM model for detailed simulations of turbulent winds with buoyancy effects under real meteorological conditions using turbulent inflow technique. Compared to the simple measurement data, especially, the 10 min averaged wind directions of the LES differ by more than 30 degrees during some period of time. However, distribution patterns of wind speeds, directions, and potential temperature are similar to the MM data. This implies that our coupling technique has potential performance to provide detailed data on contaminated area in the nuclear accidents.

1 Introduction

A significant amount of radioactive material was accidentally discharged into the atmosphere from the Fukushima Dai-ichi Nuclear Power Plant (FDNPP) from 12 March 2011. In this nuclear accident, a computer-based nuclear emergency response system, Worldwide version of System for Prediction of Environmental Emergency Dose Information (WSPEEDI-II) developed by Japan Atomic Energy Agency was used to conduct atmospheric dispersion simulations from regional to hemispheric scales (Katata et al., 2012). This system consists of a meso-scale meteorological (MM) model and Lagrangian particle dispersion model, which can provide near real-time predictions of mean air concentrations and mean surface deposition of radionuclides. However, it is difficult to estimate contaminated areas in a local-scale where turbulent motions are dominant by the influence of local terrain variability and roughness elements such as individual buildings and trees within a nuclear facility that are not explicitly resolved in MM simulation models.

For simulating wind flows and plume dispersion in a local-scale, a computational fluid dynamics (CFD) technique is commonly used. In CFD models, buildings and structures can be explicitly represented at high resolutions. In particular, the CFD simulations using large-eddy simulation (LES) are effective to capture complex behaviors of impinging, separating, and circulating flows around a bluff body. Therefore, an approach to couple an LES-based CFD model with a MM model is expected to have a potential of becoming an effective tool to provide detailed information on turbulent flows and plume dispersion in a local-scale under real meteorological conditions. For example, in the approach by Wyszogrodzki et al. (2012), the MM outputs such as pressure, wind velocity, and potential temperature calculated based on RANS simulations were directly imposed at the inflow boundaries of the LES-based CFD model. Although they provided reasonable results in comparison to field experimental data, the inflow data did not include high-frequency turbulent fluctuations appropriate to drive LES-based CFD models. Michioka et al. (2013) coupled a micro-scale LES-based CFD model with a meso-scale LES-based model and investigated spatial distributions of plume concentrations in a residential area. However, they assumed a neutral atmospheric stability condition in the micro-scale model. The application of their approach is limited to a case building-induced turbulence is dominant.

In Japan, most nuclear facilities are located at coastal complex terrains. In this case, the assumption of neutral stability is less valid and thermal stability effects should be considered. Therefore, it is important to generate thermally-stratified boundary layers with small-scale wind fluctuations in order to more faithfully reproduce meteorological conditions in the LES model from the MM outputs. In this study, we propose a calculation approach to simulation of thermally-stratified boundary layer flows by coupling an LES-based CFD model with a MM model and examine the effectiveness of the approach in comparison to measurement and MM simulation data.

2 Approach to couple between MM and CFD models

2.1 CFD model

The model used for local-scale detailed simulations is the LOHDIM-LES developed by Nakayama et al. (2014). The basic equations are the filtered continuity, Navier-Stokes, and temperature transport equations under the Boussinesq approximation. The subgrid scale (SGS) turbulent effect is represented by the standard Smagorinsky model (1963) with a constant value of 0.1. The SGS scalar flux is also parameterized by an eddy viscosity model and the turbulent Prandtl number is set to a constant value 0.71. The turbulent effects of local terrain and plant canopy are represented by the external force term and are incorporated into the Navier–Stokes equation. The terrain effects are represented by immersed boundary method proposed by Goldstein et al. (1993) as follows;

$$f_{\text{terrain},\,i} = m \int_0^t u_i(t)\,dt + n\,u_i(t)\,m < 0, n < 0, \tag{1}$$

where m and n are negative constants. The stability limit is given by $\Delta t < \frac{-n-\sqrt{(n^2-2mk)}}{m}$ where k is a constant of order 1. The plant canopy effects are expressed as follows;

$$f_{\text{canopy},\,i} = -C_{\text{d}} a(z)\, U \overline{u_i}, \tag{2}$$

where C_{d} is a drag coefficient with a constant value of 0.2. U is a wind speed. $a(z)$ is a plant area density and is determined by the forest leaf area index, thus, $\text{LAI} = \int_0^h a(z)\,dz$, where h is the canopy height.

Nudging terms for wind velocity and temperature fields are incorporated into the Navier–Stokes equation in order to maintain the mean structure of the MM model in the LES computational domain and can be expressed as the follows, respectively;

$$f_{\text{nud\_flow},\,i} = -C_{\text{nud}} \left(u_{\text{MM},\,i} - \overline{u_i} \right) \tag{3}$$

$$f_{\text{nud\_temp}} = -C_{\text{nud}} \left(\theta_{\text{MM}} - \overline{\theta} \right), \tag{4}$$

where $u_{\text{MM},\,i}$ and θ_{MM} are the wind velocity and potential temperature of the MM model, respectively. C_{nud} is a spatially dependent nudging constant (see details in Sect. 3.2).

The coupling algorithm of the velocity and pressure fields is based on the marker-and-cell method with the second-order Adams-Bashforth scheme for time integration. The Poisson equation is solved by the successive over-relaxation method. For the spatial discretization in the basic equations, a second-order accurate central difference scheme is used.

2.2 Generation of turbulent inflows from MM outputs

Mean wind directions are not always constant and often vary due to a change of weather conditions. Therefore, first, we proposed the treatment of the inflow boundary conditions in order to automatically input wind velocity data obtained by the MM model into the LES model under neutral stability conditions, depending on mean wind directions in the meteorological field (Nakayama et al., 2012, 2015). Figure 1 shows a schematic diagram of the treatment of inflow boundary conditions in the LES model. First, mean wind directions in the meteorological field are estimated and vertical planes of inflow boundaries are determined automatically depending on them. For example, when mean wind directions α of a MM model range from -315 to $0°$ or from 0 to $45°$, vertical boundary planes in the west, north, and east sides are automatically set to inflow boundaries and that in the south side is set to an outflow boundary. For the inflow boundaries, the MM outputs linearly interpolated on the spatial resolution of the LES domain. At the same time, only for the north vertical boundary plane, turbulent fluctuations generated by a turbulent inflow technique are added to the mean inflow as shown Fig. 1a.

In order to drive an LES model, a recycling method proposed by Kataoka and Mizuno (2002) is adopted. In this turbulent inflow technique, only fluctuating components are extracted at a recycling station and recycled back to the inlet

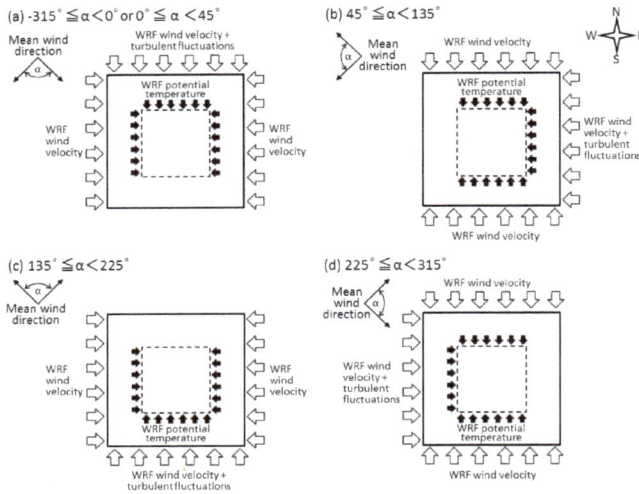

Figure 1. Schematic diagram of the treatment of inflow boundary conditions in the CFD model depending on mean wind directions in the meteorological field. The solid and dotted lines are computational regions to drive turbulent winds and simulate thermally-stratified boundary layers, respectively. The open and filled arrows indicate input of wind velocities and potential temperature obtained by a MM model, respectively.

boundary. The formulation is as follows:

$$u_{\text{inlt}}(y, z, t) = \langle u \rangle_{\text{mean}}(y, z, t) \tag{5}$$
$$+ \varphi(z)\{u_{\text{recy}}(y, z, t) - [u](y, z)\}$$
$$v_{\text{inlt}}(y, z, t) = \langle v \rangle_{\text{mean}}(y, z, t) \tag{6}$$
$$+ \varphi(z)\{v_{recy}(y, z, t) - [v](y, z)\}$$
$$w_{\text{inlt}}(y, z, t) = \langle w \rangle_{\text{mean}}(y, z, t) \tag{7}$$
$$+ \varphi(z)\{w_{\text{recy}}(y, z, t) - [w](y, z)\}.$$

Where u, v, and w are the wind components of the streamwise (x), spanwise (y), and vertical (z) directions, respectively. The suffixes of inlt and recy indicate the instantaneous wind velocity at the inlet and the instantaneous wind velocity at the recycle station, respectively. The recycle station is set at 1 km downstream position from the main inflow boundary. $[u]$, $[v]$, and $[w]$ are horizontally averaged winds over the driver domain and $\varphi(z)$ is a damping function. $\langle u \rangle_{\text{mean}}$, $\langle v \rangle_{\text{mean}}$, and $\langle w \rangle_{\text{mean}}$ are given by mean wind velocities for each component obtained by a MM model. In cases of α ranging from 45 to 135°, from 135 to 225°, and from 225 to 315°, the method to determine inflow and outflow boundaries depending on α is shown in Fig. 1b–d.

In order to produce thermally-stratified boundary layers, vertical profiles of potential temperature data obtained by a MM model are imposed at a distance of 1 km inward from each horizontal boundary. As well as the case for a wind velocity field, vertical planes of inlet and outlet boundaries are automatically determined depending on mean wind directions in the meteorological field.

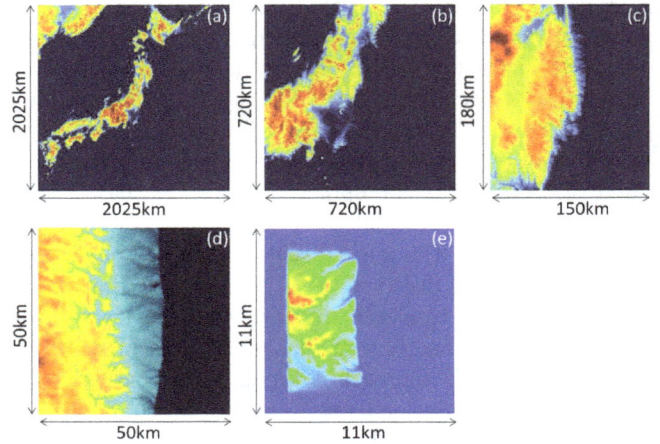

Figure 2. Computational areas of the WRF and LES models. The WRF is configured with for nested domains covering areas of (**a**) 2025 km × 2025 km at 4.5 km grid, (**b**) 720 km × 720 km at 1.5 km grid, (**c**) 150 km × 180 km at 500 m grid, and (**d**) 50 km × 50 km at 100 m grid. The LES model covers an area of (**e**) 11 km × 11 km at 20m grid.

3   Simulation settings

3.1   Meso-scale meteorological simulation

The meso-scale meteorological simulation model used here is the Weather Research and Forecasting (WRF) model, the Advanced Research WRF Version 3.3.1 (Skamarock et al., 2008) to provide the input data for the LES model. We use a nesting capability to resolve the FDNPP region at a fine grid spacing by setting two-way nested, four computational domains (with the top being at the 50 hPa level). The four domains cover areas of 2025 km by 2025 km at 4.5 km grid, 720 km by 720 km at 1.5 km grid, 150 km by 180 km at 500 m grid, and 50 km by 50 km at 100 m grid, respectively (Fig. 2a–d). The number of vertical levels is 53, with 15 levels in the lowest 1 km depth.

The terrain data used are the global 30 s data (GTOPO30) from the US Geological Survey for the outer 2 domains and the 50 m mesh digital elevation model (DEM) dataset by the Geographical Survey Institute (GSI) of Japan for the inner 2 domains. The land-use/land-cover information is obtained from the 100 mesh dataset from the Ministry of Land, Infrastructure, Transport and Tourism of Japan.

To determine the initial and boundary conditions, we use 6-hourly Mesoscale Analysis (MANAL) data of Japan Meteorological Agency (JMA), 6-hourly Final Analysis data of the US National Centers for Environmental Prediction (NCEP FNL), and daily Merged Sea Surface Temperature (MGDSST) analyses of JMA. The times of the 6-hourly MANAL and NCEP FNL are 00:00, 06:00, 12:00, and 18:00 UTC. The horizontal resolutions of MANAL and MGDSST are 10 km and 0.25 degree, respectively. Full physics processes are included in the present simulation in order to reproduce real meteorological phenomena. A physics parameterization closely relevant to the simulation of wind fields is a PBL mixing parameterization. We choose a Mellor-Yamada Level 2.5 scheme of Janjic (2002) in which mixing is done vertically between the adjacent vertical levels. A single-moment, 6-category water- and ice-phase microphysics of Hong and Lim (2006) is employed for cloud and precipitation processes in all the domains.

The case studied is the FDNPP accident on 12 March 2011. In order to simulate wind fields for this event, the time period of the WRF simulation is from 00:00 UTC 11 March 2011 to 00:00 UTC 13 March 2011. The WRF outputs during 05:00 UTC 12 March and 06:00 UTC 12 March are used for the LES model. The simulated outputs of the innermost domain at 1 min interval are used as the inputs of the LES model.

3.2 LES computational conditions

The size of the computational domain is 11.0 km by 11.0 km in the horizontal directions with the depth of 1.6 km (Fig. 2e). The total mesh number is 550 by 550 by 94 nodes. The grid spacing is 20 m in the horizontal directions and 2.5–64 m stretched in the vertical direction based on an orthogonal grid system. In the previous study, we conducted LESs of urban boundary layer flows in the urban central district by coupling with a MM model and showed reasonable results in comparison to the field experimental data of vertical profiles of wind speeds and directions (Nakayama et al., 2015). Those calculation conditions such grid resolution and computational domain were almost the same as the present ones. Therefore, it is considered that the present model set-up is reasonable to reproduce basic characteristics of the meteorological conditions in the LES model. For a wind velocity field, the inlet and spanwise boundaries are determined by the WRF wind velocity data (with 1 min interval and 100 m resolution) linearly interpolated on the spatial resolution of the LES domain with 1 min interval. At the outlet boundary, a free-slip condition is applied for each component of wind velocity. At the upper boundary, a free-slip condition for the horizontal velocity components and zero-speed condition for the vertical velocity component is imposed. For a potential temperature field, the bottom, ground surfaces, and spanwise boundaries are determined by the WRF potential temperature data (with 1 min interval and 100 m resolution) linearly interpolated on the spatial resolution of the LES domain with 1 min interval. At the outlet, a free-slip condition is imposed.

Focusing on the ground surface of the study site shown in Fig. 3a, it is found that the FDNPP is located along the coast and many forest canopies are densely situated over the land. Ground surface geometries are represented using the 50 m mesh DEM of GSI linearly interpolated on the spatial resolution of the LES model. Buffer zones with a length of 1.0 km is set in only land area and roughness blocks are placed in order to represent roughened ground surface as shown in

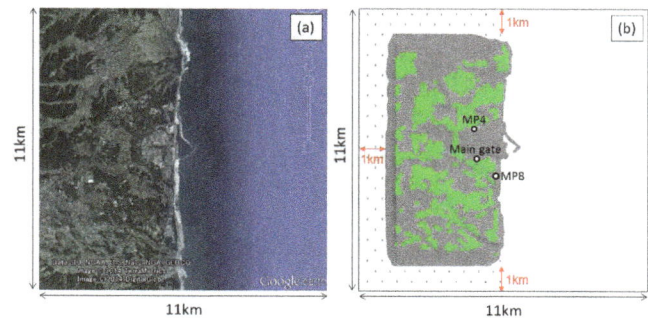

Figure 3. (a) The photograph reproduced by Google^{TM} earth graphic. (b) Configuration of the FDNPP in the LES model. The buffer zone with 1.0 km is set up from each boundary. In the land area of this buffer zone, roughness blocks are placed. Green area indicates forest canopies.

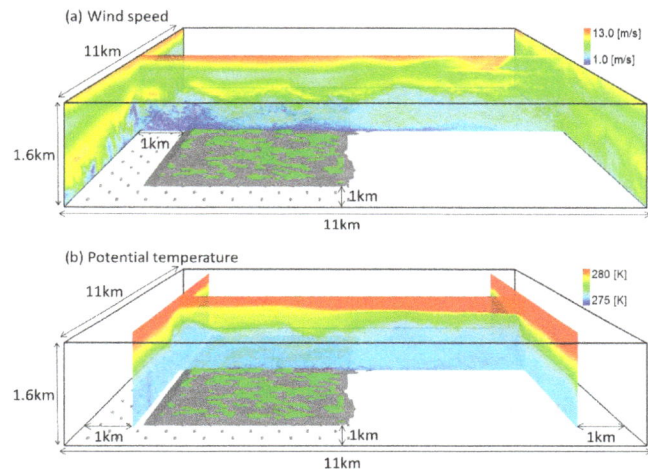

Figure 4. Instantaneous fields of (a) wind speed and (b) potential temperature.

Fig. 3b. The forest canopy is arranged based on the mixed forest in USGS 24-category land use. The LAI and canopy height are set to 4.0 and 12.0 m, respectively.

The nudging constant is set to rapidly vary from 0.0 to $0.01 \, s^{-1}$ across 750 m height using a hyperbolic tangent function for only flow field and decrease from the lateral boundaries toward the inner part using a ten-grid-point buffer zone for both flow and temperature fields. In real meteorological fields, mean wind directions are often largely different between upper and lower parts of boundary layers, which often induces numerical instabilities. Therefore, in case spatially-averaged wind directions at heights greater than 750 m height differ from those at heights less than the height by 30 degrees at the main inlet boundary, the values for each component of wind velocities are set to those at 750 m height. The time step interval is 0.05 s. The simulation period is from 05:00 UTC 12 March 2011 to 06:00 UTC 12 March 2011.

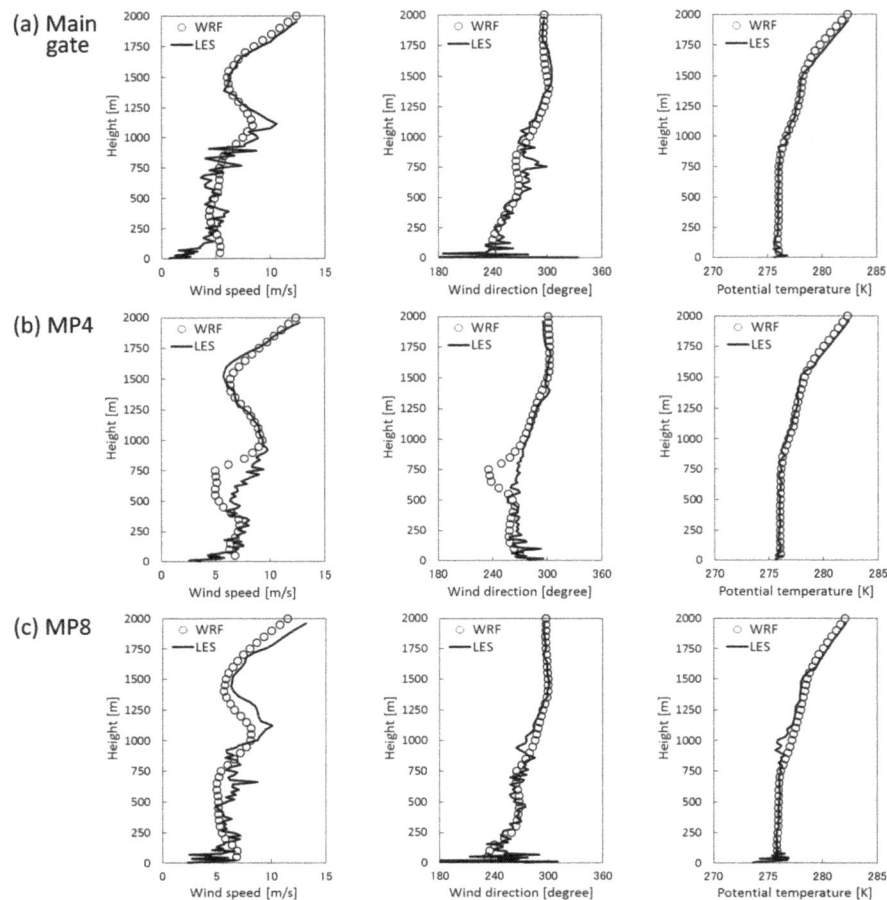

Figure 5. Vertical profiles of wind speed, wind direction, and potential temperature obtained at **(a)** Main gate, **(b)** MP4, and **(c)** MP8 at 06:00 UTC 12 March 2011. The locations of Main gate, MP4, and MP8 are shown in Fig. 3.

4 Results

Figure 4 shows instantaneous fields of (a) wind speed and (b) potential temperature. It is found that small-scale fluctuations in both wind and temperature fields are reproduced by the turbulent inflow technique. Figure 5 compares the LES results with vertical profiles of wind speeds, wind directions, and potential temperature of the WRF model obtained at (a) Main gate, (b) MP4, and (c) MP8 at 06:00 UTC 12 March 2011. The LES wind speeds and directions are found to be generally distributed along the WRF data and considerably fluctuate up to 100 m height at each point due to the turbulent effects by local terrain variability and forest canopy. However, at MP4, locally rapid variations of wind speeds and directions around 750 m are not captured in the LES model. The vertical profiles of potential temperature are similar to those of the WRF model.

Figure 6 compares the LES results with the simple measurement and WRF data of time series of wind speeds and directions. The measurement and LES data are obtained at the ground-level and the WRF data are obtained at a height of

50 m. According to the press release (28 May 2011) by Tokyo Electric Power Company (2011), all monitoring posts at the FDNPP did not work due to blackout caused by the severe earthquake. Therefore, wind speeds, wind directions, and radiation dose were measured by monitoring cars. Although these simple measurement data are not appropriate to evaluate the model performance, we use the data for a comparison. The measurement data of wind speeds vary within the range from 2.7 to 3.5 m s^{-1}. The WRF data considerably exceed the measurement data due to the difference of the measurement height. The LES 10 min values are found to be comparable to the measurement data although the instantaneous values highly fluctuate. The measurement data of wind directions vary within the range from South-southeast to South directions. The WRF wind directions vary around West-southwest direction. The LES instantaneous values highly fluctuate as well as those of wind speeds. The LES 10 min values differ by more than 30 degrees from the measurement data during some period of time.

Although the difference between the LES and measurement data are observed at a ground-level during some period

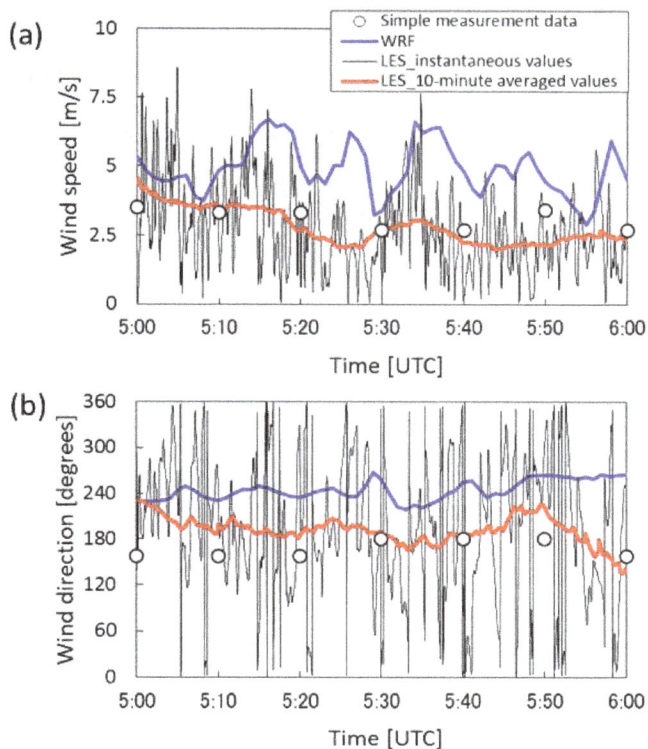

Figure 6. Time series of (**a**) wind speeds and (**b**) wind directions obtained at Main gate from 05:00 to 06:00 UTC 12 March 2011. The observed and LES data are obtained at the ground-level. The WRF data are obtained at the height of 50 m. The location of Main gate is shown in Fig. 3.

of time, it is successful in generally maintaining the simulated meteorological fields as the basic flow with buoyancy effects in the LES domain.

5   Conclusions

We proposed a calculation approach to couple the LES-based CFD model with the MM model for detailed simulations of turbulent winds with buoyancy effects under real meteorological conditions using turbulent inflow technique and examined the effectiveness in comparison to the simple measurement and MM simulation data. Inflow boundary conditions were set to automatically input wind velocity data obtained by the MM model into the LES model, depending on mean wind directions in the meteorological field. In generating thermally-stratified boundary layers from the MM outputs, first, small-scale wind fluctuations were generated at the main inlet boundary by a recycling technique. Then, the potential temperature profiles obtained by the MM model were imposed at the recycle station.

Compared to vertical profiles of the MM simulation data, it is seen that the LES wind speeds, directions, and potential temperature generally fluctuate around the MM data al-

though locally rapid variations of wind speeds and directions are not reproduced well. The 10 min averaged wind directions of the LES differ by more than 30 degrees from those of the measurement data during some period of time although the averaged wind speeds of the LES are in good agreements with them. Important issues still remain in accurately simulating local-scale turbulent winds at a ground-level under real meteorological conditions. However, our approach is successful in generally reproducing thermally-stratified boundary layer flows with turbulent fluctuations in the LES model. It can be concluded that our coupling technique has potential performance to provide detailed data on contaminated area in the nuclear accidents.

Acknowledgements. This study is partly supported by General Collaborative Research #25G-05 provided by Disaster Prevention Research Institute, Kyoto University, is also supported by JSPS KAKENHI Grant 26282107.

References

Goldstein, D., Handler, R., and Sirovich, L.: Modeling a no-slip flow boundary with an external force field, J. Comput. Phys., 105, 354–366, 1993.

Hong, S.-Y. and Lim, J.-O. J.: The WRF single-moment 6-class microphysics scheme (WSM6), J. Kor. Meteorol. Soc., 42, 129–151, 2006.

Janjic, Z. I.: Nonsingular implementation of the Mellor-Yamada level 2.5 scheme in the NCEP Meso model, NCEP Office Note, 437, 61 pp., 2002.

Kataoka, H. and Mizuno, M.: Numerical flow computation around aeroelastic 3D square cylinder using inflow turbulence, Wind Struct., 5, 379–392, 2002.

Katata, G., Ota, M., Terada, H., Chino, M., and Nagai, H.: Atmospheric discharge and dispersion of radionuclides during the Fukushima Dai-ichi Nuclear Power Plant accident. Part I: Source term estimation and local-scale atmospheric dispersion in early phase of the accident, J. Environ. Radioact., 109, 103–113, 2012.

Michioka, T., Sato, A., and Sada, K.: Large-eddy simulation coupled to meso-scale meteorological model for gas dispersion in an urban district, Atmos. Environ., 75, 153–162, 2013.

Nakayama, H., Takemi, T., and Nagai, H.: Large-eddy simulation of urban boundary-layer flows by generating turbulent inflows from mesoscale meteorological simulations, Atmos. Sci. Lett., 13, 180–186, 2012.

Nakayama, H., Leitl, B., Harms, F., and Nagai, H.: Development of local-scale high-resolution atmospheric dispersion model using large-eddy simulation Part 4: turbulent flows and plume dispersion in an actual urban area, J. Nucl. Sci. Technol., 51, 628–638, 2014.

Nakayama, H., Takemi, T., and Nagai, H.: Development of local-scale high-resolution atmospheric dispersion model using large-eddy simulation Part 5: Detailed simulation of turbulent flows and plume dispersion in an actual urban area under real meteorological conditions, submitted to J. Nucl. Soc. Technol., 2015.

Skamarock, W. C., Klemp, J. B., Dudhia, J., Gill, D. O., Barker, D. M., Duda, M. G., Huang, X., Wang, W., and Powers, J. G.: A description of the Advanced Research WRF Version 3, NCAR Tech. Note, NCAR/TN-475+STR, 1 pp., 2008.

Smagorinsky, J.: General circulation experiments with the primitive equations, Mon. Weather Rev., 91, 3, 99–164, 1963.

TEPCO, Tokyo Electric Power Company. Additional Monitoring Data at Fukushima Daiichi Nuclear Power Station, available at: http://www.tepco.co.jp/en/press/corp-com/release/11052811-e.html (last access: 04 June 2015), 2011.

Wyszogrodzki, A., Miao, S., and Chen, F.: Evaluation of the coupling between mesoscale-WRF and LES-EULAG models for simulating fine-scale urban dispersion, Atmos Res., 118, 324–345, 2012.

Quality control of 10-min soil temperatures data at RMI

C. Bertrand, L. González Sotelino, and M. Journée

Royal Meteorological Institute of Belgium, Brussels, Belgium

Correspondence to: C. Bertrand (cedric.bertrand@meteo.be)

Abstract. Soil temperatures at various depths are unique parameters useful to describe both the surface energy processes and regional environmental and climate conditions. To provide soil temperature observation in different regions across Belgium for agricultural management as well as for climate research, soil temperatures are recorded in 13 of the 20 automated weather stations operated by the Royal Meteorological Institute (RMI) of Belgium. At each station, soil temperature can be measured at up to 5 different depths (from 5 to 100 cm) in addition to the bare soil and grass temperature records. Although many methods have been developed to identify erroneous air temperatures, little attention has been paid to quality control of soil temperature data. This contribution describes the newly developed semi-automatic quality control of 10-min soil temperatures data at RMI.

1 Introduction

Of great importance in agriculture, soil temperature affects plant growth directly, e.g. in seed germination, root growth and nutrient uptake as well as indirectly in soil water and gas flow, soil structure and nutrient availability (Hillel, 1998). Soil temperature is an important parameter in energy balance applications such as land surface modeling, numerical weather forecasting, and climate prediction (e.g. Best et al., 2005). It is also important in radiative transfer applications, such as in the retrieval of land surface properties with satellite sensors, and especially in the retrieval of surface moisture with microwave sensors (e.g. de Jeu et al., 2008). Soil temperature varies in response to exchange processes that take place primarily through the soil surface. These effects are propagated into the soil profile by a complex series of transport processes, the rates of which are affected by time-variable and space-variable soil properties. At each succeeding depth, the peak temperature is dampened and shifted progressively in time (see Fig. 1). The degree of damping increases with depth and is related to the thermal properties of the soil and the frequency of the temperature fluctuation. Due to the much higher heat capacity of soil relative to air and the thermal insulation provided by vegetation and surface soil layers, soil heat anomalies of daily or weekly timescales in shallow layers near the surface do not propagate to the deeper layers. Only persistent long-term anomalies (e.g. at the in-

ter annual and decadal scale) affect temperature variations in those layers (e.g. Lachenbruch and Marshall, 1986).

While many methods have been proposed to identify erroneous air temperatures, only little attention has been paid to quality control of soil temperature data. This situation imposes a significant effort toward providing quality control and assurance (QC/QA) especially devoted to soil temperature measurements. Here we propose a semi-automatic quality control method to check 10-min grass and soil temperature records. In developing quality control tests for the automatic weather stations (AWSs) operated by the Royal Meteorological Institute of Belgium (RMI), we adapted some of the tools developed in Hu et al. (2002) and Hu and Feng (2003) to examine daily and hourly soil temperature data and expanded their approach to include additional tests introduced for some of them in quality control of air temperatures at RMI (Bertrand et al., 2013). The framework is similar to Gandin's concept of complex QA (Gandin, 1988), in that it approaches the question of the validity of a given datum from several different angles and considers errors of different types. However, in the present approach, the automated QA functions are included in a larger QA protocol involving manual inspections similarly to the method implemented at RMI for the quality control of 10-min air temperature data (Bertrand et al., 2013).

Table 1. List of the 13 RMI's Automatic Weather Stations performing at least one soil temperature record and the associated measurements.

AWS			Air temperature measurement (s)							QC	Coordinates	
Code	Name	Climate zone	Grass	Bare soil 0 cm	Soil −5 cm	Soil −10 cm	Soil −20 cm	Soil −50 cm	Soil −1 m	group	Lat. (° N)	Lon. (° E)
6414	Beitem	1		X	X	X	X	X		3	50.91	3.12
6431	Zelzate	1		X						5	51.18	3.81
6434	Melle	1	X	X	X	X	X	X		2	50.98	3.83
6438	Stabroek	1	X	X	X	X	X	X		2	51.33	4.37
6439	Sint Katelijn Waver	2	X	X		X	X	X		4	51.08	4.53
6447	Uccle	1	X	X	X	X	X	X		2	50.80	4.36
6455	Dourbes	3	X	X	X	X	X	X		2	50.10	4.60
6459	Ernage	3	X	X	X	X	X	X		2	50.58	4.69
6464	Retie	2	X	X	X	X	X	X		2	51.22	5.03
6472	Humain	4	X	X	X	X	X	X	X	1	50.19	5.26
6477	Diepenbeek	2	X	X	X	X	X	X		2	50.92	5.45
6484	Buzenol	4		X	X	X	X	X		3	49.62	5.59
6494	Mont Rigi	5		X	X	X	X	X		3	50.51	6.07

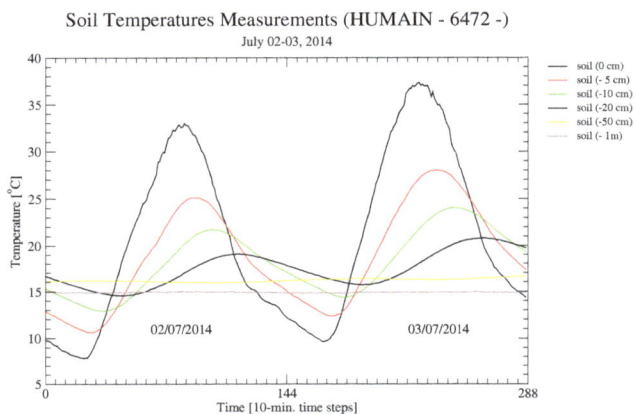

Figure 1. 10-min soil temperature at various depths vs time of the day for 2–3 July 2014, at the Humain station (RMI's AWS 6472, see Table 1).

2 Soil temperature records at RMI

Automatic weather stations operated by RMI range from basic climatological stations to fully equipped synoptic stations performing a complete set of meteorological observations. Soil temperature measurements are performed in 13 of them to provide soil temperature profile in different regions across Belgium for agricultural management as well as for climate research. At each station, soil temperature can be measured at up to 5 different depths (from 5 to 100 cm) in addition to the bare soil and grass temperatures records (see Table 1). Due to the large heterogeneity within the RMI's AWSs, five groups based on the recorded temperature parameters are distinguished for the automated data QC and the stations are gathered in five climatic zones.

3 Complex automatic QA

RMI's AWSs are built around a programmable data logger that measures the sensors, then processes, stores and transmits the data to the central database (DB) in Uccle, Brussels. Once converted to digital values a first processing is performed on the raw data at the data logger level allowing calculation of 10-min temperatures values from the 5-s measurements. To ensure that gross errors are trapped before being further transmitted in the central DB a first basic QC is performed on all temperature values once acquired centrally. Automated procedures monitor the data to make sure they are collected and that the system performance is acceptable. After an existence test, a module checks for physical limits and flags the data violating these limits (erroneous when data lie outside physical limits and suspect when lying outside basic long-term climatological extremes that do not take into account the time of year and location). A list of missing and flagged data is automatically produced after each control cycle and transmitted to the AWS network maintenance team for further intervention. Note that values flagged as erroneous fail immediately and do not require further testing.

Second, each night automated QA procedures check the previous day 10-min temperatures values for more subtle errors. Implementation of the complex automated QA is diagrammed in Fig. 2. Daily temperature values are first checked using annual variation expectancies envelopes and the shape of the diurnal variation of the soil temperature at the different depths are compared to ensure consistency between them (see Sect. 3.1 for details). Individual 10-min records are further scrutinized for plausibility (using adjusted limits to reflect climatic conditions more precisely than in the first near real time range test), internal consistency, temporal consistency and spatial consistency (see Sect. 3.2 for details).

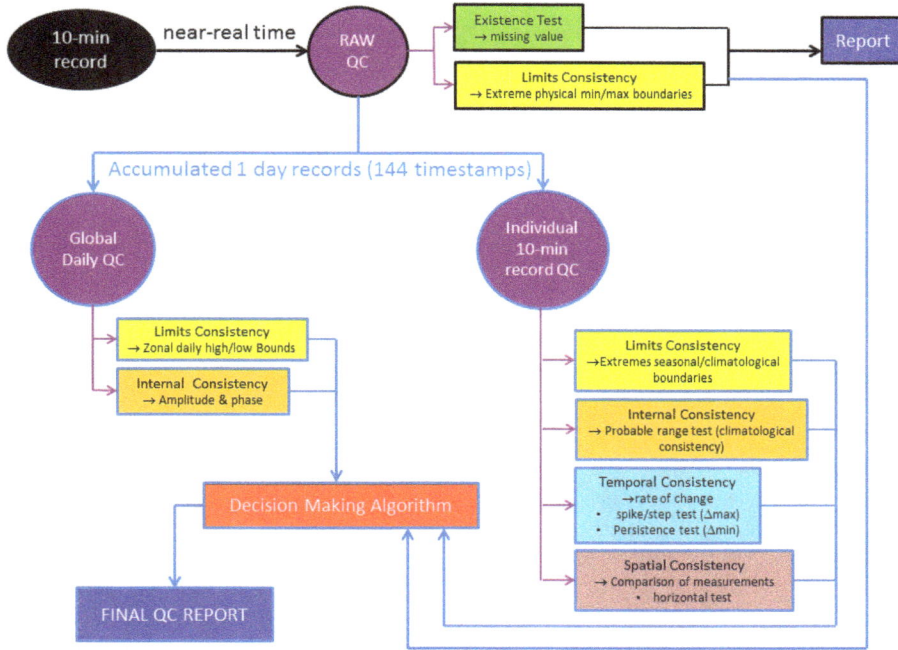

Figure 2. Complex QA of 10-min grass/soil temperatures records.

At the end of the checks, when all 10-min daily temperature time series recorded in a given station have been analyzed, a decision algorithm (that is applicable to all variables and all sites) interprets the scores obtained at each of the individual tests and attributes a final flag (i.e. erroneous, suspect or valid) to each particular data given the weight of evidence. At the end of the process, a report is automatically generated for each AWS and sent to the QC staff.

3.1 Global daily QC

In nature, soil temperature varies continuously in response to the ever changing meteorological regime acting upon the soil-atmosphere interface. That regime is characterized by a regular periodic succession of days and nights, and of summers and winters. The regular diurnal and annual cycles are perturbed by irregular episodic phenomena as cloudiness, cold wave, warm waves, rain storms or snow storms, and periods of drought. In addition to these external influences, there are the soil's own changing properties (i.e. temporal changes in reflectivity, heat capacity, and thermal conductivity as the soil alternately wets and dries, and the variation of all these properties with depth), as well as the influences of geographic location and vegetation. While the thermal regime of soil profiles is very complex, a simple mathematical representation of the fluctuating thermal regime in a soil profile is obtained by assuming that at all depths in the soil the temperature oscillates as a pure harmonic (sinusoidal) function of time around an average value (van Wijk and de Vries, 1963).

To rapidly identify data outside the variation range of soil temperatures at each depth, daily values (i.e. computed from the 10-min measurements) are compared to lower and upper bounds for soil temperature at given depths. Similarly to the LIM test in Hu and Feng (2003), lower and upper bounds for a given temperature data series (i.e. grass and soil temperatures) are constructed for each of the five climate zones by retrieving the highest and lowest daily values on each calendar day of the year from 9 years (2005–2013) of manually quality controlled historical data. Assuming that annual soil/grass temperature variations follow a sinusoidal curve, envelopes of annual variation of these extreme temperatures were then defined using wave functions of the form:

$$T_{L/U}(z,d) = T_{Lo/Uo}(z) + A_{Lo/Uo}(z) \cdot \sin(\omega_o d) \quad (1)$$

where, T_L and T_U are the lower and upper bounds of soil temperature variations, respectively, d is the day of year, ω_o is the angular frequency, which is 2π times the actual frequency (i.e. $2\pi/365$ in case of an annual forcing), z is depth, and $T_{Lo/Uo}(z)$ and $A_{Lo/Uo}(z)$ are the annual mean of these extreme soil temperatures and their amplitude of variations, respectively. To include the extreme values within the derived expectancy envelopes from Eq. (1), the boundaries are adjusted as follows (see Fig. 3 for an illustration): data satisfying

$$T_L(z,d) - 2 \le T(z,d) \le T_U(z,d) + 2$$

succeed the limits consistency test, and data failing it are soft flagged if they are less than 10 % outside the range delimited by the adjusted boundaries and hard flagged otherwise.

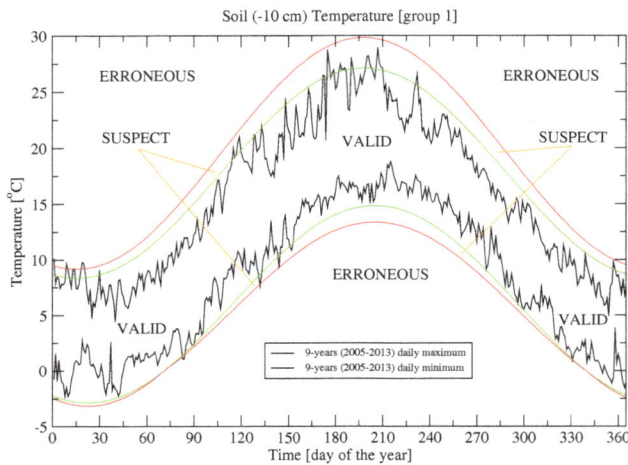

Figure 3. Example of annual variation expectancy envelopes used in the Global Daily QC Limits Consistency test. Illustrated limits apply to the daily mean soil temperature recorded at -10 cm for stations of climate zone 1 (see Table 1)

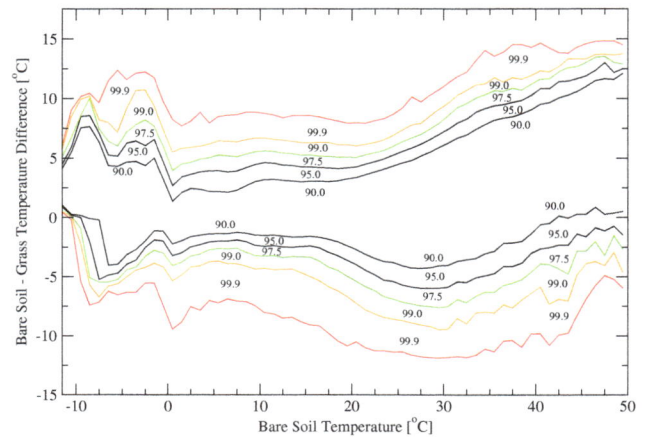

Figure 4. Probable bare soil–grass temperature differences as a function of the bare soil temperature.

To check the reduction in amplitude of the diurnal temperature cycle and the phase shift of the temperature maximum and minimum with depth (see Fig. 1), the recorded diurnal temperature cycle at any depth z, and time t, is modeled as follows:

$$T(z,t) = \overline{T}(z) + A(z) \cdot \sin(\omega t - \phi(z)) \qquad (2)$$

where, $\overline{T}(z)$ is the daily mean temperature at depth z, ω is the angular frequency (diurnal forcing), $A(z)$ is the amplitude of the temperature wave at depth z, and $\phi(z)$ is the phase constant at depth z (aligns soil temperature variation with the forcing). The different $A(z)$ and $\phi(z)$ which are functions of z but not of t are then compared and temperature data series which do not respect the damping and retarding of the temperature waves with depth fail the internal consistency test.

Note that these two tests are denoted as global because they cannot discern which observations within the daily time series are responsible for the offense.

3.2 Individual 10-min record QC

The question of the validity of a given datum is approached using a number of additional tests applied to each particular 10-min temperature record. To verify whether the values are within acceptable range limits depending on the climatic conditions of the measurement site, individual data values are compared with upper and lower seasonal bounds derived for each temperature parameters and zones from several years of previous manually controlled data. The check provides information as to whether the values are erroneous or suspect. To minimize the possibility of a false positive identification, the algorithm does not report an anomaly in case where the

majority of the recorded parameters in a given site at time t are flagged as being either to cool or to warm. Grass temperature and bare soil temperature are further examined for climatological consistency.

Bare soil temperature is measured by a PT100 sensor in contact with the ground on a horizontal surface fully exposed to the open sky. Similarly, grass temperature is measured by a PT100 sensor, fully exposed to the open sky, suspended horizontally over an area covered with short cropped turf and in contact with the tips of grass blades. With the advent of automation and the lack of daily attention by an observer or caretaker, this set up has proved limitations (e.g. lack/lost of contact between the temperature probe and the ground surface or the grass blades, probe fully covered by grass, ...). The probable range test provides a more stringent constraint than simple valid maximum/minimum limit test by requiring consistency among temperature parameters as well as consistency with historical data. Basically, the differences between bare soil and grass temperatures as a function of the bare soil temperature are compared to probable difference determined from several years of previous data. Contours in Fig. 4 indicate which combinations of grass and bare soil temperatures fall within a given percentile of joint probability density. Following a review of the values that fall outside the 99.9 % boundary, the 99.9th percentile was selected as the boundary of acceptability. Combinations are hard flagged when falling outside the 99.9 % boundary and soft flagged when falling between the 99.0 and 99.9 % boundaries. Because two comparisons (involving three parameters) are necessary to unambiguously identify which parameter is problematic, similar joint probability densities were established involving the soil temperature at -10 cm. This last parameter has been chosen for the probable range test as it is systematically recorded in stations where both grass and bare soil temperatures are measured (see Table 1).

To examine the temporal consistency of the data, two tests involving the rate of change of the variables from a preceding acceptable level are applied: the spike/step (ΔMax) test and the persistence (ΔMin) test. Both, the maximum and minimum probable changes for each analyzed parameters (i.e. grass and soil temperatures) are based on the 99.9th percentile change for several years of previous data. Because the rate of change for deep soil temperatures can be very small, the persistence test does not apply to soil temperature below -20 cm. Values are checked for 10-min, 1, 2, 3 and 6 h time steps. To minimize the possibility of a false positive identification, the data must fail in at least 3 of the 5 tested time steps prior to be flagged as suspect or erroneous. Moreover, because in case of extreme meteorological conditions, unusual variability in the air temperature may occur, grass and bare soil temperatures data may be flagged as suspect, although correct. To prevent from this, the algorithm does not report a spike/step anomaly for grass and bare soil temperatures if both temperatures at the same site fail the spike/step test. Similarly, the algorithm does not report a persistence anomaly for the grass and bare soil temperatures in case of snow cover. Below 0 °C in snow free condition, a persistence anomaly is reported based on the assumption that freezing conditions can affect multiple sensors. Note that for stations where only one of the grass or bare soil temperatures is recorded, the -5 cm soil temperature value is used (when applicable) in place of the lacking parameter to adjust the ΔMax and ΔMin tests.

Finally, horizontal comparisons of the same measurement at different stations are performed for all recorded temperature parameters. As for the quality control of 10-min air temperature data implemented at RMI (Bertrand et al., 2013), the horizontal check works in two steps. First, an outlier detection is performed on both the station data being quality controlled and the data of the surrounding stations using the daily 10-min temperature time series of each stations.

Let $T_{i,t}$ be a 10-min temperature record at station i ($i = 1, M$ with $M \leq 13$) at time t ($t = 1, N$ with $N \leq 144$) in a given day. $Y_{i,t} = T_{i,t} - \overline{T_i}$ with $\overline{T_i} = \frac{1}{N} \sum_{t=1}^{N} T_{i,t}$ the daily mean temperature at station i. $Z_{i,t} = Y_{i,t} - \overline{Y_t}$ with $\overline{Y_t} = \frac{1}{M} \sum_{i=1}^{M} Y_{i,t}$ the stations' mean at time t. We test whether or not the $Z_{i,t}$ values fall within the confidence interval defined by:

$$\overline{Z_t} - C\sigma_{Z_t} \leq Z_{i,t} \leq \overline{Z_t} + C\sigma_{Z_t} \tag{3}$$

where $\overline{Z_t} = \frac{1}{M} \sum_{i=1}^{M} Z_{i,t} = 0$, $\sigma_{Z_t} = \sqrt{\frac{1}{M-1} \sum_{i=1}^{M} (Z_{i,t} - \overline{Z_t})^2}$ is the estimated standard deviation at time t, and C is an adjustment parameter function of the considered soil/grass temperature parameter. Values $T_{i,t}$ that do not satisfy the relation in Eq. (3) are considered as outliers.

If an outlier is detected for the station being quality checked, then the data fails the horizontal consistency test.

Table 2. Overall performance of the automatic QC. The evaluation is performed over the full month of November 2014. A total of 298 512 (100 %) 10-min records including all soil/grass temperature parameters recorded within the RMI AWS network have been analyzed.

| | 10-min records | | |
QC$_{AUTO}$	Manual QC TRUE	Manual QC FALSE	Total
Valid	97.71 %	1.60 %	99.31 %
Suspect	0.17 %	0.09 %	0.26 %
Erroneous	< 0.01 %	0.43 %	0.43 %
Total	97.88 %	2.12 %	100 %

Otherwise, the algorithm tests on a 10-min basis whether the analyzed station value, $T_{i,t}$, falls inside a confidence interval formed from surrounding stations data that were not classified as outliers. Measurements that fail the test are soft or hard flagged depending upon the departure of the data from the confidence interval. Note that the outlier check in Eq. (3) can lead to false positives if one or several of the comparison measurements are spatial outliers able to influence the stations mean, $\overline{Y_t}$, in such a way that the measurement under the test is erroneously flagged as an outlier while being valid. In such cases either the decision algorithm at the end of the checking process identifies the false positives as valid based on the scores obtained at the other tests involved in the automated QA system or they will be reviewed during the manual follow up (see Sect. 4) and set to valid if justified.

In developing quality control methods for the US Department of Agriculture (USDA) Natural Resources Conservation Service (NRCS) Soil Moisture-Soil Temperature (SM-ST) network, Hu et al. (2002) established a soil heat diffusion model to screen and identify erroneous soil temperature data. Because such kind of model was found to perform well only in sunny and clear days situations, in our case, modeled data are not used to examine the soil temperatures records. Instead a soil model is used to assist the QC staff in their corrections and estimations (see Sect. 4).

3.3 Automated QC performance

Quality assurance consists of procedures or rules against which data are tested. Each procedure will either detect the data as being valid, suspect or erroneous. False positives (i.e. type I error) increase the burden on the manual QC, and false negatives (i.e. type II error) reduce the quality of the data. One month of data (e.g. November 2014) has been used to determine the overall performance of the automated QA system. Independent manual QC applied on the recorded 10-min soil temperatures during the same month has been considered as reference for the evaluation. Table 2 presents a general overview of the performance of the newly developed

Table 3. Quantitative evaluation of the different tests involved in the automated QC of the 10-min soil/grass temperatures records (QC1 = physical limits test, QC2 = Min–Max range test, QC3 = spike/step − persistence test, QC4 = spatial horizontal test and QC5 = probable range test). The evaluation is performed over the full month of November 2014. A total of 298 512 (100 %) 10-min records including all soil/grass temperature parameters recorded within the RMI AWS network have been analyzed (v = valid, s = suspect, e = erroneous, and nt=no check).

QC TESTS		10-min records	
		Manual TRUE 292 187 = 100 %	Manual FALSE 6325 = 100 %
QC1	v	100 %	100%
	s		
	e		
	nt		
QC2	v	98.15 %	71.40 %
	s	1.81 %	26.12 %
	e	0.04 %	2.34 %
	nt		0.14 %
QC3	v	99.36 %	89.38 %
	s	0.51 %	7.67 %
	e	0.13 %	2.81 %
	nt		0.14 %
QC4	v	91.45 %	96.13 %
	s	7.07 %	1.44 %
	e	< 0.01 %	2.43 %
	nt	1.48 %	
QC5	v	99.96 %	78.95 %
	s	0.04 %	1.56 %
	e	< 0.01 %	19.34 %
	nt		0.15 %
QC$_{AUTO}$	v	99.83 %	75.38 %
	s	0.17 %	4.49 %
	e	< 0.01 %	20.13 %
	nt		

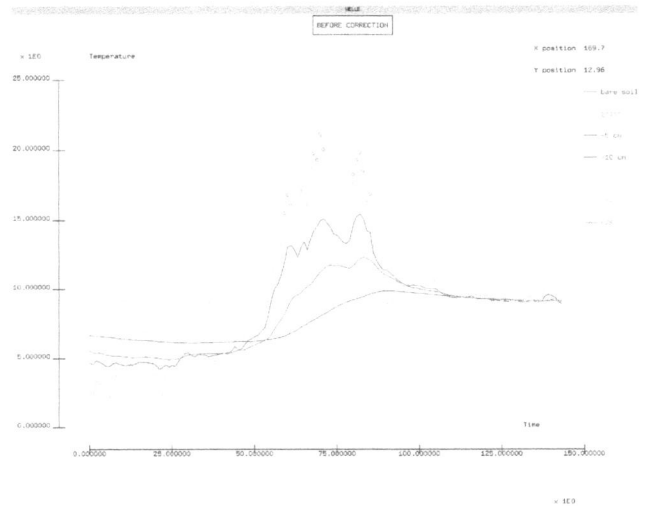

Figure 5. Visualization of the automated QC applied on the 10-min soil/grass temperatures records performed on 22 November 2014 at the Melle station (AWS 6434, see Table 1). Erroneous grass temperature data are indicated by a orange circle on the green curve.

complex QA system while Table 3 provides a quantitative evaluation of the various tests involved in the data checking. It is worth pointing out that both tables refer to the 10-min data tests (daily tests cannot identify which observations are responsible for the offense) and that the results could differ for a given temperature parameter or a station type (i.e. QC group in Table 1). Table 2 indicates that type I errors generated by the automated QA system are very low (less than 0.01 % of the true 10-min records were detected as erroneous by the algorithm). By contrast the percentage of type II errors is very large (more than 75 % of the false 10-min records were found as valid by the automated QC as indicated in Table 3). However, this apparent very bad performance of the automated QA system in term of type II errors has to be handle with caution. First, it often occurred that while

the algorithm effectively detected 10-min erroneous/suspect measurements in a daily parameter time series, the operators corrected more records than the ones found problematic by the system. As an example, Fig. 5 indicates that the 10-min grass temperature records were found erroneous by the automated QA system 14 times (orange circles on the green curve) on 22 November 2014 at the Melle station. After visualization of the station grass temperatures time series on 22 November 2014, 33 corrections were performed by the operator on the 10-min records (i.e. the full time segment where problematic measurements were detected by the algorithm was manually corrected). Second, Tables 2 and 3 only deal with the 10-min tests and do not account for the daily tests. It is worth pointing out that during the month of November 2014, the grass temperature measurements performed in the Stabroek station were found systematically wrong by the QC staff during 28 days. Over this time period, the grass temperature parameter in this station was detected as erroneous (suspect) 23 (4) times on a daily basis while the 10-min tests did not necessarily reported any erroneous/suspect measurements for this parameter as illustrated in Fig. 6.

When accounting for both daily and 10-min automated QC results, the algorithm succeeded to identify the stations, days and parameters on which corrections were made by the QC staff. A detailed analysis of the type II errors revealed that they mainly concern grass temperature measurements. We strongly suspect that the grass temperature database (and to a lesser extend the bare soil temperature database) used to derive the tests was not validated as it should have been. Typically, probable range test aims at detecting problematic situations as the one illustrated in Fig. 6 for the grass tem-

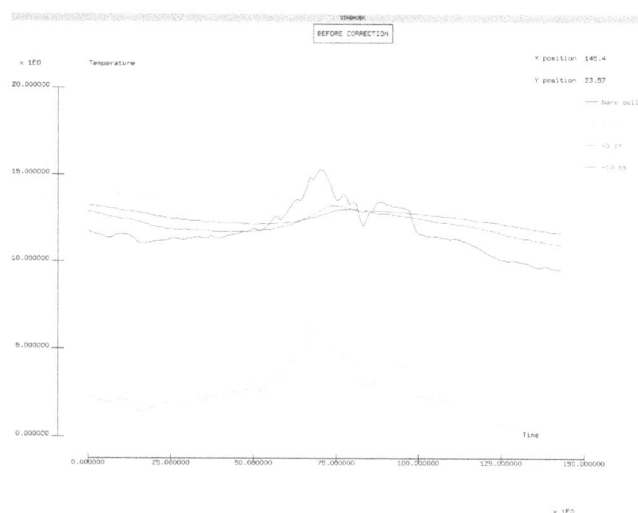

Figure 6. Visualization of the automated QC applied on the 10-min soil/grass temperatures records performed on 3 November 2014 at the Stabroek station (AWS 6438, see Table 1). No erroneous or suspect data were found in the daily 10-min temperatures time series.

perature 10-min daily time series. Because erroneous data have been involved when defining the boundaries used in the automated tests, probable differences between bare soil and grass temperatures given in Fig. 4 are certainly too permissive. This drawback has a direct impact on the detection performance of problematic grass, bare soil and −10 cm soil temperatures records by the probable range test. As an example, Table 3 indicates that this test (i.e. QC5) produced type II errors in about 79 % of the cases while this category of test has proven to be one of the most efficient in the detection of erroneous 10-min air temperature records performed by the RMI's AWSs.

4 Manual QA

Each day, the QC staff analyses the preceding day 10-min temperature records in the light of the assigned quality flags from the automated QA system. Results of the automated QA system can be graphically plotted on the operator terminal screen as illustrated in Figs. 5 and 6. In that case, all the analyzed station 10-min soil/grass temperatures records of the inspected day are illustrated in a graphic window and erroneous or suspect data are indicated in the corresponding parameter daily time series (e.g. orange circles on the green line in Fig. 5). Visual inspection of all records flagged by the automated decision making algorithm is done to distinguish instrumental problems from plausible behaviors. It is the human decision whether or not a value is accepted. When errors are verified or visually detected, faulty records are eliminated and "trouble tickets" are issued where needed to the maintenance team so that sensors can be replaced or repaired. More than simply deleting erroneous measurements, human

operators supply corrections and estimations (i.e. when values are missing) where possible. They are supported in this task by automated procedures. As an example, assuming that the thermal properties are constant with depth, soil temperature at any depth below the ground surface (i.e. $0 < z < \infty$) can be estimated using a soil heat diffusion model (van Wijk and de Vries, 1963).

The correction/estimation process is fully interactive, operators directly visualize on screen the corrections they applied on the parameters time series (the graphic window displaying the station temperatures time series being automatically updated after each modification). They have the opportunity to visualize different corrections on the problematic time series in order to determine the most appropriate in their specific case. When the correction/estimation process is completed, all modifications introduced by the operator are automatically implemented in the central RMI database. Note that the original parameters values are kept in the database and still accessible by the QC staff if required.

5 Conclusions

Automation of the RMI's AWSs data quality control is ongoing. After the automated quality control of 10-min air temperature data (Bertrand et al., 2013), automated quality assurance procedures devoted to 10-min grass/soils temperature records have been operationally implemented to support the QC staff in their work. The purpose of this automated data screening is to objectively identify abnormal data values for subsequent review by an experienced data analyst. The review is necessary to determine whether an anomaly results from a problem with instrumentation or whether it accurately reflects unusual meteorological conditions. Validation exercises have revealed that the complex automatic QA system is able to correctly identify problematic parameters in a particular station on a given day. However, automated tests applied to 10-min temperature records produce a very high percentage of type II error. In depth analysis of type II errors indicates that because the database of grass temperature records (and to a lesser extent bare soil temperature records) used to derive the boundaries involved in the automated tests was not validated as it should be, the probable range test fails to perform correctly. To overcome such a limitation an extensive validation of our historical records of 10-min grass and bare soil temperatures will be undergone as soon as possible. Once available, the new validated database will be used to refine the automated tests in general and in particular the probable range test involving the grass, bare soil and -10 cm soil temperatures. This forthcoming version of the algorithm will be evaluated using test data days from a whole year as the use of one single month of data could have masked sensitivities of the automated QA system to seasonal variations.

References

Bertrand, C., Gonzalez Sotelino, L., and Journée, M.: Quality control of 10-min air temperature data at RMI, Adv. Sci. Res., 10, 1–5, doi:10.5194/asr-10-1-2013, 2013.

Best, M. J., Cox, P. M., and Warrilow, D.: determining the optimal soil temperature scheme for atmospheric modeling applications, Bound.-Lay. Meteorol., 114, 111–142, 2005.

de Jeu, R. A. M., Wagner, W., Holmes, T. R. H., Dolman, A. J., van de Giesen, N. C., and Friesen, J.: Global soil moisture patterns observed by space borne microwave radiometers and scatterometers, Surv. Geophys., 29, 399–420, 2008.

Gandin, L. S.: Complex quality control of meteorological observations, Mon. Weather Rev., 116, 1137–1156, 1988.

Hillel, D.: Environmental soil physics. Academic press, London, 771 pp., 1998.

Hu, Q. S. and Feng, S.: A daily soil temperature dataset and soil temperature climatology of the contiguous United states, J. Appl. Meteorol., 42, 1139–1156, 2003.

Hu, Q. S., Feng, S., and Schaefer, G.: Quality control for USDA NRCS SM-ST Network soil temperatures: a method and a dataset, J. Appl. Meteorol., 41, 607–619, 2002.

Lachenbruch, A. H. and Marshall, B. V.: Changing climate: geothermal evidence from permafrost in the Alaskan Arctic, Science, 234, 689–696, 1986.

van Wijk, W. R. and de Vries, D. A.: Periodic temperature variations in homogeneous soil, in: Physics of plant environment, edited by: van Wijk, W. R., North-Holland Publ. Co., Amsterdam, 102–143, 1963.

7

UV and global irradiance measurements and analysis during the Marsaxlokk (Malta) campaign

J. Bilbao[1], R. Román[1], C. Yousif[2], D. Mateos[1], and A. de Miguel[1]

[1]University of Valladolid, Spain, Atmosphere & Energy Laboratory, Faculty of Sciences, Valladolid, Spain
[2]University of Malta, Institute for Sustainable Energy, Marsaxlokk, Malta

Correspondence to: J. Bilbao (juliab@fa1.uva.es)

Abstract. A solar radiation measurement campaign was performed in the south-eastern village of Marsaxlokk ($35°50'$ N; $14°33'$ E; 10 m a.s.l), Malta, between 15 May and 15 October 2012. Erythemal solar radiation data (from a UVB-1 pyranometer), and total horizontal solar radiation (global and diffuse components) from two CM21 pyranometer were recorded. A comparison of atmospheric compounds from ground measurements and satellites shows that TOC (total ozone column) data from the Ozone Monitoring Instrument OMI, TOMS and DOAS algorithms correlate well with ground-based recorded data. The water vapour column and the aerosol optical depth at 550 nm show a significant correlation at the confidence level of 99 %. Parametric models for evaluating the solar UV erythemal (UVER), global (G) and diffuse (D) horizontal irradiances are calibrated, from which aerosol effects on solar irradiance are evaluated using the Aerosol Modification Factor (AMF). The AMF_{UVER} values are lower than AMF_G, indicating a greater aerosol effect on UVER than on global solar irradiance. In this campaign, several dust event trajectories are identified by means of the HYbrid Single-Particle Lagrangian Integrated Trajectory (HYSPLIT) model and by synoptic conditions for characterizing desert dust events. Hence, changes in the UV index due to atmospheric aerosols are described.

1 Introduction and objectives

The UV erythemal (UVER) radiation affects human health (short-term effects include erythema or sunburn and long-term effects include photo-ageing/skin cancer), damages aquatic life, and affects plants as well as the conservation and durability of materials, in addition to impacting global energy balance and climate change (UNEP, 2010). Yet, not all its effects are harmful, with the synthesis of vitamin D being one of the beneficial effects of UV (Webb, 2006; Fioletov et al., 2009). Different action spectra are used to quantify these effects. McKinlay and Diffey (1987) established the erythemal action spectrum, which represents the spectral response of human skin to UV radiation being able to trigger erythema (or sunburn). Solar radiation weighted using this spectrum is called erythemal radiation (UVER; 280–400 nm) (Webb et al., 2011; CIE, 1998).

Various authors have proposed new models for estimating solar irradiance (e.g. Mateos et al. (2010) who assessed ultraviolet and global cloud modification factors in central Spain). The attenuation of UVER by different atmospheric components has been analysed by De Miguel et al. (2011); the results confirm that small aerosols scatter more solar UV wavelengths than larger ones.

Román et al. (2013) described the effects of a desert dust episode at Granada (Spain) on global, direct and diffuse spectral UV irradiance, and reported that the attenuation of direct UV was about 50 %, while the diffuse irradiance increased by up to 40 %. Authors such as Pace et al. (2006) have studied the influence and identification of different aerosol types at Lampedusa Island (Central Mediterranean); their observations were combined with air mass trajectories in order to identify different types of particles and to determine their mean optical properties.

Also it is known that aerosols are atmospheric particles that can affect incoming solar radiation directly by absorption and scattering, and indirectly by acting as cloud condensation nuclei (modifying cloud microphysical properties).

Both effects contribute to cooling the Earth's surface and simultaneously warming the lower atmosphere (IPCC, 2007).

The aerosol radiative forcing in the Central Mediterranean has also been studied (Gómez-Amo et al., 2011); the results show the effects of different aerosols on the radiative budget during a dust event within the boundary layer. The sensitivity of shortwave radiative fluxes to changes in the vertical distribution of aerosols and a case of Sahara dust layer above urban aerosols have also been studied (Gómez-Amo et al., 2010; Kaskaoutis et al., 2010a, b).

The present work seeks to explore the influence of aerosols on solar irradiance in the area of Malta. In the study, parametric models for evaluating solar UVER, global (G) and diffuse (D) horizontal irradiances under cloudless conditions were calibrated, from which the aerosol Modification factor (AMF) was evaluated. The models are a function of the solar zenith angle (SZA), the total ozone column (TOC), and the aerosol optical depth at 550 nm (AOD_{550}). The models were validated through a graphical comparison of measured and estimated data and by different statistical indices (Bilbao et al., 2014). Atmospheric ground reference data were compared to those retrieved from satellites; the agreement was assessed. In this campaign, several dust event trajectories were identified by means of the HYbrid Single-Particle Lagrangian Integrated Trajectory (HYSPLIT) model and by the synoptic conditions (Bilbao et al., 2014). Changes in the UV index due to atmospheric aerosols were thus described.

In the following sections, the site description, instrument and data collection are detailed, the methodology is explained, and the observations and results are analyzed in relation with the influence of aerosols on UVER, G and D irradiances.

2 Site and instrumentation

2.1 Measurement site

The Universities of Malta and Valladolid (Spain) conducted a solar radiation measurement campaign, which took place at the Institute for Sustainable Energy in the south-eastern village of Marsaxlokk (35°50′ N; 14°33′ E; 10 m a.s.l.), Malta. Measurements were recorded between May and October 2012. The campaign involved two European Institution Groups: the Institute for Sustainable Energy at the University of Malta, and the Atmosphere and Energy Laboratory at the University of Valladolid, Spain. Solar radiation instruments used for this study were located on the rooftop of the Institute for Sustainable Energy, which had an obstruction-free horizon.

2.2 Ground-based instruments

Global and diffuse horizontal irradiance were recorded using two CM21 (Kipp & Zonen) pyranometers, one of which was equipped with a shadow-band. Diffuse solar irradiance data

were corrected following the method proposed by Batlles et al. (1995) and Perez et al. (1990) that take into account geometric and atmospheric (clearness index) corrections. The sampling rate of the global and diffuse irradiance pyranometers was 10 s although measurements were averaged and recorded at 1 and 10 min. As a result, 10 min solar irradiance measurements were obtained. The CM21 (Kipp & Zonen) instruments have a flat spectral response from 305 to 2800 nm and the cosine effect is below 3 % for a solar elevation above 10°. The CM21 sensors were regularly calibrated by comparison with a reference sensor at the Kip & Zonen manufacturer; the differences obtained were below 2 % (Bilbao et al., 2014).

The UVER measurements were recorded using a Yankee Environmental Systems (YES) UVB-1 radiometer, which has a spectral range between 280 and 400 nm and a spectral sensitivity which resembles the erythemal action spectrum. The UVB-1 sensor is designed to operate continuously and autonomously in the field (Esteve et al., 2009). The cosine response is greater than ±5 % for solar zenith angles (SZA) below 60°. This instrument is calibrated by the standard rules in the National Institute for Aerospace Technology (INTA) in Spain. The calibration consists of measuring the spectral response of the sensor indoors and comparing with a Brewer MKIII spectroradiometer outdoors. After this process, a double input matrix with the calibration factors depending on SZA and the total ozone column (TOC) was obtained using a radiative transfer model (Hülsen and Gröbner, 2007). The error given by the calibration matrix is below 9 % for a SZA of less than 70°. The output-voltage signal of the sensor is converted into units of erythemal irradiance (W m^{-2}) by multiplying the signal voltage by the calibration factor obtained from the matrix, taking into account the average SZA during the 10 min measure and the daily TOC. The experimental uncertainty of the sensor according to the results is in the 4.6–7 % range (De Miguel et al., 2012).

A Solar Light Microtops-II manual Sun photometer was used throughout the campaign to measure TOC and AOD at 1020 nm (AOD_{1020}), and water vapour column. The Microtops-II Sun photometer is equipped with five optical collimators, with a field of view of 2.5°, to perform direct radiation measurements at the following nominal wavelengths: 305, 312, 320, 936, and 1020 nm. The instrument is calibrated every two years at the Mauna Loa Observatory (Hawaii) by Solar Light Company (Glenside, PA), further details are available in De Miguel et al. (2011). Furthermore, an inter-comparison of ozone data between the Microtops-II and the Brewer spectrometer was performed at Lampedusa island (in the Central Mediterranean) over five days. The differences observed between the TOC estimations of the two instruments were below 2 % (Mateos et al., 2014b). More details concerning solar irradiance sensors are given in De Miguel and Bilbao (2005).

Solar sensors were connected to a Campbell CR10X Data Logger, which was programmed to take measurements each

10 s from which 10 min average values were computed and stored. There are 6 readings for every 10 min rintervals in an hour which were added together providing a single hourly reading. Hourly and daily irradiances were then evaluated from the 10 min average values (Bilbao and de Miguel, 2013). Data were transmitted continuously from data-logger to the Atmosphere and Energy Laboratory via web. Measurements were collected in Greenwich Mean Time (GMT).

2.3 Satellite-borne instruments

TOC, aerosol optical depth measured at 550 nm wavelength (AOD_{550}), and precipitable water column (w), were retrieved from a satellite sensor (MODIS OMI). During the campaign, whenever the Microtops TOC measurements were not available because of cloudy conditions, data from the Ozone Monitoring Instrument (OMI; onboard Aura satellite) were used. If TOC from OMI was not available, data from the Global Ozone Monitoring Experiment-2 instrument (GOME-2; onboard MetOp-A satellite) were used. These satellite-based data were obtained from the Aura Validation Data Centre (AVDC: http://avdc.gsfc.nasa.gov) as overpass files. The TOC from OMI retrieved by the TOMS (OMITOMS) and DOAS (OMIDOAS) algorithms, and the TOC retrieved from GOME-2 (GOME2) were compared with Microtops measurements, and showed a high correlation between ground based and satellite-based TOC for the three databases. OMITOMS was seen to be the most similar TOC data series to ground measurements, followed by OMIDOAS and then GOME-2. The comparison can be found in Bilbao et al. (2014). The correlations obtained between the sun photometer and satellite TOC retrieved daily values are shown and compared in Bilbao et al. (2014), with RMSE values being below 2.5 %. In addition, the correlation coefficient between the OMI and Microtops-II TOC daily data of 0.95, the frequency distribution of the differences showed a standard deviation of 6 DU and also a RMSE of 2.2 % were obtained during a two month campaign in Trisaia campaign, Southern Italy (Mateos et al., 2014a).

Water-vapour column data were also obtained from noon measurements of the MODIS instrument (Moderate Resolution Imaging Spectroradiometer) on board the Terra satellite, using MODIS infrared channels. Comparisons of this MODIS product against ground-based water-vapour measurements in a Mediterranean region show that the combined uncertainty of MODIS water vapour increases from 0.38 to 0.52 cm when w rises from 0.5 to 3 cm. This may be found in Román et al. (2014).

The AOD_{550}, was also obtained from the MODIS instrument on board the Terra satellite and the results were considered as constant daily values. The uncertainty in the AOD_{550} daily values from MODIS was evaluated by the expression proposed by Kaufman et al. (1997). The Ångström exponent values were also observed from daily AERONET (http://aeronet.gsfc.nasa.gov/new_web/index.html) data at

the nearest Italian station of Lampedusa (35.52° N; 12.63° E; 45 m a.s.l.) (Meloni et al., 2003) and from MODIS. AOD_{1020} was calculated using the Ångström exponent and the AOD_{550} from MODIS. Comparisons of Microtops II AOD_{1020} values and calculated AOD_{1020} show a difference of 0.06 % (Bilbao et al., 2014).

The MODIS data used in this work were downloaded from the GIOVANNI application (http://disc.sci.gsfc.nasa.gov/giovanni/overview/index.html) as a spatial the average value in a 0.2° × 0.2° square with centre at the Institute for Sustainable Energy, Marsaxlokk, Malta.

3 Methodology

3.1 Modelling

Parametric models for evaluating solar UVER and horizontal global and diffuse irradiances under cloudless conditions were used; they are called the UVER, Global and Diffuse Model, respectively. The UVER model is based on an analytical expression for clear-sky conditions proposed by Madronich (2007). The Global model was derived from the analytical relation between the horizontal global solar irradiance, G, and the cosine of SZA. It has been observed (Bilbao et al., 2014) that G is a linear function of SZA, while D shows a power function with SZA. The inputs to the UVER model consist of SZA and TOC values. The input to the Global model is SZA and to the Diffuse model the SZA and the AOD. The algorithms of the models are detailed in Table 1.

Data from 15 May to 31 July 2012 were used to estimate the model coefficients and data from 1 August to 15 October 2012 were used for model validation. Statistical indices such as the mean bias, root mean square errors, square fits and correlation coefficients, were used for model validation.

3.2 Statistical estimators used

The accuracy of the different models was assessed by means of two widely used statistics: the root mean square error (RMSE) and mean bias error (MBE). The following expressions for RMSE and MBE, as a percentage of the measured average value, are used:

$$\text{RMSE}(\%) = \frac{100}{x_{\text{ave}}} \left[\sum_{i=1}^{n} (x_{i\,\text{sim}} - x_{i\,\text{mea}})^2 / n \right]^{0.5} \qquad (1)$$

$$\text{MBE}(\%) = \frac{100}{x_{\text{ave}}} \left[\sum_{i=1}^{n} (x_{i\,\text{sim}} - x_{i\,\text{mea}}) / n \right], \qquad (2)$$

where n is the number of data points, $x_{i\,\text{sim}}$ is the ith simulated value, $x_{i\,\text{mea}}$ is the ith measured value, and x_{ave} is the mean of the measured values. The estimators were calculated for each models.

During the campaign, different atmospheric conditions were observed, particularly events with high aerosols due to

Table 1. UVER, global and diffuse solar horizontal irradiance calibrated models under cloudless conditions, calibration coefficients and their expanded uncertainty, during the period May to October 2012 at Marsaxlokk, Malta.

Clear sky models	Solar irradiance	Coefficients
UVER	$\text{UVER} = a(\cos \text{SZA})^b \left(\frac{\text{TOC}}{300} \right)^c$	$a = 0.270 \pm 0.002 \, \text{W m}^{-2}$ $b = 2.417 \pm 0.002$ $c = -0.78 \pm 0.02$
Global	$\text{G} = a(\cos SZA)^b$	$a = 1032 \pm 14 \, \text{W m}^{-2}$ $b = 1.279 \pm 0.002$
Diffuse	$\text{D} = a(\cos \text{SZA})^b \left(\frac{\text{AOD}_{550}}{0.05} \right)^c$	$a = 90.2 \pm 0.9 \, \text{W m}^{-2}$ $b = 0.451 \pm 0.009$ $c = 0.274 \pm 0.006$
Aerosol modification factor	$\text{AMF} = \frac{\text{G}_{\text{measured}}}{\text{G}_{\text{clear sky}}}$	

Table 2. Estimation of solar UVER, global and diffuse irradiance values by means of linear regressions for clear-sky conditions in Marsaxlokk, Malta in the period August to October 2012.

	Estimated $= a \times$ measured $+ b$		
	a	$b \, (\text{W m}^{-2})$	r
UVER	0.972 ± 0.004	0.0023 ± 0.0004	0.995
Global	1.001 ± 0.004	-7 ± 4	0.995
Diffuse	1.004 ± 0.016	13.4 ± 1.6	0.921

desert dust. The agreement between the measured solar irradiance data and the estimated values was also assessed by determining the percentages of the estimated values that agree with the ± 10, ± 20 and $\pm 30\%$ of the measured reference data, which are denoted as W_{10}, W_{20} and W_{30} (%) (Tanskanen et al., 2007).

3.3 AMF

In order to evaluate the aerosol effect, the AMF was defined as the relation between the measured solar irradiance and the cloudless sky one under the same atmospheric conditions.

$$\text{AMF} = \frac{\text{SR}_{\text{aerosol}}}{\text{SR}_{\text{aerosol-free}}}, \tag{3}$$

where $\text{SR}_{\text{aerosol}}$ and $\text{SR}_{\text{aerosol-free}}$ represent solar irradiance under the corresponding conditions. AMF is the ratio of clear-sky solar irradiance with the atmosphere containing aerosols to that solar irradiance recorded under cloudless conditions and an atmosphere without aerosol, for the same SZA and TOC values. In this work, we compare two adjacent days that verify these conditions (26th with aerosols and 25th aerosol free).

3.4 Air mass trajectories

The HYSPLIT model http://ready.arl.noaa.gov/HYSPLIT. php was used to evaluate air mass backward trajectories in order to describe the influence of aerosols on solar irradiance. The input data for a backward trajectory evaluation are: measurement station, considered as the final position of the backward trajectory ($35°50'$ N; $14°33'$ E); 72 h duration, concluding at 12:00 UTC; the day analyzed the trajectories at three heights above sea level, 500, 1500 and 3000 m. The dates analyzed are 25, 26, 27 August 2012.

3.5 Atmospheric component correlations

Satellite atmospheric component data were compared with ground-based data; frequency histograms were generated before analyzing their influence on solar irradiances. As observed by Koepke et al. (2002), AOD_{550} depends on humidity. In this paper, the relation between AOD_{1020} and watervapour column was evaluated, and the correlation and its significance was also calculated. In addition, the comparison of TOC measurements from the Microtops and from the OMI-TOMS, OMI-DOAS and GOME-2 sensors was carried out and their uncertainty frequency distributions were plotted so as to verify the agreement between the two data series. The linear fit of calculated AOD_{1020} data from MODIS versus AOD_{1020} data from Microtops was performed. The correlation coefficient value and the frequency distribution of the difference of its values were evaluated so as to study the agreement between MODIS and Microtops data.

3.6 UVI evaluation

The UV index (UVI) was also evaluated in the measurement campaign. To calculate UVI, it is necessary to calibrate the UVER sensor and to have the calibration matrix available, using the facilities at the INTA Laboratory, in our case. The raw signal of the UVB-1 pyranometer is multiplied by a calibration coefficient in order to convert it to erythemal UV-B irradiance. The calibration coefficient depends on SZA and TOC, this coefficient being obtained each time by interpolation in a matrix calibration. Once the raw signal is converted into erythemal UV-B irradiance ($W\,m^{-2}$), this amount is multiplied by $40\,m^2\,W^{-1}$ in order to convert it into UVI as indicated in WHO (2002). The influence of aerosol desert dust on UVI was also studied. This evaluation also serves as an important vehicle to raise public awareness and to alert people vis-à-vis the need to adopt protective measures when exposed to UV radiation.

4 Results

After examining the campaign data series, it was observed that on cloudless days, the diffuse horizontal irradiance increases with increasing atmospheric water vapour. The linear fit between AOD_{550} from MODIS and w from MODIS shows a correlation coefficient of 0.51, with a confidence interval of (0.160–0.751), at a 99 % confidence level, and a p value of 0.00041. In addition, w uncertainty for the Microtops and MODIS data were evaluated and a standard deviation of the residuals of 0.5 cm was obtained. The linear fit between TOC correlation coefficients between 0.95 and 0.86, while the frequency distribution of the differences is in the range from 4.5 to 7.6 DU and a standard deviation value of 6 DU was obtained; this result can be used as uncertainty of satellite-derived TOC. A similar evaluation can be found in Román et al. (2014).

4.1 Estimated vs. ground-based measurements

In this study, three different models for UVER, global and diffuse irradiance under cloudless conditions were calibrated. The data set for the calibration models contain 10 min clear sky data for May, June and July 2012, a total of 684 10 min values. For the evaluation period (August, September and October), a total of 650 data points were used. The solar irradiance clear-sky model expressions and coefficient values are shown in Table 1. The UVER model fits as a SZA cosine and TOC power function. The solar global horizontal irradiance fits as a function of cos(SZA) cosine, and the solar diffuse horizontal irradiance is a function of cos(SZA) and AOD_{550}. Similar results have been obtained by Bilbao et al. (2014) when studying the effects of atmospheric components on UVER and horizontal irradiances. A model relating clear sky UV-B irradiance with SZA and TOC was developed by Dubrovsky (2000) at two stations in the Czech Republic.

Model validations have been evaluated using scatterplots. Figure 1 consists of three scattered graphs, and the diagonal line represents the ideal match between the estimated and measured values. Table 2 shows the linear regression equations and the correlation coefficient, r, between measured and estimated values. It is observed that UVER and the global solar models show the highest correlation coeffi-

cients of 0.995. The diffuse model obtains a lower correlation coefficient of 0.921.

Table 3 shows the model performance, namely mean measured and mean estimated values, the RMSE, MBE deviations and the W_i (%) variation coefficients between the measured and the estimated values. The W_i coefficients for UVER indicate that 86.7 % of the data exhibit a variation within ± 10 % in respect of the measured data, and 96.9 % of the data a variation within ± 20 %. From Table 3, it can be deduced that 94.6 % of the global solar data show a variation within ± 10 % compared to the measured data. A similar validation coefficient analysis was used by Mateos et al. (2013) and Tanskanen et al. (2007).

4.2 Case study: desert aerosols

Figure 2 shows the 10 min evolution and daily value of UVER and UVI 10 min evolution and UVI daily maximum values from the 22 to 28 August 2012. 25 and 26 August were classified as cloudless days; however, recorded UVER and calculated UVI decreased on 26 August. In addition, TOC dropped from 296 to 286 DU starting with on 23 August, AOD_{550} increased on 26 August only, and UVI increased on 25 August and decreased again on 26 August. While TOC does not show any change (< 1 DU), AOD shows a difference of two units, from 0.10 on 25 to 0.3 on 26 August. Although 27 August was not classified as a cloudless day, it can be observed that the mean daily UVER and AOD values are similar to those for 25 August.

As regards the UVI values, 25 August shows a maximum with 9.8, which drops to 8.6 on 26 August and increases to 9.7 on 27 August. It is also observed that cloudless days with high AOD and water vapour value of 3.8 ± 0.5 cm coincide with a decrease in UVI and UVER, and an increase in daily diffuse horizontal irradiance, which value was $5.05 \pm 0.02\,MJ\,m^{-2}$ on 26 August 2012.

In addition to this, it is known (Bilbao et al., 2014) that UVER irradiance does not depend on w because w has no absorption bands in the UV range. It was obtained that w reduces G irradiance between -2.44 and $-4.53\,\%\,cm^{-1}$, and D irradiance increasing with w between $42.15\,\%\,cm^{-1}$ and $20.30\,\%\,cm^{-1}$ from low to high SZA angles. It has been observed that D irradiance does not increase with w when AOD_{550} is fixed (Bilbao et al., 2015). In addition, the increase in the D component is due to the dust episodes in Malta whose synoptic conditions also increase w and it makes possible the hygroscopic growth of aerosols and as consequence the increases of D irradiance depends on aerosols. Aerosols cause a UVER reduction between 28 and 52 % and a G irradiance reduction between 13 and 41 %. These results show that w impact on irradiances in Malta measurement station is a magnitude lower compared to aerosols, as it can be seen at Bilbao et al. (2014). Under this situation, and in order to study the aerosol impact, it was decided to study air mass trajectories.

Table 3. Statistical estimators for the comparison of measured and estimated 10 min UVER, global and diffuse solar irradiance at Marsaxlokk, Malta during the period August to October 2012.

	Mean (Wm^{-2})		MBE (%)	RMSE (%)	W_{10} (%)	W_{20} (%)	W_{30} (%)
	Measured	Estimated					
UVER	0.115	0.114	0.22	6.56	86.7	96.9	100.0
Global	602.6	595.7	−0.90	4.52	94.6	98.8	99.6
Diffuse	110.6	103.7	−0.66	19	68.0	80.7	92.7

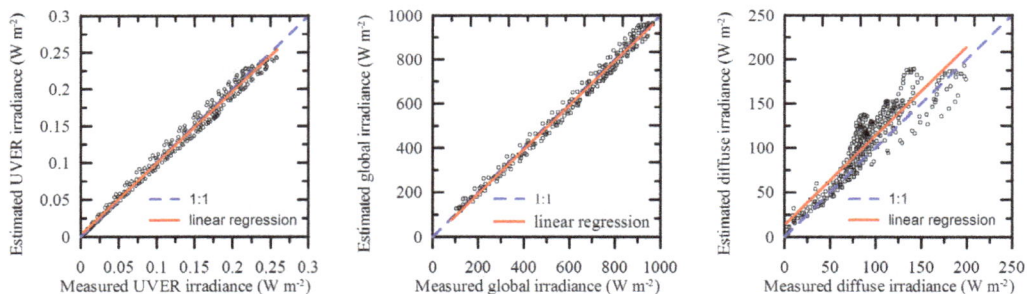

Figure 1. Comparison of 10 min measured and estimated UVER (left), global (centre) and diffuse (right) solar irradiances under cloudless conditions during the period May to October 2012 at Marsaxlokk. Red line is the regression fit. Blue line is the 1 : 1 line.

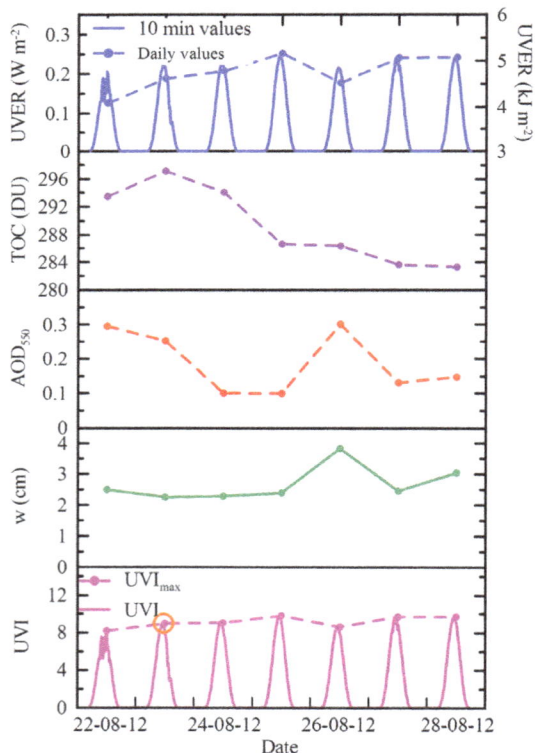

Figure 2. 10 min and mean daily values of UVER and UVI, and TOC, AOD$_{550}$ and w from satellites at Marsaxlokk in the period 22–28 August 2012.

4.3 Air-mass trajectories

Figure 3 shows the daily backward trajectories on 25, 26 and 27 August 2012. It can be observed that on 25 August, the air masses come from the eastern Mediterranean, that the situation changes on 26 August when the air masses come from the African continent. The air mass at 3000 m crosses Africa at very low altitude and might explain the high aerosol content. The situation changes on 27 August when the air masses come from the Iberian Peninsula. As a result, it is concluded that 26 August was characterized by a dust event from Africa.

4.4 AMF

AMF was calculated so as to determine the impact of AOD on horizontal solar radiation. For these evaluations, measurements on 25 August (cloudless sky) were used as reference measurements for an aerosol-free atmosphere. The AMFs were evaluated for erythemal UVER (AMF$_{UVER}$), solar global (AMF$_G$), and diffuse (AMF$_D$) solar components.

An aerosol intrusion from Africa was described on 26 August 2012. In addition, we evaluate the AMF, defined previously, using 25 August measurements as a reference for an aerosol-free atmosphere. In this way, erythemal UVER, global horizontal and diffuse (AMF$_{UVER}$), (AMF$_G$) and (AMF$_D$) were calculated, respectively. First, from the data series it is observed that on 26 August, Fig. 2, the water-vapour column increases to a value of 4.0 cm and the diffuse irradiation to a value of 5.3 MJ m^{-2} day^{-1} due to a desert dust intrusion.

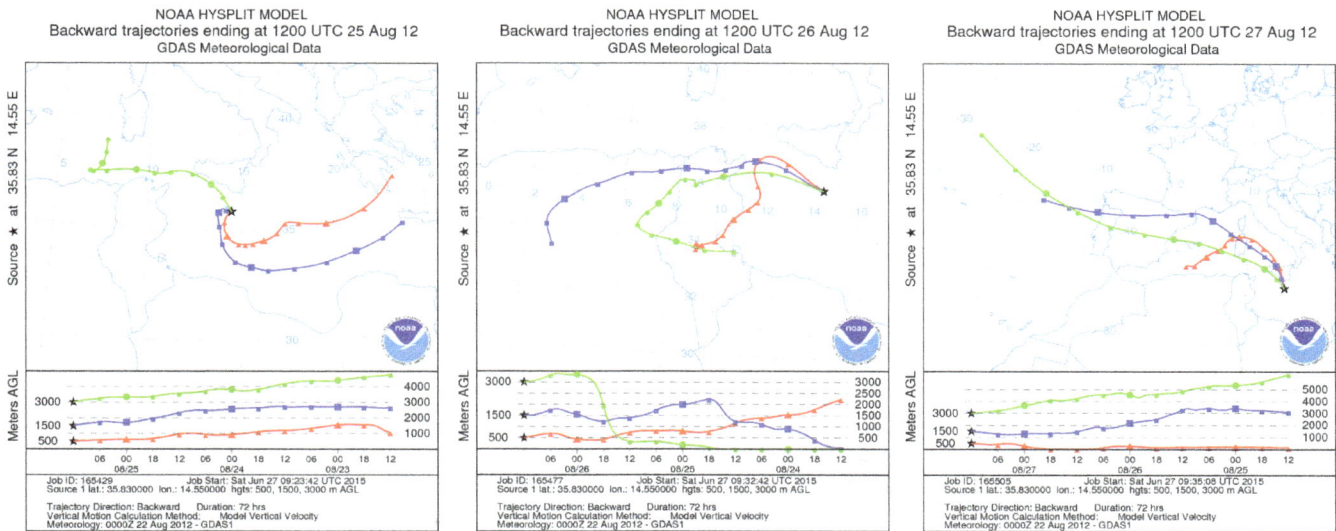

Figure 3. Air-mass backward trajectories at 500, 1500 and 3000 m a.g.l. obtained from NCEP/NCAR reanalysis data and HYSPLIT model at Marsaxlokk in the period 25–27 August 2012.

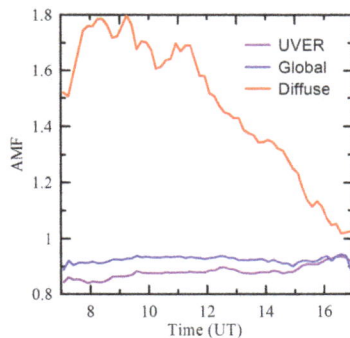

Figure 4. Daily evolution Aerosol Amplification Factors, AMF_{UVER} (purple) AMF_G (blue) and AMF_D (red), on 26 August 2012 at Marsaxlokk, Malta.

The AMFs were evaluated for 26 August. The influence of aerosol events on UVER, global and diffuse solar irradiances is compared in Fig. 4. The results show that AMF_{UVER} (purple) values increase from 0.8 to 0.9 after four o'clock GMT. The AMF_G does not show changes during the day; its value is around 0.9. On 26 August, the AMF_D shows high values in the morning and lower in the afternoon, and AMF_D values are higher than one throughout the whole day. The AMF_{UVER} shows lower values than AMF_G, indicating a more important aerosol effect on erythemal than on global solar irradiance, similar result having been obtained by Bilbao et al. (2014). In the afternoon, AMF_D decreases, probably due to a fall in the aerosol concentration. This result may be confirmed taking into account the air mass origin in Fig. 3 (27 August). The desert dust event observed in Malta with MODIS data of AOD_{550} and air-mass back trajectories is also corrobo-

rated with the evolution (not shown here) of aerosol properties (AOD), Ångström exponent, among others at the nearby site of Lampedusa island (e.g., Meloni et al., 2003).

5 Conclusions

A solar radiation measurement campaign was carried out in the south-eastern village of Marsaxlokk (35°50′ N; 14°33′ E; 10 m a.s.l), Malta, between 15 May and 15 October 2012. Observations of erythemal (UVER; 280–400 nm) and global and diffuse horizontal irradiances (G; D; 305–2800 nm), TOC, water-vapour column, AOD and Ångström exponent were performed during the campaign, in south-east Malta.

The agreement of satellite-derived data with ground-based measured data was studied so as to validate satellite data. The TOC daily values from OMI were compared with Microtops measurements and the results showed a high correlation coefficient of 0.95. The frequency distribution of the differences showed a standard deviation of 6 DU. The daily water vapour column measured by Microtops and retrieved from MODIS was compared and the frequency distribution of the differences showed a standard deviation of 0.5 cm. A comparison between AOD_{1020} from MODIS and from Microtops showed a correlation coefficient of 0.68 and an RMSE of 0.6. The frequency distribution of said values have a standard deviation of 0.06.

Three different clear-sky models for solar UVER, global and diffuse were calibrated taking into account that solar irradiance values depend on SZA, TOC and aerosols. The RMSE values obtained ranged between 6.6 and 4.5 % for UVER and global solar, while for the diffuse irradiance of 19 %. It was observed that the statistical estimators give better results for solar global irradiance followed by UVER and diffuse.

Aerosol effects have been considered as a case study of dessert aerosol influence. The AMFs were evaluated on 26 August. The influence of the aerosol events on UVER, global and diffuse solar irradiances was compared. The results showed that on 26 August 2012, the AMF_{UVER} and AMF_G changed during the day and their values were around 0.9. The AMF_D showed high values in the morning and lower in the afternoon. The AMF_{UVER} displays lower values than AMF_G, pointing to a more important effect of aerosols on erythemal solar irradiance than on global. A similar result was obtained by Bilbao et al. (2014). The evolution of the AMF_D values was due to an increase in AOD observed by daily backward air-mass trajectories. In the afternoon, both AMF_D and AMF_{UVER} decreased, probably due to a drop in aerosol concentration. This result may be confirmed taking into account the air-mass origin in Fig. 3 (27 August).

Acknowledgements. The authors gratefully acknowledge the support of the Spanish Research and Economy Ministry through Projects CGL2010-25385 and CGL2010-12410-E. The authors also would like to thank the European Space Agency for satellite products, as well as the principal investigator D. Meloni for making the Lampedusa AERONET data available. The authors also thank the NOA Research Reanalysis department for the data obtained.

References

Batlles, F. J., Olmo, F. J., and Alados-Arboledas, L.: On shadow-band correction methods for diffuse irradiance measurements, Sol. Energy, 54, 105–114, 1995.

Bilbao, J. and Miguel, A.: Contribution to the study of UV-B solar radiation in Central Spain, Renew. Energy, 53, 79–85, 2013.

Bilbao, J., Román, R., and De Miguel, A.: Total ozone column, water vapour and aerosol effects on erythemal total solar shortwave irradiance in Marsaxlokk, Malta, Atmos. Environ., 99, 508–518, doi:10.1016/j.atmosenv.2014.10.005, 2014.

CIE (Commission Internationale de l' Eclairage): Erythema Reference Action Spectrumand Standard Erythema Dose, CIE S007E-1998, CIE Central Bureau, Vienna, Austria, 1998.

De Miguel, A. and Bilbao, J.: Test reference year generation from meteorological and simulated solar radiation data, Sol. Energy, 78, 695–703, doi:10.1016/j.solener.2004.09.015, 2005.

De Miguel, A., Mateos, D., Bilbao, J., and Román, R.: Sensitivity analysis of ratio between ultraviolet and total shortwave solar radiation to cloudiness, ozone, aerosols and precipitable water, Atmos. Res., 102, 136–144, doi:10.1016/j.atmosres.2011.06.019, 2011.

De Miguel, A., Bilbao, J., Román, R., and Mateos, D.: Measurements and attenuation of erythemal radiation in Central Spain, Int. J. Climatol., 32, 929–940, doi:10.1002/joc.2319, 2012.

Dubrovsky, M.: Analysis of UV-B irradiances measured simultaneously at two stations in the Czech-Republic, J. Geophys. Res., 105, 4907–4913, doi:10.1029/1999JD900374, 2000.

Esteve, A. R., Martinez-Lozano, J. A., Marin, M. J., Estelles, V., Tena, F., and Utrillas, M. P.: The influence of ozone and aerosols on the experimental values of UV erythemal radiation at ground level in Valencia, Int. J. Climatol., 29, 2171–2182, doi:10.1002/joc.1847, 2009.

Fioletov, V. E., McArthur, L. J. B., Mathews, T. W., and Marrett, L.: On the relationship between erythemal and vitamin D action spectrum weighted ultraviolet radiation, J. Photochem. Photobiol. B Biol., 95, 9–16, doi:10.1016/j.jphotobiol.2008.11.014, 2009.

Gómez-Amo, J. L., DiSarra, A., Meloni, D., Cacciani, M., and Utrillas, M. P.: Sensitivity of shortwave radiative fluxes to the vertical distribution of aerosol single scattering albedo in the presence of a desert dust layer, Atmos. Environ., 44, 2787–2791, 2010.

Gómez-Amo, J. L., Pinti, V., Di Iorio, T., Di Sarra, A., Meloni, D., Becagli, S., Bellantone, V., Cacciani, M., Fua, D., and Perrone, M. R.: The June 2007 Saharan dust event in the central Mediterranean: Observations and radiative effects in marine, urban, and sub-urban environments, Atmos. Environ., 45, 5385–5393, 2011.

Hülsen, G. and Gröbner, J.: Characterization and calibration of ultraviolet broadband radiometers measuring erythemally weighted irradiance, Appl. Opt., 46, 5877–5886, 2007.

IPCC: Climate change 2007-the physical science basis: Working group I contribution to the fourth assessment report of the IPCC (Vol. 4), edited by: Solomon, S., Cambridge University Press, 129–235, 2007.

Kaskaoutis, D. G., Nastos, P. T., Kosmopoulos, P. G., and Kambezidis, H. D.: The combined use of satellite data, air-mass trajectories and model applications for monitoring dust transport over Athens, Greece, Int. J. Remote Sens., 31, 5089–5109, doi:10.1080/01431160903283868, 2010a.

Kaskaoutis, D. G., Kosmopoulos, P. T., Kambezidis, H. D., and Nastos, P. T.: Identification of the Aerosol Types over Athens, Greece: The Influence of Air-Mass Transport, Adv. Meteorol., 2010, 15 pp., doi:10.1155/2010/168346, 2010b.

Kaufman, Y. J., Tanré, D., Remer, L. A., Vermote, E. F., Chu, A., and Holben, B. N.: Operational remote sensing of tropospheric aerosol over land from EOS moderate resolution imaging spectroradiometer, J. Geophys. Res., 102, 17051–17076, 1997.

Koepke, P., Reuder, J., and Schwander, H.: Solar UV radiation and its variability due to the atmospheric components, Recent Res. Dev. Photochem. Photobiol., 6, 11–34, 2002.

Madronich, S.: Analytic formula for the clear-sky UV index, Photochem. Photobiol., 83, 1537–1538, 2007.

Mateos, D., De Miguel, A., and Bilbao, J.: Empirical models of UV total radiation and cloud effect study, Int. J. Climatol., 30, 1407–1415, doi:10.1002/joc.1983, 2010.

Mateos, D., Bilbao, J., Kudish, A. I., Parisi, A. V., Carbajal, Di Sarra, G., Román, R., and De Miguel, A.: Validation of OMI satellite erythemal daily dose retrievals using ground-based measurements from fourteen stations, Remote Sens. Environ., 128, 1–10, doi:10.1016/j.rse.2012.09.015, 2013.

Mateos, D., Pace, G., Meloni, D., Bilbao, J., Di Sarra, A., de Miguel, A., Casasanta, G., and Min, Q.: Observed influence of liquid cloud microphysical properties on ultraviolet surface radiation, J. Geophys. Res.-Atmos., 119, 2429–2440, doi:10.1002/2013JD020309, 2014a.

Mateos, D., Di Sarra, A., Bilbao, J., Meloni, D., Pace, G., de Miguel, A., and Casasanta, G.: Spectral attenuation of global and diffuse UV irradiance and actinic flux by clouds, Q. J. Roy. Meteorol. Soc., 141, 109–113, doi:10.1002/qj.2341, 2014b.

Meloni, D., Marenco, F., and Di Sarra, A.: Ultraviolet radiation and aerosol monitoring at Lampedusa, Italy, Ann. Geophys. Italy, 46, 373–383, 2003.

McKinlay, A. F. and Diffey, B. L.: A Reference Action Spectrum for Ultraviolet Induced Erythema in Human Skin, Commission Internationale de l'Eclairage (CIE), 6, 17–22, 1987.

Pace, G., di Sarra, A., Meloni, D., Piacentino, S., and Chamard, P.: Aerosol optical properties at Lampedusa (Central Mediterranean). 1. Influence of transport and identification of different aerosol types, Atmos. Chem. Phys., 6, 697–713, doi:10.5194/acp-6-697-2006, 2006.

Perez, R., Ineichen, P., and Seals, R.: Modelling daylight availability and irradiance components from direct and global irradiance, Sol. Energy, 44, 271–289, 1990.

Román, R., Antón, M., Valenzuela, A., Gil, J. E., Lymani, H., de Miguel, A., Olmo, F. J., Bilbao, J., and Alados-Arboledas, L.: Evaluation of the desert dust effects on global, direct and diffuse spectral ultraviolet irradiance, Tellus B, 65, 19578, doi:10.3402/tellusb.v65i0.19578, 2013.

Román, R., Bilbao, J., and de Miguel, A.: Uncertainty in water vapor column, aerosol optical depth and Angström Exponent, and its effect on radiative transfer simulations in the Iberian Peninsula, Atmos. Environ., 89, 556–569, doi:10.1016/j.atmosenv.14.02.027, 2014.

Tanskanen, A., Lindfors, A., Määttä, A., Krotkov, N., Herman, J., Kaurola, J., Koskela, T., Lakkala, K., Fioletov, V., Bernhard, G., McKenzie, R., Kondo, Y., O'Neill, M., Slaper, H., den Outer, P., Bais, A. F., and Tamminen, J.: Validation of daily erythemal doses from Ozone Monitoring Instrument with ground-based UV measurement data, J. Geophys. Res., 5, D24S44, doi:10.1029/2007JD008830, 2007.

UNEP (United Nations Environment Programme): Environmental Effects of Ozone Depletion and its Interactions with Climate Change: 2010 Assessment, ISBN 92-807-2312-X, 2010.

Webb, A. R.: Who, what, where and when-influences on cutaneous vitamin D synthesis, Prog. Biophys. Mol. Biol., 92, 17–25, 2006.

Webb, A. R., Slaper, H., Koepke, P., and Schmalwieser, A. W.: Know your standard: clarifying the CIE erythema action spectrum, Photochem. Photobiol. 87, 483–486, 2011.

WHO: Global Solar UV Index: A Practical Guide, World Health Organization: Geneva, Switzerland, ISBN 92-4-1590076, 2002.

8

Ten years water and energy surface balance from the CNR-ISAC micrometeorological station in Salento peninsula (southern Italy)

P. Martano[1], C. Elefante[2], and F. Grasso[1]

[1]CNR-Istituto di Scienze dell'Atmosfera e del Clima – UOS Lecce, Via Monteroni, 73100 Lecce, Italy
[2]Ripartizione Informatica, Università del Salento, Viale Gallipoli 49, 73100 Lecce, Italy

Correspondence to: P. Martano (p.martano@isac.cnr.it)

Abstract. Data of surface-atmosphere energy and water transfer from a ten years (2003–2013) period of activity of the ISAC-Lecce micrometeorological station (http://www.basesperimentale.le.isac.cnr.it) have been analyzed: to the authors' knowledge this is the first decadal data set of surface-atmosphere transfer in Salento peninsula. The surface energy budget shows a tendency to a positive bias possibly due to several reasons that require more investigations. Some suitable indices related to the surface water balance, such as the precipitation intensity, the aridity index and the ground water infiltration fraction have been calculated. Possible trends of these annual averages in the decadal period are considered, also taking into account the statistical uncertainty associated to measurement errors and missing data. The results indicate a significant increasing in the precipitation intensity together with an experimental evidence of increasing of the ground water infiltration in the measurement area, that is in agreement with recent estimations for the whole Salento peninsula. On the other hand, recent studies show that seawater intrusion and salinization of the deep underground aquifer keep increasing in the same period.

1 Introduction

Semi-arid coastal regions are common in Mediterranean areas. The correct assessment of the surface water and energy balance can be difficult for such regions, due the extreme changes in the surface moisture conditions that require direct measurements. Indeed reliable estimates of evapotranspiration generally based on the potential evaporation concept (Allen et al., 1998) are not easy to obtain over surfaces that are often in very dry conditions. In this context evapotranspiration constitutes one of the less known components of the surface water budget especially for climatic studies, and long term time series of measurements are often not easy to be found. Long term measurements of the surface water balance can be also helpful for the management of the often scarce fresh water resources. Indeed coastal groundwater aquifers are often the only fresh water supply for the local populations, and imbalance between drainage and recharge by surface infiltration can lead to loss of pressure and seawater intrusion in the aquifer, significantly degrading water

quality (Custodio, 2010; Delle Rose et al., 2000). Salento peninsula, in the south-east of the Italian peninsula (Fig. 1) is a semiarid region characterized by an almost plane surface with a small orographic relief in the south-west part (200 m height). The non-urbanized lands are generally covered by olive groves, vineyards or arable lands, mostly non-irrigated, while few remaining natural extensions are characterized by pine forests or mixed Mediterranean shrubs. The precipitation climatology is typically Mediterranean with warm dry summers and quite mild wet winters. The average annual precipitation is about 650 mm but with a certain variation within tens of kilometres due to the particular geographical shape of the southern end. This, together with the presence of the low orographical relief, triggers air lifting and a convergence zone, thus enhancing local precipitation in this area (Fig. 1). The average annual temperature is about 16–17 °C, with a spatial variation of about 1 °C from the east to the west coast that is not directly exposed to the mistral winds. Indeed the wind regime has two main components: south-west, prevailing in cyclonic conditions, and north-west, that

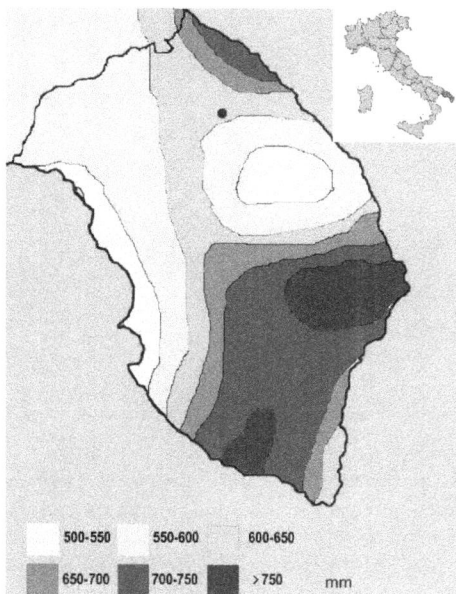

Figure 1. Salento peninsula with averaged 1921–1996 precipitation distribution (mm) (adapted from http://www.supermeteo.it). The dot indicates the measurements site.

are quite common in this area due to the channelling effect of the Otranto channel, and are prevailing in anti-cyclonic conditions. Hydrology is characterized by ephemeral stream systems and karstic geological characteristics that allow the presence of a deep good quality ground water aquifer, the major source of freshwater for anthropic use (more than 80% of the total fresh water necessity in the area). Monitoring and modelling studies of the aquifer balance in the last decades (Margiotta and Negri, 2005; Portoghese et al., 2005; Romanazzi and Polemio, 2013) showed a decreasing aquifer pressure with increasing salty seawater intrusion. Thus, efforts have been recently made to specifically modelling the groundwater flow in the fractured karst aquifer (Giudici et al., 2012; De Filippis et al., 2013). The aim of this work is to present a data set of long term experimental measurements of the surface energy and water balance as measured by the CNR-ISAC micrometeorological station in Salento peninsula (the only available long term data set in Salento, to the authors' knowledge), and to give an experimental support to the hydrological modelling for the local water resources management.

2 Data processing

2.1 Data collection

The CNR-ISAC micrometeorological base is placed in a suburban area 5 km south-west from Lecce (Fig. 1), within the Salento University campus (Lat: 40°20′12″ Lon: 18°7′17″) characterized by mixed typical local vegetation (pines, olive

groves and Mediterranean shrubs) with an increasing content of buildings and non-natural surfaces. Buildings and trees have an average height of about 10 m and with estimated roughness length z_0 and displacement height d of about 0.5 and 7 m respectively (Martano, 2000). A 16 m high mast is equipped with a fast response eddy covariance system and standard meteorological sensors for routinely collecting half-hour averaged data of air temperature and humidity. The mast height allows a flux footprint fetch of the order of several hundreds of meters (Hsieh et al., 2000), thus taking contributions from the campus and some immediately surrounding vegetated areas. An ancillary Campbell meteorological station collects measurements of air temperature, air humidity and net radiation and soil surface measurements of temperature, soil moisture and soil heat flux. Data are averaged and stored either in a data logger (meteorological station) or in a dedicated netbook for fast response data processing (eddy covariance system), that outputs half-hour averaged turbulent flux data in the streamline coordinate system (McMillen, 1988). Before the final storage in the web database as 30 min averages, data undergo a quality control that eliminates out of range measurements generated by possible malfunctioning of the sensors. A site map and a more detailed description of the station are available elsewhere (http://www.basesperimentale.le.isac.cnr.it, Martano et al. 2013).

2.2 Time series post processing

The 30 min averaged data time series have been post processed trying to minimize the possible uncertainties and errors coming from the measurement procedure, such as the eddy covariance method for the turbulent fluxes of heat and water vapor (Foken and Wichura, 1996). Massman (2000) spectral corrections and mass flux Webb corrections (Webb et al., 1980) have been applied. In addition an estimation of the uncertainty to be associated to the annual cumulative and/or average value has been attempted. It has been defined as the maximum between the uncertainty associated to the measurement procedure (instrumental uncertainty for slow response measurements and statistical uncertainty for eddy covariance measurements), and the uncertainty associated to gaps of data lacking in the time series (however, annual time series with more than 1 month total data gap have not been used). The statistical uncertainty E_{ec} associated to the eddy covariance fluxes $< w'q' >$ was estimated as the standard deviation of the instantaneous measurements $w'q'$ in the averaging time (30 min).

An evaluation of the uncertainties associated to the presence of gaps of dN data in the (annual) time series of N (30 min averaged) measurements is proposed as (if $dN \ll N$):

$E_{ave} = \sigma_N (dN/N)^{1/2}$ for average values and $E_{cum} = Q dN/N$ for cumulant values (total sum) where

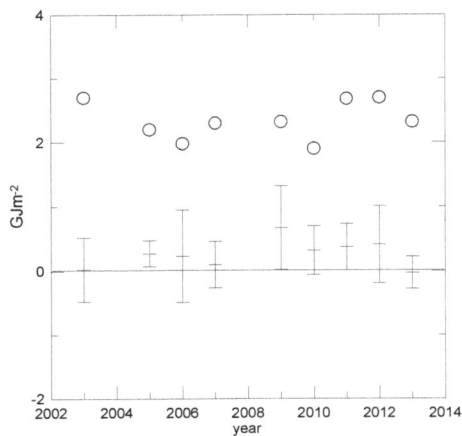

Figure 2. Total annual surface energy balance ($R_n - G - LE - H$) compared with the measured annual net radiation (R_n, circles).

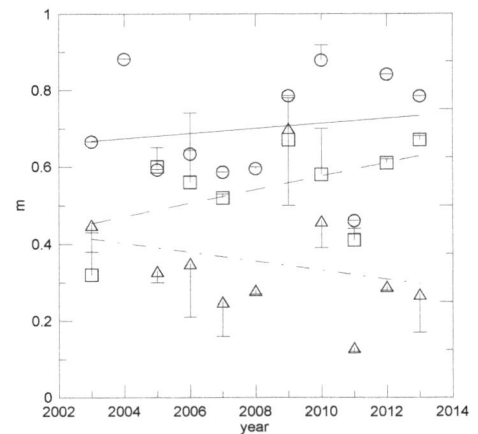

Figure 3. Annual averages of precipitation (m, circles), water infiltration fraction (squares) and aridity index (triangles) with estimated maximum uncertainties. The continuous (P) dashed (WIF) and dash-dotted (AI) lines are least squares linear regressions.

σ_N is the standard deviation of the (annual) time series and Q its sum.

Besides directly measured quantities some other derived indices are used in the following analysis:

 – Evaporative Fraction $EVF = LE/(LE + H)$ where LE and H are respectively the latent and sensible heat fluxes.

 – Aridity Index $AI = P/ET_0$ where P and ET_0 are the precipitation and the potential evapotranspiration calculated by the Penman–Monteith formula (Allen et al., 1998).

 – Water Infiltration Fraction $WIF = (P - Ev)/P$ where Ev is the actual (measured) evapotranspiration and runoff is neglected, that is a very reasonable assumption in the measurement area, and quite generally also in Salento peninsula (Portoghese et al., 2005).

 – Precipitation Intensity $PI = <P>_{wd}$ where $<...>_{wd}$ indicates the time series average for wet days only (days in which $P > 1$ mm).

In addition the North Atlantic Oscillation Index NAO (difference between the measured sea level atmospheric pressure in Lisbon and Reykjavik) has been computed for the same periods by web data (http://www.cpc.ncep.noaa.gov/products/precip/CWlink/pna/nao.shtml).

3 Results and discussion

The annual averages of the energy balance ($R_n - G - LE - H$, where R_n is the net radiation and G the soil heat flux at 2 cm depth) are shown in Fig. 2. Although the budget appears to be generally closed within the quite large uncertainties, a tendency to a positive imbalance is also apparent. Several reasons can contribute: from the

divergence of the soil heat flux between the surface and 2 cm depth, to the tendency of the eddy correlation system to underestimate the turbulent fluxes due to the finite averaging time (Cava et al., 2008). Even possible uncertainties in the calibration of the net radiometer after suffering strong rain events with dome breaking cannot be excluded, and all will require a separate analysis. Figure 3 shows some trends for the water balance: total precipitation, water infiltration fraction and aridity index. Here and in the next figure two tests have been performed to check the decadal trends. In the first the sign of the trend has been checked after deleting the first and the last annual average in the series, while in the second the statistical uncertainty of the regression slope was calculated to verify whether a change of sign of the slope were possible within the slope uncertainty. The trend is considered significant if both tests give negative result (no change in the slope sign), marginally significant if only one of them gives negative result, and not significant if both give positive result. In Fig. 3 the trends are marginally significant and positive for both total precipitation and water infiltration fraction, while the AI is decreasing, indicating a probable decrease in the year-averaged surface soil moisture, as AI is quite well correlated with local soil moisture measurements when available (correlation coefficient 0.85 for annual averages). The apparent disagreement between these last results (increasing precipitation and infiltration, and decreasing surface soil moisture) is perhaps less surprising when observing Fig. 4, that shows a significant positive trend for the total precipitation intensity. It is possible that the increasing concentration of precipitation events in an otherwise semi-arid region with karstic geologic features and generally negligible slopes and runoff is likely to increase infiltration more than evaporation. This is because the soil surface becomes wet within short time intervals during the year, while it is going to dry up quickly in a few days after rain, preventing strong long term evaporation. The NAO negative trend also shown in Fig. 4 suggests that the locally measured increasing precipitation intensity could be linked to a regional trend of increasing precipitation associated to the decreasing NAO phase during the decade of observations (Willems, 2013). A recent statistical downscaling analysis also found a scenario of increasing precipitation intensity, over this region for the next decades, especially in the dry season (Palatella et al., 2010). These results suggest a potentially increasing availability of water recharge for the groundwater aquifers during the past decade in the measurement area. Although the extension of this experimental result to the whole Salento peninsula would require more measurement sites, this possibility is suggested by the following remarks. Indeed the measurement site is far from the maximum of the precipitation distribution (Fig. 1) and characterized by Mediterranean arboreus vegetation that tends to enhance and stabilize evapotranspiration with respect to bare soil or dry shrub Mediterranean areas (Scanlon et al., 2006). Thus the annual difference between measured precipitation and evapotranspiration is not expected to overestimate the average over the Salento peninsula and the measured positive trend of the surface infiltration also appears as an experimental confirmation of recent estimations for the whole peninsula (Portoghese et al., 2013). This trend can be constrasted with a recent analysis of the available data from monitoring wells by Fidelibus and Tulipano (2014), showing that the freshwater column thickness of the local deep aquifer keeps decreasing during the last decade, together with an increasing height of the freshwater-seawater interface under Salento peninsula. In synthesis, the water surface budget measurements presented here give an experimental evidence of an increasing availability of potential groundwater recharge by infiltration, still not affecting the increasing salinization and marine intrusion in the Salento deep aquifer in the last decade.

Acknowledgements. The authors wish to acknowledge the PON 2007-13 I-AMICA (OR1) project (http://www.i-amica.it), and the CNR GIIDA project (http://www.dta.cnr.it/content/view/2735/244/lang,en/), for partially supporting the micrometeorological base and the database, and the HyMeX project (http://www.hymex.org) for the scientific support.

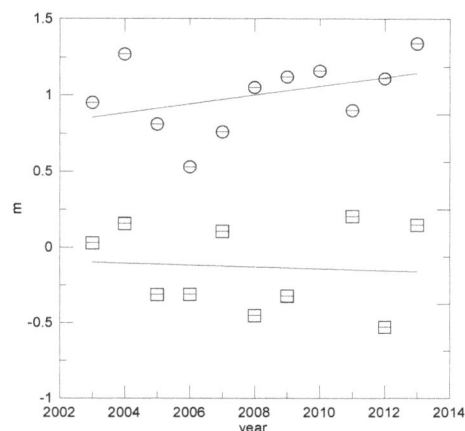

Figure 4. Annual averages of precipitation intensity (m, circles) and NAO (squares). The straight lines are least squares regressions.

References

Allen, R. G., Pereira, L. S., Raes, D., and Smith, M.: Crops Evapotranspiration: Guidelines for computing crop Water Requirements, FAO Irrigation and Drainage Paper 56, FAO, Rome, Italy, 300 pp., 1998.

Cava, D.,Contini, D., Donateo, A., and Martano, P.: Analysis of short-term closure of the surface energy balance above short vegetation, Agr. Forest Meteorol., 148, 82–93, 2008.

Custodio, E.: Coastal aquifers of Europe: an overview, Hydrogeol. J., 18, 269–280, 2010.

De Filippis, G., Giudici, M., Margiotta, S., Mazzone, F., Negri, S., and Vassena, C.: Numerical modelling of the groundwater flow in the fractured and karst aquifer of the Salento peninsula (Southern Italy), Acque Sotterranee, 1, 17–28, 2013.

Delle Rose M., Federico A., Fidelibus C.. A computer simulation of groundwater salinization risk in Salento peninsula (Italy), in: Risk Analysis II, edited by: Brebbia, C. A., Wessex Institute of Tecnology, Southampton, 465–475, 2000.

Fidelibus, M. D. and Tulipano, L.: Monitoring seawater intrusion by means of long-term series of EC and T logs (Salento coastal karstic aquifer, Southern Italy), SWIM 23rd SaltWater Intrusion Meeting, 16–20 June 2014, Husum, Germany, 2014.

Foken, T. and Wichura, B.: Tools for quality assessment of surface-based flux measurements, Agr. Forest Meteorol., 78, 83–105, 1996.

Giudici, M., Margiotta, S., Mazzone, F., Negri, S., and Vassena, C.: Modelling hydrostratigraphy and groundwater flow of a fractured and karst aquifer in a Mediterranean basin (Salento peninsula, southeastern Italy), Environ. Earth Sci., 67, 1891–1907, 2012.

Hsieh, C. I., Katul, G., and Chi, T.: An approximate analytical model for footprint estimation of scalar fluxes in thermally stratified atmospheric flows, Adv. Water Resour., 23, 765–772, 2000.

Margiotta, S. and Negri, S.: Geophysical and stratigraphical research into deep groundwater and intruding seawater in the mediterranean area (the Salento Peninsula, Italy), Nat. Hazards Earth Syst. Sci., 5, 127–136, doi:10.5194/nhess-5-127-2005, 2005.

Martano, P.: Estimation of Surface Roughness Length and Displacement Height from Single-Level Sonic Anemometer Data, J. Appl. Meteorol., 39, 708–715, 2000.

Martano, P., Elefante, C., and Grasso, F.: A database for long term atmosphere-surface transfer monitoring in Salento Peninsula (Southern Italy), Dataset Papers Geosci., 2013, 946431, doi:10.7167/2013/946431, 2013.

Massman, W. J.: A simple method for estimating frequency response corrections for eddy covariance systems, Agr. Forest Meteorol., 104, 185–198, 2000.

McMillen, R.: An eddy correlation technique with extended applicability to non-simple terrain, Bound.-Lay. Meteorol., 43, 231–245, 1988.

Palatella, L., Miglietta, M. M., Paradisi, P., and Lionello, P.: Climate change assessment for Mediterranean agricultural areas by statistical downscaling, Nat. Hazards Earth Syst. Sci., 10, 1647–1661, doi:10.5194/nhess-10-1647-2010, 2010.

Portoghese, I., Uricchio, V., and Vurro, M.: A GIS tool for hydrogeological water balance evaluation on a regional scale in semi-arid environments, Comput. Geosci., 31, 15–27, 2005.

Portoghese, I., Bruno, E., Dumas, P., Guyennon, N., Hallegatte, S., Hourcade, J. C., Nassopoulos, H., Pisacane, G., Struglia, M. V., and Vurro, M.: Impacts of Climate Change on Freshwater Bodies: Quantitative Aspects, in: Regional Assessment of Climate Change in the Mediterranean, Advances in Global Change Research 50, Ch. 9, edited by: Navarra, A. and Tubiana, L., Springer, Dordrecht, 2013.

Romanazzi, A. and Polemio, M.: Modelling of coastal karst aquifers for management support: a case study of Salento (Apulia, Italy), Ital. J. Eng. Geol. Environ., 1, 65–83, 2013.

Scanlon, B. R., Keese, K. E., Flint, A. L., Flint, L. E., Gaye, C. B., Edmunds, W. M., and Simmers, I.: Global synthesis of groundwater recharge in semiarid and arid regions, Hydrol. Process., 20, 3335–3370, 2006.

Webb, E. K., Pearman, G. I., and Leuning, R.: Correction of flux measurements for density effects due to heat and water vapour transfer, Q. J. Roy. Meteorol. Soc., 106, 85–100, 1980.

Willems, P.: Adjustment of extreme rainfall statistics accounting for multidecadal climate oscillations, J. Hydrol., 490, 126–133, 2013.

9

Urban warming in villages

J. Lindén[1], **C.S.B. Grimmond**[2], and **J. Esper**[1]

[1]Department of Geography, Johannes-Gutenberg University, Mainz, Germany
[2]Department of Meteorology, University of Reading, Reading, UK

Correspondence to: J. Lindén (j.linden@geo.uni-mainz.de)

Abstract. Long term meteorological records (> 100 years) from stations associated with villages are generally classified as rural and assumed to have no urban influence. Using networks installed in two European villages, the local and microclimatic variations around two of these rural-village sites are examined. An annual average temperature difference (ΔT) of 0.6 and 0.4 K was observed between the built-up village area and the current meteorological station in Geisenheim (Germany) and Haparanda (Sweden), respectively. Considerably larger values were recorded for the minimum temperatures and during summer. The spatial variations in temperature within the villages are of the same order as recorded over the past 100+ years in these villages (0.06 to 0.17 K/10 years). This suggests that the potential biases in the long records of rural-villages also warrant careful consideration like those of the more commonly studied large urban areas effects.

1 Introduction

Changes in the surface energy balance of urban areas caused by, for example, increased thermal admittance of urban materials, limited radiative and advective cooling (due to urban morphology), lowered evapotranspiration cooling (due to sealed surfaces and reduced vegetation coverage), and additional anthropogenic heat release, tend to cause increased temperatures in urban areas compared to surrounding rural environments (e.g. Arnfield, 2003). This well documented urban heat island (UHI) effect is generally most pronounced in larger settlements with dense, tall built structures and sparse vegetation (e.g. Oke, 1982), but observable UHI effects are found in towns (e.g. Magee et al., 1999; Steeneveld et al., 2011) and small villages (Hinkel et al., 2003).

To assess long-term temperature trends, meteorological (met) stations are classified according to potential urban influence, e.g. "associated with urban area" or "rural", where "rural" is considered to have no significant urban warming bias (e.g. Hansen et al., 2010). However, metadata from long-term "rural" stations often reveal that these stations are located in or near villages. The influences on met stations associated with large cities (e.g. populations > 100 000) are well studied but the potential bias in small built-up urban areas,

such as villages (e.g. populations < 15 000, note this is not a precisely defined term), are not.

The objective of this study is to assess local and microclimatic variations around two long-term, rural-village, met stations, relative to their long-term records. The analysis uses air temperature sensors installed which are representative of the current and past met stations locations, as well as the general area.

2 Methods

A network of air temperature sensors was installed around a long-term Deutsche Wetterdienst (DWD) station in Geisenheim, Germany and a Swedish Meteorological and Hydrological Institute (SMHI) station, Haparanda in Sweden (Table 1). In the Global Historical Climatological Network (GHCN) database (http://www.ncdc.noaa.gov/ghcnm/) both are classed as rural. These villages currently have 11 000 and 5000 inhabitants, respectively (Hessische Statistische Landesamt, 2014; Statistiska Centralbyrån, 2014). Since the sites were initiated both villages have more than trebled their population (Table 1) and size of their built area increased, although the urban centres remains similar in structure and density. These two GHCN sites were selected based on their location, length of record, size of population, climate zone

Table 1. Description of examined met stations and associated village. Source of data are the meta-data archives of DWD (Deutscher Wetterdienst, Klima und Umwelt, Climate Data Centre, Frankfurter Straße 135, 63067 Offenbach, Germany) and SMHI (SMHI, Folkborgsvägen 17, 601 76 Norrköping, Sweden).

Station/ village name	Location	Altitude (m a.s.l.)	Climate	Station initiation year	Village pop. current/ initiation year	Description of built structure	Topography
Geisenheim	50.0° N 8.0° E	95	Temperate maritime	1882	11 000 3000	Dense village centre, mainly stone houses and impervious surfaces	hilly, up to 8° slope
Haparanda	65.8° N 24.1° E	4	Sub-Arctic	1859	4900 1000	Open structure, building material often wood, abundant vegetated surfaces	flat

Table 2. Description of temperature sensor network.

	Location	Site description/sensor placement (period station was there)	Elevation (m)
Geisenheim	Village centre (cen)	Open-set mid-rise built structure/on free standing post near central square	92
	River (riv)	Low plants, scattered trees, water/on post next to small road	83
	Residential (res)	Open set low-rise built structure/on street post	124
	Park (park)	Scattered trees/in park near the 1st location of the met station (1888–1915)	100
	Vineyard (Vin/met.stn.)	Scattered trees/on post in vineyard 155 m away from met station (1977–2006), and 180 m current met station (2006–ongoing)	116
Haparanda	Village centre (cen)	Open-set low-rise built structure/on post in vegetated yard surrounded by wooden 2-floor buildings, 40 m away from met station location (1859–1942)	5
	River (riv)	Low plants, scattered trees, water/on post, ~ 150 m from met station location (1942–1977)	2
	Residential 1 (res1)	Open set low-rise built structure/on tree in garden, 3 m away from met station location (1980–2005)	10
	Residential 2 (res2)	Open set low-rise built structure/on post of met station (2005–2010) and 230 m away from station (1977–2005 – parallel stations)	6
	Met station (met.stn)	Scattered and dense trees/ on fence surrounding current met station (2010–ongoing)	9

and logistical possibilities following advice (personal communication from SMHI and DWD staff). The different characteristics (e.g. regional climates, built form, materials) are summarized in Table 1.

Temperature sensors (HOBO Pro v2 U23-001 in radiation shields RS1, Onset Computer Corporation, Bourne, MA 02532, USA) were installed at multiple sites in the village and vicinity (Table 2, Fig. 1) in mid-2013. Analysis of the meta-data files in the DWD and SMHI archives was used to identify past met-station locations so they, or comparable locations, could be instrumented. The sensors record a sample every 30 min. Analysis of the pre-deployment 22 day inter-instrument comparison, over a −4 to 18° C range, found the mean difference to be < 0.05 K, with < 5 % exceeding ±0.1 K (max 0.3 K). These inter-sensors differences are not removed so these values provide a measurement error. The data analyzed in this paper are for 321 and 352 days (data collection is on-going) in Geisenheim and Haparanda.

The Geisenheim station has been re-located multiple times, but due to incomplete metadata, the exact location is only known for the first and the last two locations. For the remaining three station locations, only minute-precision coordinates and short site descriptions are available. Sensors were deployed in the known sites, as well as in representative area types for the village and surroundings (Fig. 1). Although the current met station is situated to avoid urban influences (Behrendt et al., 2011) the nearby vineyard was used as a reference as unfortunately our sensor deployed at the current met site had to be removed for part of the year. The vineyard temperatures was found not to differ significantly based on analysis of concurrent data. In Haparanda the well documented previous met station locations were all identified for sensor placements (Fig. 1). The current met station is situated to avoid urban influence (Andersén, 2010) so can be used as a reference for analyzing the urban influence.

To examine potential urban effects, differences in measured maximum, average and minimum tem-

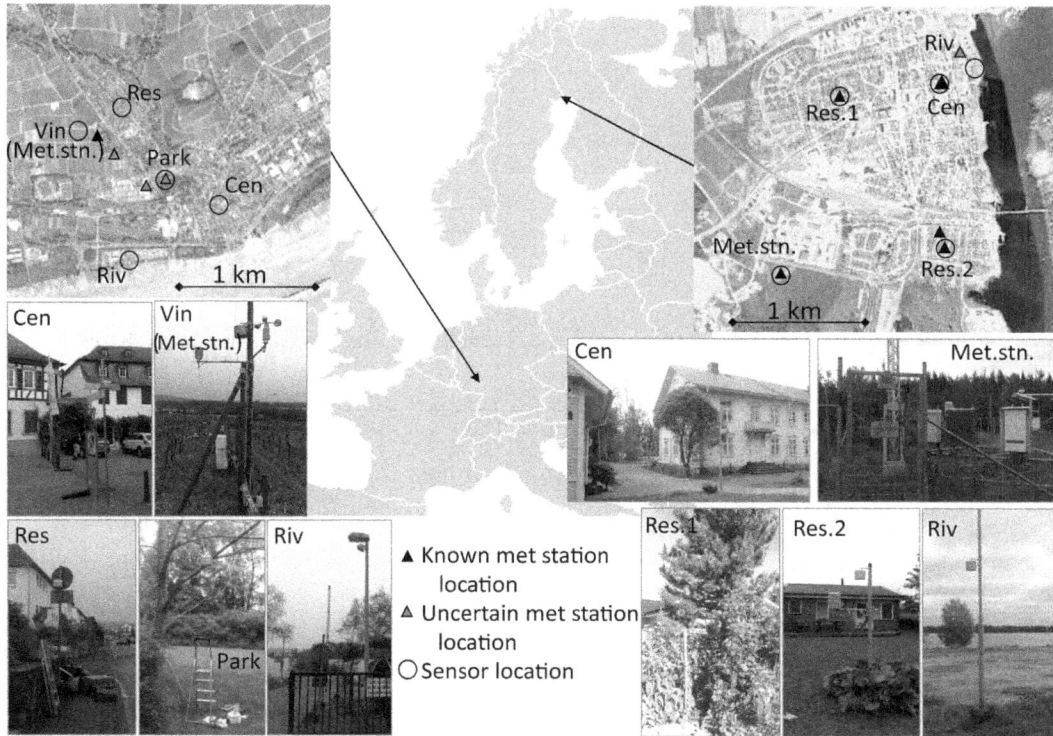

Figure 1. Location of the met stations and temperature sensors in Geisenheim (left panels) and Haparanda (right panels): previous (triangles) and current (black location known, grey location uncertain or has changed substantially since station was located there), circles mark sensor locations. Photos show locations of sensors used in this study. Satellite images from Google earth.

perature ($\Delta T_{\text{max/avg/min}}$), between the met station site and the remaining sites, were calculated for each day (e.g. $\Delta T_{\text{max/avg/min(site)}} = T_{\text{max/avg/min (site)}} - T_{\text{max/avg/min (met.stn)}}$). A t test was used to assess if the differences are significant. The DWD and SMHI station temperature data are used to calculate the long-term temperature trends. Elevation differences (up to 41 m) in Geisenheim will influence the temperatures, i.e. higher elevation sites are adiabatically cooler that lower sites (if all other characteristics are constant). However, since other factors influence each site causing differing effects whether max/avg/min temperature are considered (Blandford et al., 2008; Lindén et al., 2015), this was not accounted for in this study.

3   Results and discussion

3.1   Sensor network $_1T$

The daily differences of the averages between each site and the met station site (ΔT_{avg}) by season (Fig. 2) are positive and mostly statistically significant, indicative of a warming influence in both villages. This finding supports the expectation that the current met station locations are cooler than previous sites, given they were located in the respective villages previously. However, this study did not establish whether the current met sites have any urban influence and if locations further away from the villages would have shown lower temperatures.

The warming influence is most pronounced in the two village centres, where significantly increased daily average temperatures were found in all seasons (median yearly $\Delta T_{\text{avg(cen)}} = 0.4/0.6$ K for Haparanda/Geisenheim). The surrounding residential areas have a smaller, but still significant, warming influence (median yearly $\Delta T_{\text{avg(res)}} = 0.1/0.2$ K) revealing patterns similar to those found in large urban structures: increased urban warming with increased building density and sealed surfaces. However, the overall magnitude of temperature residuals is smaller than that found in larger settlements (e.g. Svensson and Eliasson, 2002; Unger, 2004; Offerle et al., 2006).

Evidence of a moderating influence from waterbodies were found in both villages, particularly during summer, when higher minimum and lower maximum temperatures were measured near the rivers. In Haparanda, the influence in T_{min} was similar between the river and the village centre, though this could be an effect of the proximity of the river station site to the village centre, located upwind in general wind direction (Bergström, 2007). The slightly warmer temperatures found, especially in summer, in the urban park in Geisenheim could be a consequence of the proximity to the

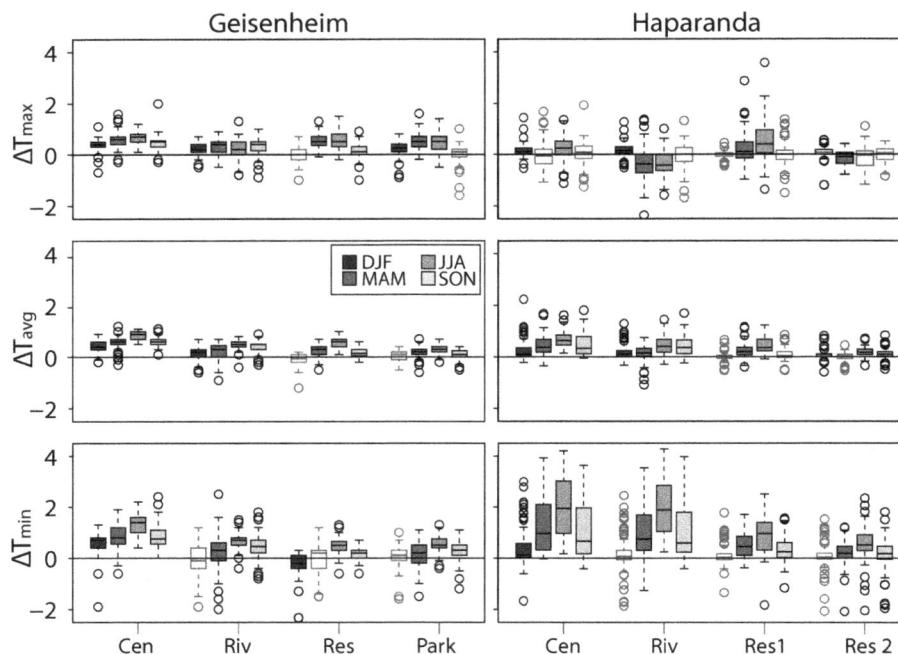

Figure 2. Boxplots of seasonal ΔT_{max} (top panels), ΔT_{avg} (middle panels) and ΔT_{min} (bottom panels), between the previous station locations/area typical environments and the current location of the met stations in Haparanda (left panels) and Geisenheim (right panels). Boxes show 25–75 percentile, with line across at median, whiskers extend to approximately 5–95 percentile and circles show outliers. Open, grey boxes are shown when sites do not differ significantly from the reference station or when difference is less than 0.1 K (due to accuracy limitations of the loggers). Site names see Table 2.

urban centre, as well as of prevented ventilation and nocturnal radiative cooling by the deciduous trees canopy (e.g. as shown by Spronken-Smith and Oke, 1998).

Temperature differences are larger for ΔT_{min} (median yearly $\Delta T_{min(cen)} = 0.8/0.8$ K) and smaller for ΔT_{max} (median yearly $\Delta T_{max(cen)} = 0.1/0.5$ K). This is a common result that is attributed the generally more stable nocturnal boundary layer preventing mixing of air from different areas, supporting site-specific nocturnal cooling (e.g. Krueger and Emmanuel, 2013). More unstable conditions caused by solar heating enhance vertical mixing which can generate horizontal winds that mitigate spatial air temperature differences. The timing of the maximum/minimum were generally consistent between sites in each village area on most days.

Substantial seasonal differences include more pronounced temperature residuals in summer than winter. The enhanced summertime solar radiation generates spatially heterogeneous heating of built and vegetated surfaces during the daytime. If strong surface heating occurs, the influence of site-specific nocturnal cooling is more important. The larger summer temperature differences (median $\Delta T_{min(cen)} = 1.9/1.4$ K), are thus likely a result of the stronger solar influence compared to winter (median $\Delta T_{min(cen)} = 0.1/0.6$ K). The seasonal pattern is especially pronounced in Haparanda. Given the high latitude (65.8° N) of that site, the mid-winter sun rises only around 3° above

the horizon, while the sun does not set for a few weeks in summer.

Although the general patterns are the same in the two villages, several differences are found in the data. The larger ΔT in the Geisenheim centre is likely due to the taller buildings and denser urban centre with more sealed surfaces to house the greater population of Geisenheim. Furthermore, the variability in ΔT is smaller in Geisenheim. This may be a consequence of non-urban effects such as the differing relief causing cold air drainage, particularly during calm and clear weather situations, which prevents the development of strong spatial temperature differences (Bigg et al., 2014). If elevation was accounted for the warming bias in the village centre, park and river in Geisenheim would further increase, while that in the residential area would decrease.

3.2 Potential urban warming bias in data from the examined stations

Sources for inhomogeneity in climate data are many, for example, relocations of measurement sites, changes in surroundings, sheltering, exposure and instrumentation, calculation methods and observation practices (Aguilar et al., 2003). The urban influences documented in this study in Geisenheim and Haparanda are of the same order of magnitude as the temperature trends recorded in the long term records of the stations (Table 3). This suggests that station relocations in villages could potentially cause substantial bias in the data recorded and should be taken into consideration when homogenizing data.

Analysis of previous met station locations in the two examined villages (Fig. 1) show that at no point had the Geisenheim station been located in the village centre. Thus it is less likely to have been biased by urban effects. The station was originally placed in the centrally located park, which was slightly warmer than the current met station location. However, the north-west part of the park has been converted into a parking lot since the station was located there, which limits the possibility of accurately determining the exact bias for this location. Incomplete meta-data for some prior Geisenheim station sites only benefits from the knowledge they were always outside the most densely built area.

In Haparanda, the station was located in the village centre for the first 83 years of operation, then moved to the riverside (where several minor moves took place), then to two residential locations, before its current location outside the residential area. Historical maps show that the village centre is still very similar to 1924, which suggest the temperature differences from this study can be assumed to be close to those for the siting 90 years ago, although it is important to recognise that changes in heat sources (wood burning to central heating) and house insulation will have affected heat emissions by the local residents. Historically, the Haparanda station likely does contain an urban warming bias, primarily in data from the first station location (in the urban centre). The second location (river side) also shows a bias, likely a combination of warming from the nearby urban area and the river. The residential areas would have had a smaller but still significant bias, particularly in T_{min}. Given that the current met station appears to have less urban influences compared to all previous sites, correcting for such biases in the long record would result in an increased warming trend through the past 100+ years. Such a correction would, however, need to consider the population increase since the mid-19th century and other effects in building emissions and properties. As the general knowledge on how to avoid urban biases in station record has increased greatly in recent decades, it is likely that more rural-village stations have undergone transitions similar to the Haparanda station with the possibility that they include urban effects, likely stronger in earlier records. This should be considered when using such data for analysis of long-term climate trends.

4 Conclusions

In this paper it is shown that the urban effects in villages can be sufficient to significantly modify temperatures, thus potentially causing a warming bias in rural-village met stations. The effect is largest in minimum temperatures, and during summer, and influenced by latitude (stronger seasonal dif-

Table 3. Linear temperature trend in the complete raw data set from DWD Geisenheim (1882–2014) and SMHI Haparanda (1859–2014), and median of the daily temperature difference between the village centre and the site of the current met station for the measurement period.

	Temperature trend, K $[10\,\mathrm{yr}]^{-1}$	Median $\Delta T_{(cen)}$, K
Geisenheim		
T_{max}	0.06	0.5
T_{avg}	0.11	0.6
T_{min}	0.17	0.8
Haparanda		
T_{max}	0.15	0.1
T_{avg}	0.10	0.4
T_{min}	0.17	0.7

ferences at higher latitude) and relief (less variability in the data in sloping terrain compared to flat). Urban influences of similar order were found in Geisenheim (Germany) and Haparanda (Sweden) potentially causing substantial biases in the temperature trends from these stations. Thus the classification of stations to indicate rural or village may provide a key flag to the interpretation of data in sets such at the GHCN.

Acknowledgements. We are grateful to the Hermann Mächel at the DWD (Climate Data Centre, Frankfurter Straße 135, Offenbach, Germany) and Sverker Hellström at the SMHI (Enheten för statistik och information, Folkborgsvägen 17, Norrköping, Sweden) for supplying the data and meta-data used in this study. The help of local authorities in finding suitable locations and getting permissions to install sensors is also greatly appreciated.

References

Aguilar, E., Auer, I., Brunet, M., Peterson, T. C., and Wieringa, J.: Guidelines on climate metadata and homogenization, WMO/TD No. 1186, WMO, World Climate Data and Monitoring Programme (WCDMP) series, available online: http://www.wmo.int/datastat/documents/WCDMP-53_1.pdf, 2003.

Andersén, M.: Allmän beskrivning av OBS2000-stationerna, Swedish Meteorological and Hydrological Institute (SMHI), Norrköping, Sweden, 2010

Arnfield, A. J.: Two decades of urban climate research: A review of turbulence, exchanges of energy and water, and the urban heat island, Int. J. Climatol., 23, 1–26, 2003.

Behrendt, J., Penda, E., Finkler, A., Heil, U., and Polte-Rudolf, C.: Beschreibung der Datenbasis des NKDZ, Referat "Nationale Klimaüberwachung", Deutscher Wetterdienst, Deutsche Wetter Dienst (DWD), Offenbach, Germany, available at: http://www.dwd.de/bvbw/generator/DWDWWW/Content/ Oeffentlichkeit/KU/KU2/KU21/datenbasis/german/kollektive_ _beschreibung,templateId=raw,property=publicationFile.pdf/ kollektive_beschreibung.pdf, 2011.

Bergström, H.: Wind resource mapping of Sweden using the MIUU-model, Wind Energy Report WE2007. Department of Earth Sciences, Uppsala University, Sweden. available at: https://www.energimyndigheten.se/Global/Omoss/Vindkraft/ Wind3.PDF, 2007.

Bigg, G. R., Wise, S. M., Hanna, E., Mansell, D., Bryant, R. G., and Howard, A.: Synoptic climatology of cold air drainage in the Derwent Valley, Peak District, UK, Meteorol. Appl., 21, 161–170, 2014.

Blandford, T. R., Humes, K. S., Harshburger, B. J., Moore, B. C., Walden, V. P., and Ye, H.: Seasonal and synoptic variations in near-surface air temperature lapse rates in a mountainous basin, J. Appl. Meteorol. Clim., 47, 249–261, 2008.

Hansen, J., Ruedy, R., Sato, M., and Lo, K.: Global surface temperature change, Rev. Geophys., 48, RG4004, doi:10.1029/2010RG000345, 2010.

Hessische Statistische Landesamt: Bevölkerung, Gebiet, http:// www.statistik-hessen.de/, last access: 01 December 2014.

Hinkel, K. M., Nelson, F. E., Klene, A. F., and Bell, J. H.: The urban heat island in winter at Barrow, Alaska, Int. J. Climatol., 23, 1889–1905, 2003.

Krueger, E. and Emmanuel, R.: Accounting for atmospheric stability conditions in urban heat island studies: The case of Glasgow, UK, Landsc. Urban Plan., 117, 112–121, 2013.

Lindén, J., Esper, J., and Holmer, B.: Using Land Cover, Population, and Night Light Data for Assessing Local Temperature Differences in Mainz, Germany, J. Appl. Meteorol. Clim., 54, 658–670, 2015.

Magee, N., Curtis, J., and Wendler, G.: The urban heat island effect at Fairbanks, Alaska. Theor. Appl. Climatol., 64, 39–47, 1999.

Offerle, B., Grimmond, C. S. B., Fortuniak, K., Klysik, K., and Oke, T. R.: Temporal variations in heat fluxes over a central European city centre, Theor. Appl. Climatol., 84, 103–115, 2006.

Oke, T. R.: The energetic basis of the urban heat island, Q. J. Roy. Meteorol. Soc., 108, 1–24, 1982.

Spronken-Smith, R. A. and Oke, T. R.: The thermal regime of urban parks in two cities with different summer climates, Int. J. Remote Sens., 19, 2085–2104, 1998.

Statistiska Centralbyrån: Befolkningsstatistik, http://www.scb.se/, last access: 01 December 2014.

Steeneveld, G. J., Koopmans, S., Heusinkveld, B. G., van Hove, L. W. A., and Holtslag, A. A. M.: Quantifying urban heat island effects and human comfort for cities of variable size and urban morphology in the Netherlands, J. Geophys. Res.-Atmos., 116, D20129, doi:10.1029/2011JD015988, 2011.

Svensson, M. K. and Eliasson, I.: Diurnal air temperatures in built-up areas in relation to urban planning, Landsc. Urban Plan., 61, 37–54, 2002.

Unger, J.: Intra-urban relationship between surface geometry and urban heat island: review and new approach, Clim. Res., 27, 253–264, 2004.

10

The benefits of emergency rescue and reanalysis data in decadal storm damage assessment studies

P. Jokinen, A. Vajda, and H. Gregow

Finnish Meteorological Institute, P.O. Box 503, 00101 Helsinki, Finland

Correspondence to: P. Jokinen (pauli.jokinen@fmi.fi)

Abstract. Studying changes in storm-induced forest damage in Finland has not been possible previously due to the lack of continuous, long series of impact data. We overcome this by combining emergency rescue data from the Finnish rescue services "PRONTO" (2011-) with ERA-Interim reanalysis data of wind gusts and soil temperatures to define exceedance thresholds for potential forest damage days. These thresholds were applied as a proxy for the period 1979–2013 in order to study the spatial and decadal characteristics of forest damage in Finland due to windstorms.

The results indicated that the area most impacted by potential forest damage was the south-western part of Finland along the coast, with 1–10 damaging storm cases per year. A decadal examination highlighted a lull period in the number of potential forest damage days during the 1990s compared to the 1980s and 2000s, albeit no trend was evident.

The inclusion of emergency rescue data allowed us for the first time to estimate the spatial distribution and decadal variations of potential forest damage days due to windstorms in Finland. The results achieved will encourage further development of thresholds for potential forest damage by including additional data sources and applying them to future climate scenarios.

1 Introduction

Extratropical storms can cause extensive socioeconomic impacts, by damaging infrastructure, disrupting society for several days or weeks, and even causing fatalities. Europe has experienced several high-impact storms in the past decade, such as "Kyrill" in 2007 (Levinson and Lawrimore, 2008), "Xynthia" in 2010 (Blunden et al., 2011) and "Dagmar" in 2011 (Blunden and Arndt, 2012). In fact, storms are the main natural hazard affecting Europe in terms of insured losses (Munich-Re, 2013).

Recent studies have indicated a decrease in the total number of extratropical cyclones, but an increase in the number of extreme cyclones over western and central Europe that enhance damage and economical losses (e.g. Donat et al., 2011; Zappa et al., 2013; Pinto et al., 2013). Thus, information about changes in storm impacts on infrastructure, society and economy in present and future climates is essential when planning adaptation and mitigation strategies for future adverse weather.

Storm-induced economic losses include damage to the forestry sector and infrastructure losses caused by fallen trees over power lines that then lead to power outages. In this study we will concentrate on the forest damage risks due to extratropical windstorms. Our study area is Finland, which is the most heavily forested country in Europe (FAO, 2010).

There are several factors that contribute to the risk of fallen trees, e.g. forest management, topography and tree type (Schmoeckel and Kottmeier, 2008). However, at higher latitudes, an important large scale factor in addition to strong winds is the soil conditions (Peltola et al., 1999). Limitations to root development, soil moisture and soil temperature, i.e. whether the ground is frozen or not, influence the vulnerability of forests to windthrows. Frozen soil anchors trees solidly to the ground and thereby makes them less liable to windthrow during storms relative to unfrozen soil conditions.

Our study aims to estimate recent changes in potential forest damage days by combining recent high quality forest damage observations with atmospheric and soil conditions from a reanalysis dataset. This will additionally demonstrate the benefits of including and combining new types of datasets within the larger context of improving and diversifying climate services.

2 Data and methods

ERA-Interim reanalysis data was used to create thresholds for wind gusts and soil temperatures on the days when forest damage occurred in Finland. These thresholds were then used as a proxy to utilise 35 years of ERA-Interim reanalysis data in order to obtain the decadal and spatial variation in potential forest damage days.

2.1 Rescue data and fallen trees

The statistics system of the Finnish rescue services, "PRONTO", stores information on various emergency rescue related incidents in Finland such as first response operations and building fires, that is when e.g. ambulances or the fire department are called to a location (Pronto, 2015). In addition, it includes cases related to natural disasters categorized according to different causes. Strong wind events were selected from this dataset, and it is estimated that at least 75 % of these wind-related events were due to fallen trees that either fell over roads or downed power lines (A.-J. Punkka, personal communication, 10 December 2014). Thus, they were considered to be a good indicator of forest (tree) damage. The term "forest damage" is used although it is acknowledged that not all reported wind-related events were necessarily due to forest damage.

This dataset provided information on the place and time of forest damage caused by storms since the coordinates and times were included for each event reported to the emergency response centres. The rescue dataset starts in 2011, with data up until summer of 2014 used in this study. Since wind speeds related to small-scale summertime convective events are not captured by the coarse resolution reanalysis data, they were not included in our study. Therefore, only the months September to April were considered when defining the thresholds.

2.2 Reanalysis data

ERA-Interim reanalysis data (Dee et al., 2011) was used to define exceedance thresholds for wind gusts and soil temperatures on the days when forest damage occurred in Finland. ERA-Interim output covers the time period from 1979 onwards at a spatial resolution of $0.7°$, which is approximately 80 km at the equator. Since the main contributors to possible forest damage conditions are wind speed and soil state (unfrozen conditions induce high risk), we utilise the ERA-Interim 12 h forecasts of 10 m maximum wind gusts, and temporally-averaged soil temperature data provided at four levels: 0–7, 7–28, 28–100 and 100–255 cm.

2.3 Combining the datasets

The reanalysis maximum wind gust and average soil temperatures were extracted from two days prior to and including the day of the reported damage time in the "PRONTO" dataset by using the closest reanalysis grid point to the location of observed impact. Histograms of these variables were created that depicted the conditions leading to forest damage based on the overlapping time period of the two datasets (January 2011–April 2014). A combination of these thresholds was used as a proxy when selecting similar weather conditions in the ERA-Interim dataset from 1979 to 2013. The days when wind gusts at 10 m and soil temperatures at the four mentioned depths met or exceeded the defined thresholds were considered to be potential forest damage days.

3 Results

3.1 Forest damage thresholds

The histogram of ERA-Interim peak wind gusts corresponding to the days with observed wind damage is shown in Fig. 1. Although most damage reports given to the rescue services are logged almost immediately after the wind damage has occurred, there can sometimes be a much longer lag time between event and logging of the report, with no exact knowledge of when the damage occurred (e.g. people travelling to their summer cabins and finding damage that had occurred during the previous months). This is thought to explain some of the weaker wind speed values in the histogram. We therefore selected $19 \, \mathrm{m \, s^{-1}}$ as the threshold for wind gusts as a clear jump in the number of damage events was observed above this value.

The histograms for the soil level temperatures at different levels depicted in Fig. 2 reveal that forest damage can occur even though the near surface layer temperature (0–7 cm) is slightly below freezing. However, on moving to deeper soil levels, the bulk of the histogram shifts to higher temperatures. These results suggest that trees can be uprooted when the near-surface level, down to a depth of 28 cm, is frozen, but not when the deeper soil layers (28–100 cm) are frozen anchoring trees more securely to the ground; there were only few cases reported for frozen deep soil layers. This is in agreement with the earlier findings of Gregow (2013) suggesting that frozen soil conditions at a depth of 30 cm substantially reduced the risk of uprooting.

The combined threshold levels based on the wind gust and soil temperature histograms are summarized in Table 1.

Table 1. The forest damage probability thresholds defined for the ERA-Interim dataset.

ERA-Interim parameter	Threshold
Wind gusts	$\geq 19\,\mathrm{m\,s^{-1}}$
0–7 cm soil temperature	$\geq -2\,°\mathrm{C}$
7–28 cm soil temperature	$\geq -1\,°\mathrm{C}$
28–100 cm soil temperature	$\geq 1\,°\mathrm{C}$
100–289 cm soil temperature	$\geq 3\,°\mathrm{C}$

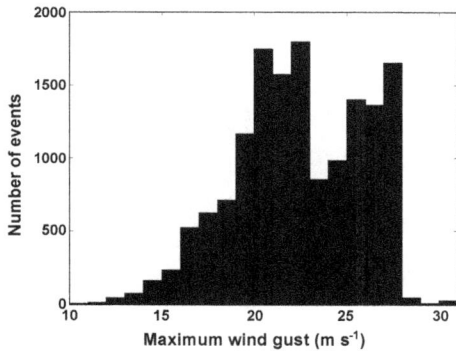

Figure 1. Histogram of the ERA-Interim maximum simulated wind gusts leading to observed forest damage.

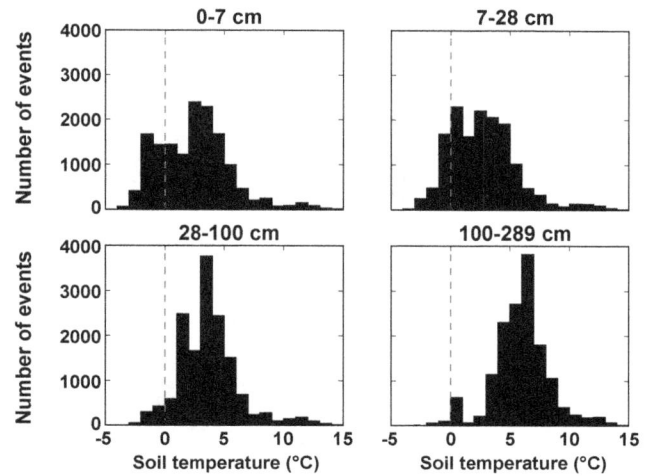

Figure 2. Histograms of ERA-Interim temperatures for four different soil levels leading to observed forest damage. The red dashed line highlights the 0 °C temperature.

3.2 Potential forest damage days

The average number of annual potential forest damage days in 1979–2013 is shown in Fig. 3. It indicates that the highest frequency is in the south-western parts of Finland along the coasts with values ranging between 1 and 10 per year. The lowest frequencies were found in the northern inland areas where forest damage conditions occurred on average less than once per year. This difference is most likely due to the longer duration of frozen ground in the north during the windy winter season and to the lower intensity of wind gusts in the more continental stable conditions. The spatial map in Fig. 3 also covers areas around Finland, although we must stress that the forest damage criteria was calibrated for Finland only and therefore values shown elsewhere are only suggestive; the thresholds would presumably require modification to accommodate different land surface and tree type.

Decadal maps in Fig. 3 as well as a time series of the areal average in southern and central parts of Finland (Fig. 4) indicate that the highest frequencies of potential forest damage conditions occurred in the mid-1980s and close to the year 2010 with a clear lull period during the 1990s. If the gust threshold only was considered, the number of damage days in the 1990s would have seen a peak. Thus the drop in the 1990s was due to lower soil temperatures on stormy days. There was also considerable variability from year to year, but no statistically significant trend existed for the time period covered in this study.

4 Discussion and conclusions

In this work, the benefits of combining emergency rescue data with reanalysis data were studied within the context of potential forest damage days in Finland. Although the calibration period for this study was relatively short, from 2011 to 2014, it did include six high-impact named storms in the study area (Tapani, 26 December 2011; Hannu, 27 December 2011; Antti, 30 November 2012; Eino, 17 November 2013; Oskari, 1 December 2013; Seija, 13 December 2013). Naturally, the thresholds will become more accurate as more data becomes available in the future. Furthermore, other weather factors such as wind direction and storm duration could be included in the list of parameters (Kupfer et al., 2008).

Future improvements in this method would include using the detailed impact data to calibrate the threshold values separately for each area (grid point) to account for the fact that forest type, and therefore damage conditions, vary from one location to another. For example, using Finnish thresholds would most likely not lead to similar damage in Germany. Additionally, the thresholds were specifically calibrated for ERA-Interim reanalysis data, which does not necessarily represent the thresholds that would be derived based on actual wind gust observations.

This study shows a large annual variation in the number of potential forest damage days in Finland, with the highest risk in the south-western parts of the country. The lowest number of damage-inducing storms was in the 1990s; this agrees well with the amount of forest damage compensation paid to forest owners due to wind damage, with higher amounts in the 1980s and 2000s compared to the 1990s (Ylitalo, 2010). This demonstrates that the inclusion of soil thresholds in the

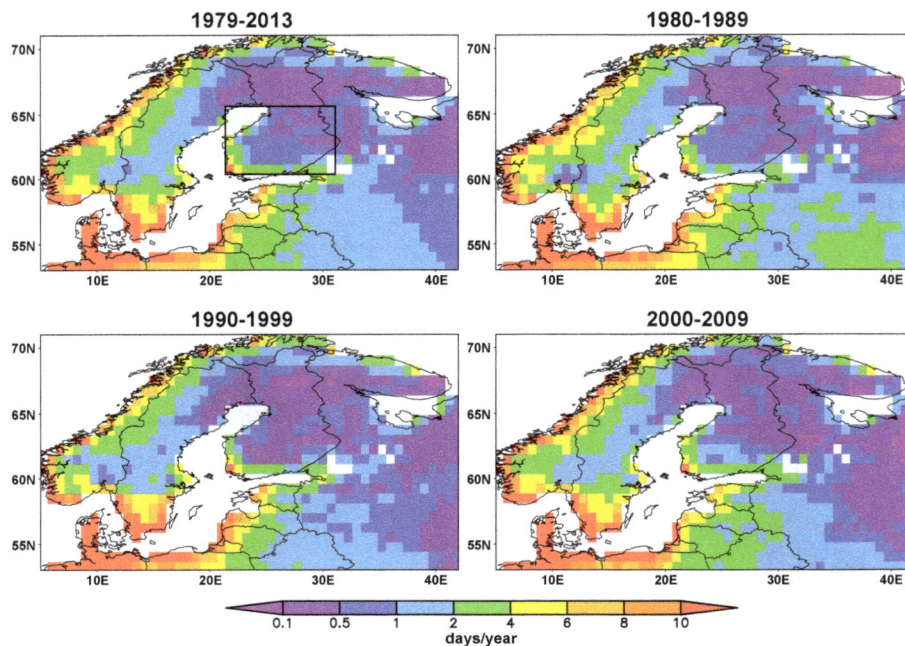

Figure 3. Potential forest damage days per year (days when the threshold levels are met or exceeded) in ERA-Interim during 1979–2013 (upper left panel) and decadal averages. The black box indicates the region over which the areal average in Fig. 4 is calculated.

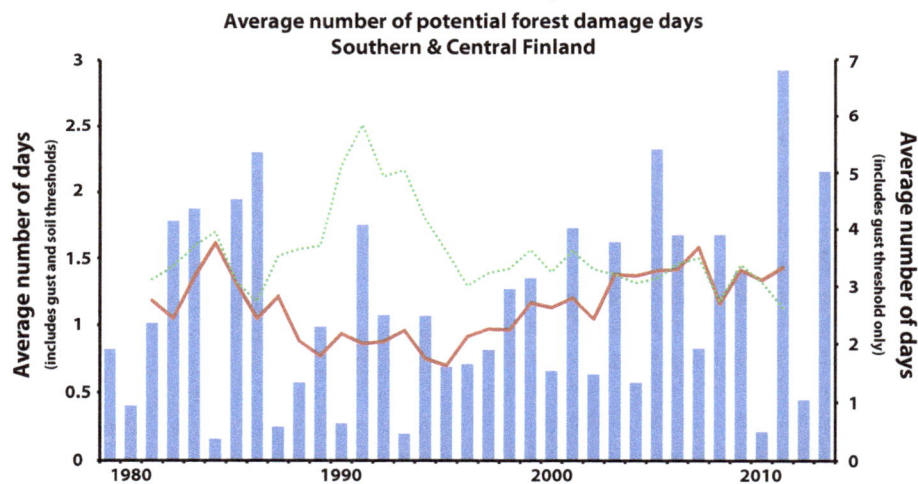

Figure 4. Blue columns are the average number of potential forest damage days in southern and central Finland based on ERA-Interim when wind gust and soil thresholds are met or exceeded. Vertical axis is on the left, and the red line is the 5-year running mean. The green dashed line is a 5-year running mean that uses the wind gust threshold only (vertical axis on the right). The values are calculated for the area depicted in Fig. 3.

assessment is a major improvement over using a wind-gust threshold only.

The most significant Finnish forest damage and rescue operations due to windstorms occurred during the Pyry-storm, 31 October 2001, and Janika-storm, 15 November 2001 (total forest damage 7.3 Mm^3). Additionally, the Tapani-storm (also called Dagmar) on 26 December 2011 caused 3.5 Mm^3 in forest damage (Gregow, 2013). Both these years were

above average in our assessment of the number of potential forest damage days, although it should be noted that our methodology does not attempt to estimate the damage severity of individual storms.

This study has evaluated the feasibility of using emergency rescue data in conjunction with reanalysis data in storm damage studies. The collaboration in data sharing within the public sector has been fruitful and found to boost the effectiveness of weather and climate services in Finland. Since the aim of National Hydrological and Meteorological Services is also to serve the private sector, more cooperation in data sharing is required in this regard; for example, data from energy or telecommunication companies would be very valuable in storm-related impact research. As the data from private companies is rarely available, more effort is needed to collect valuable data from multiple sources, including the general public, in order to gain the most comprehensive view on past and future storm impacts.

Based on the present results we plan to develop the method further and apply it to future climate scenarios. With an accurate forest damage model we can then estimate how forest damage conditions in Finland are expected to change in the upcoming decades, providing useful information for adaptation and mitigation strategies.

Author contributions. Pauli Jokinen has prepared the manuscript and conducted the analyses. Andrea Vajda and Hilppa Gregow have supervised, commented and participated in writing.

Acknowledgements. This work was financially supported by the EU FP7 CORE-CLIMAX project. We are grateful to Ewan O'Connor for proofreading the manuscript. We also thank Carin Nilsson and one anonymous reviewer for their helpful comments.

References

Blunden, J. and Arndt, D.: State of the climate in 2011, B. Am. Meteorol. Soc., 93, S1–S282, doi:10.1175/2012BAMSStateoftheClimate.1, 2012.

Blunden, J., Arndt, D. S., and Baringer, M. O.: State of the climate in 2010, B. Am. Meteorol. Soc., 92, S1–S236, doi:10.1175/1520-0477-92.6.S1, 2011.

Dee, D. P., Uppala, S. M., Simmons, A. J., Berrisford, P., Poli, P., Kobayashi, S., Andrae, U., Balmaseda, M. A., Balsamo, G., Bauer, P., Bechtold, P., Beljaars, A. C. M., van de Berg, L., Bidlot, J., Bormann, N., Delsol, C., Dragani, R., Fuentes, M., Geer, A. J., Haimberger, L., Healy, S. B., Hersbach, H., Hólm, E. V., Isaksen, L., Kllberg, P., Köhler, M., Matricardi, M., McNally, A. P., Monge-Sanz, B. M., Morcrette, J.-J., Park, B.-K., Peubey, C., de Rosnay, P., Tavolato, C., Thépaut, J.-N., and Vitart, F.: The ERA-Interim reanalysis: Configuration and performance of the data assimilation system, Q. J. Roy. Meteorol. Soc., 137, 553-597, doi:10.1002/qj.828, 2011.

Donat, M. G., Leckebusch, G. C., Wild, S., and Ulbrich, U.: Future changes in European winter storm losses and extreme wind speeds inferred from GCM and RCM multi-model simulations, Nat. Hazards Earth Syst. Sci., 11, 1351–1370, doi:10.5194/nhess-11-1351-2011, 2011.

FAO: Global Forest Resources Assessment 2010, Food and Agriculture Organization of the United Nations, Rome, Italy, available at: http://www.fao.org/forestry/fra/fra2010/en/ (last access: 10 January 2015), 2010.

Gregow, H.: Impacts of strong winds, heavy snow loads and soil frost conditions on the risks to forests in northern Europe, Ph.D. thesis, contributions 94, Finnish Meteorological Institute, Finland, 178 pp., 2013.

Kupfer, J., Myers, A., McLane, S., and Melton, G.: Patterns of Forest Damage in a Southern Mississippi Landscape Caused by Hurricane Katrina, Ecosystems, 11, 45–60, doi:10.1007/s10021-007-9106-z, 2008.

Levinson, D. H. and Lawrimore, J. H.: State of the climate in 2007, B. Am. Meteorol. Soc., 89, S1–S179, doi:10.1175/BAMS-89-7-StateoftheClimate, 2008.

Munich-Re: Topics Geo Natural Catastrophes: Analyses, assessments, positions, Munich Re Publications, Munich, Germany, 64 pp., 2013.

Peltola, H., Kellomäki, S., and Väisänen, H.: Model computations of the impact of climatic change on the windthrow risk of trees, Climatic Change, 41, 17–36, 1999.

Pinto, J. G., Bellenbaum, N., Karremann, M. K., and Della-Marta, P. M.: Serial clustering of extratropical cyclones over the North Atlantic and Europe under recent and future climate conditions, J. Geophys. Res.-Atmos., 118, 12476–12485, doi:10.1002/2013JD020564, 2013.

Pronto: http://www.pelastusopisto.fi/fi/tutkimus-_jatietopalvelut/tutkimus-_ja_kehittamispalvelut/tilastotpronto, last access: 10 January 2015.

Schmoeckel, J. and Kottmeier, C.: Storm damage in the Black Forest caused by the winter storm "Lothar" – Part 1: Airborne damage assessment, Nat. Hazards Earth Syst. Sci., 8, 795–803, doi:10.5194/nhess-8-795-2008, 2008.

Ylitalo, E. (Ed.): Finnish Statistical Yearbook of Forestry 2010, Finnish Forest Research Institute, Sastamala, Finland, 2010.

Zappa, G., Shaffrey, L. C., Hodges, K. I., Sansom, P. G., and Stephenson, D. B.: A multi-model assessment of future projections of north atlantic and european extratropical cyclones in the CMIP5 climate models, J. Climatol., 26, 5846–5862, 2013.

11

Wind variability in a coastal area (Alfacs Bay, Ebro River delta)

P. Cerralbo[1], M. Grifoll[1,2], J. Moré[3], M. Bravo[3], A. Sairouní Afif[3], and M. Espino[1,2]

[1]Maritime Engineering Laboratory, Polytechnic University of Catalonia (LIM/UPC), c./Escar, 6, 08039 Barcelona, Spain
[2]International Centre of Coastal Resources Research, Polytechnic University of Catalonia (CIIRC/UPC), c./Jordi Girona, 1–3, 08034 Barcelona, Spain
[3]Servei Meteorològic de Catalunya (SMC), c./Berlín, 38–46, 08029 Barcelona, Spain

Correspondence to: P. Cerralbo (pablo.cerralbo@upc.edu; pablocerralbo@gmail.com)

Abstract. Wind spatial heterogeneity in a coastal area (Alfacs Bay, northwestern Mediterranean Sea) is described using a set of observations and modelling results. Observations in three meteorological stations (during 2012–2013) along the coastline reveal that wind from the N–NW (strongest winds in the region) appears to be affected by the local orography promoting high wind variability on relatively short spatial scales (of the order of few kilometres). On the other hand, sea breezes in late spring and summer also show noticeable differences in both spatial distribution and duration. The importance of wind models' spatial resolution is also assessed, revealing that high resolution (= 3 km) substantially improves the results in comparison to coarse resolution (9 km). The highest-resolution model tested (400 m) also presents noticeable improvements during some events, showing spatial variability not revealed by coarser models. All these models are used to describe and understand the spatial variability of the typical wind events in the region. The results presented in this contribution should be considered on hydrodynamic, ecological and risk management investigations in coastal areas with complex orography.

1 Introduction

In the open sea and ocean, wind variability responds mostly to mesoscale structures like cyclones and anticyclones, as well as more permanent structures such as easterly (polar) and westerly (middle latitudes) winds. But when orographic constraints appear, such as oceanic islands (Chavanne et al., 2002) or mountains in coastal areas (Jiang et al., 2009), the wind presents high spatial variability, showing important curl gradients and becoming less predictable. In coastal areas, several examples of high spatial variability due to topographic constraints have been described (e.g. Herrera et al., 2005; Boldrin et al., 2009). In recent years, the application of numerical models in both the atmosphere and oceans has contributed to improving the understanding and description of this variability (Schaeffer et al., 2011). Moreover, modelling studies have revealed that the model resolution is a key factor for the correct representation of wind patterns, which

could be essential in a correct prediction of flood episodes (Brecht and Frank, 2014) and could allow for the correct application of hydrodynamic modelling (Signell et al., 2005; Bignami et al., 2007). Several authors have described wind variability in lakes (e.g. Venäläinen et al., 2003), whilst only few studies focused on the wind description in a small-scale domain such as small estuaries or coastal areas. Considering the importance of wind on hydrodynamics, water mixing, waves and air quality, this contribution seeks to fulfill this gap, presenting an example of wind variability in a small-scale coastal area through observations and modelling results. A small bay, Alfacs Bay, in the northwestern Mediterranean Sea was selected.

This contribution is organized as follows. First, a short description of the study area (Sect. 2) and the set of observations is presented (Sect. 3). In this section the spatial and temporal variability observed is described. Then, the numeri-

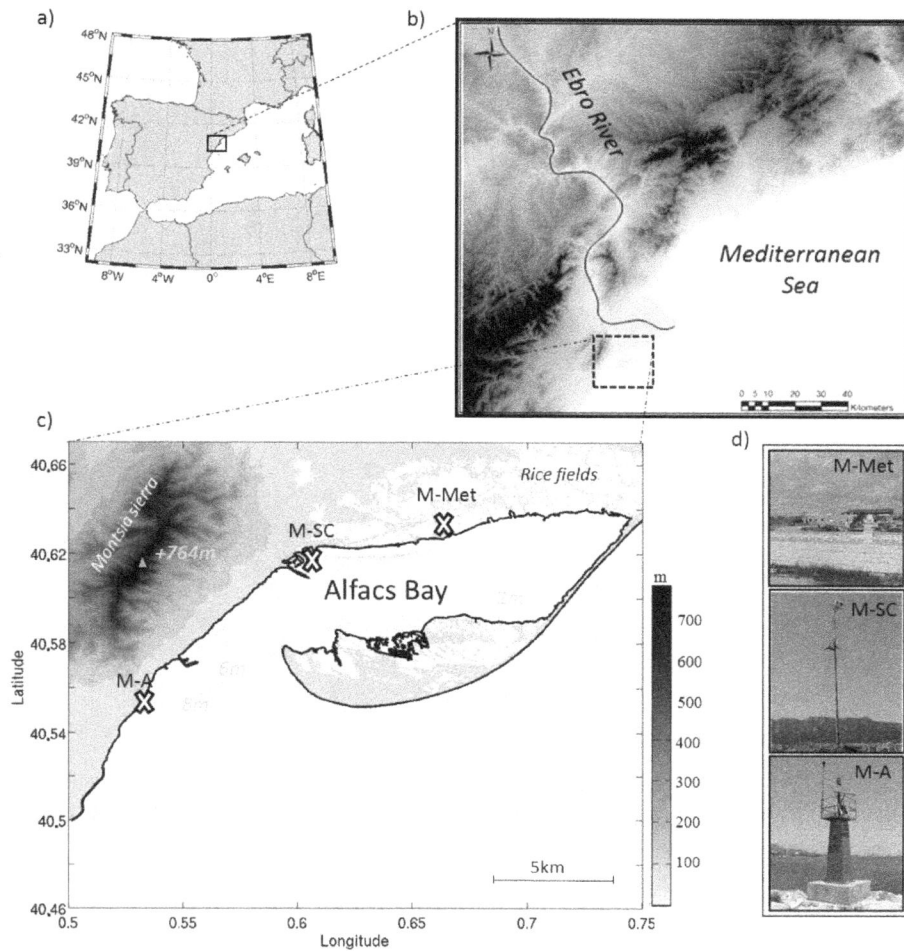

Figure 1. (a) Location map. **(b)** Ebro River delta region, with Alfacs Bay delimited by dashed square. **(c)** Study area showing meteorological stations (white crosses): M-A for Les cases d'Alcanar station, M-SC for Sant Carles de la Ràpita harbour station and M-Met for Meteocat station. Colour bar indicates altimetry above mean sea level. The bathymetry is also shown (isobaths, each 2 m). **(d)** Meteorological stations pictures.

cal weather prediction model implementation and outputs are shown, compared to the observed winds (Sect. 4). Finally, results for selected wind events and patterns observed are discussed.

2 Study area

The Ebro Delta (NE coast of Spain) forms two semi-enclosed bays, Fangar and Alfacs (to the north and south, respectively). The dimensions of Alfacs Bay are 16 km from head to mouth (Fig. 1c), 4 km wide and a mouth connection to the open sea of about 2.5 km. The bay is surrounded by rice fields to the north – which spill around 10 m^3 s^{-1} of freshwater 10 months per year to the bay (Serra et al., 2007) – and a sand beach enclosing it on the eastern side, which can suffer breaching processes under severe storm conditions (Gracia et

al., 2013). Serra de Montsià, with maximum altitudes around 700 m, closes the bay on the west side (Fig. 1).

The synoptic winds on the Catalan coast are affected by orographic constraints, such as the blocking winds of the Pyrenees that promote tramuntana (N) and mistral (NW) winds over some areas, and the wind channelling due to river valleys (Sánchez-Arcilla et al., 2008). Northerly winds in the region are mainly produced by high pressures over the Azores and lows over the British Isles and Italy; other synoptic situations could also lead to strong winds from the NW in the Ebro Delta (Martín Vide, 2005). Winds in the bay have been characterized as having a northwestern and southwestern predominance, with the strongest ones coming from the NW (channelized by the Ebro River valley; see Fig. 1b), being also the most common strongest winds on the Catalan coast during autumn and winter (Bolaños et al., 2009). On the other hand, some authors have reported the high spatial

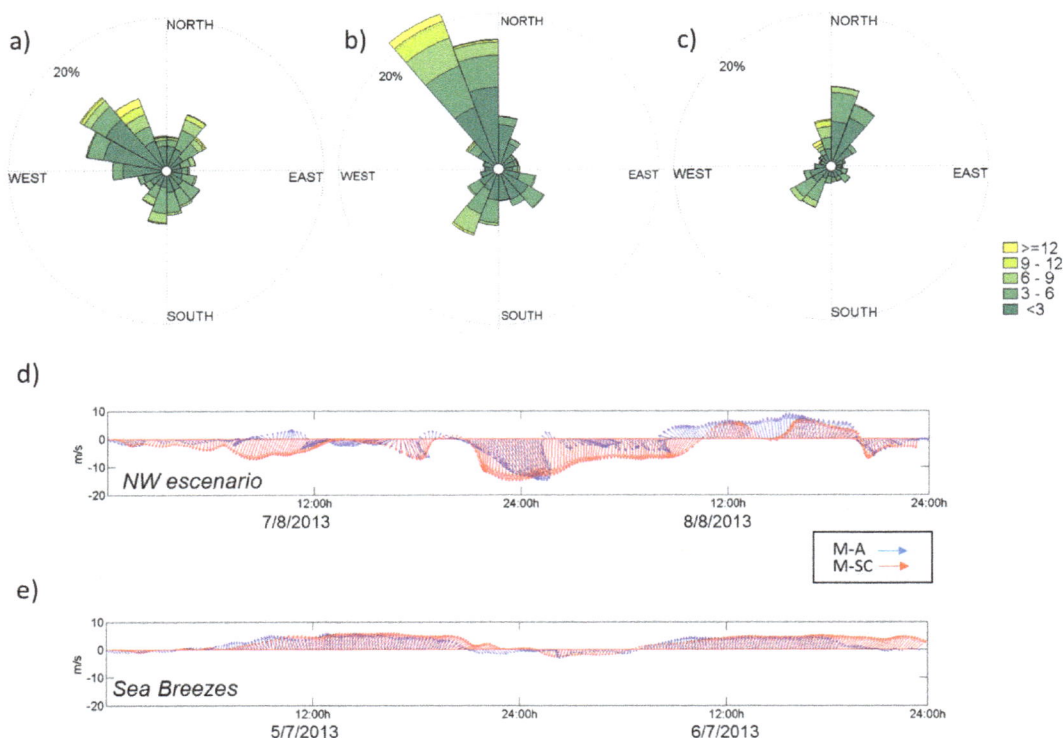

Figure 2. Wind roses for M-A (**a**), M-Met (**b**) and M-SC (**c**) during the period 2012–2013. Velocities – colour bar in (**c**) – are in $\mathrm{m\,s^{-1}}$. Dotted line indicates frequency of 20 %. Wind time series for northwesterlies and sea breeze events in (**d**) and (**e**), respectively. For (**d**) and (**e**), a Lanczos filter of 2 h has been applied to the 10 min wind data.

heterogeneity of the wind fields inside the bay (Camp, 1994), in agreement with observed events during field campaigns by the authors of this manuscript, in which winds from the northwest were blowing inside the bay whilst in the mouth of the bay the wind was almost calm.

3 Observations

Atmospheric data – wind, atmospheric pressure, solar radiation and humidity – were obtained from three fixed land stations: Alcanar (M-A); Sant Carles (M-SC) from Xarxa d'Intruments Meteorològics de Catalunya (XIOM) described in Bolaños et al. (2009); and Alfacs-Meteocat (M-Met), which belongs to the automatic weather stations network of the Meteorological Service of Catalunya (http://www. meteocat.cat). Pictures of the stations are found in Fig. 1d. Both M-A and M-SC are at 10 m above the ground, while M-Met measures at 2 m. In order to compare wind intensities from all stations, we have adapted the measurements at 2 m to the standard height of 10 m. The method adapted in Herrera et al. (2005) from Oke (1987) is used to compute the velocities at 10 m (w_{10}) from the observed values (w_h), following $w_{10} = w_h \dfrac{\log\left(\frac{10}{z_s}\right)}{\log\left(\frac{h}{z_s}\right)}$, where h represents the measurement height (2 m). Following Agterberg and Wieringa (1989), we

have considered a typical roughness (z_s) for plains with low vegetation (rice fields) of ≈ 0.03 m. The roughness variability as a function of the wind direction is not considered.

Observations from June 2012 to June 2013 at three meteorological stations show noticeable differences (Fig. 2), confirming high variability among them. This period has been chosen for this analysis because it is the only one with data from all three stations. In M-SC (Fig. 2c) a bimodal behaviour with the most common winds from the southwest and north–northeast are shown. These winds are in agreement with data acquired at the same station for about 16 years (not shown), indicating high representation for the period selected. The highest intensities correspond to N–NW winds ($> 9\,\mathrm{m\,s^{-1}}$). For M-A the directions are more scattered, the most common winds being from W-NW to NW. The highest-intensity winds still come from the NW–NE, but the purely northern winds are less common. Finally, winds from M-Met show also a clear bimodal behaviour, with winds from the NW and SW–SE being the most common. However, M-Met is altered by the human buildings on the south side, which would alter on the wind measurements. Among the three stations most of the differences are clearly seen in land winds due to possible effects of the Serra de Montsià mountain range, showing high heterogeneity in wind fields in short distances. To better understand the wind variability, directional

Table 1. Summary of the main characteristics of the three different model configurations used in this study.

Model	Domain	Nominal resolution	Lead time	Outputs
WRF9	Iberian Peninsula	9 km	72 h	3 h
WRF3	NE Spain	3 km	48 h	1 h
CALMET	SW Catalonia	0.4 km	48 h	1 h

scatter plots for stations with data at 10m height are shown in Fig. 3a. The corresponding mean velocity for the two stations (M-A and M-SC) is also shown in colours. This figure shows that winds from the SW–SE in M-A are rotated 20–30° clockwise in M-SC, as well as that winds from the NW–N in M-A seem to be rotated a bit clockwise in M-SC (also seen in Fig. 2). On the other hand, winds from W to NW in M-SC are not observed as in M-A; M-SC concentrates more on the N–NE winds. All these data illustrate that Serra de Montsià (Fig. 1) could act as a physical barrier to some type of northerly winds, redirecting them. These effects are probably most clear on mistral winds (NW), which are not completely reproduced by M-SC (Fig. 2d) but oriented to the north. Another meteorological station was planned to be operating between these two stations, but in the end it was no possible to deploy it. The winds from the SW–SE correspond mainly to see breezes in spring and summer (Bolaños et al., 2009). The time and spatial evolution of sea breeze differences between stations are also observable (Fig. 2e). Weak nocturnal offshore winds (NW) rotate and increase in intensity until the maximum of around 5 m s^{-1} and S–SW direction in M-A is reached. Similar behaviour is observed in M-SC, but the nocturnal winds come from the NE and the rotation is very noticeable, as breezes arise in the afternoon from the SW. In M-Met the behaviour is almost the same as M-A. All these data reflect that even during sea breezes the orography and probably variation of land uses (rice fields in the delta's plain versus brush forest in Serra de Montsià) affect the direction, intensity and durability of winds. Winds from the S–SE (not related to sea breezes) are also probably affected by topography (Fig. 3), but there are not enough observations to investigate the spatial variability in this case. The lack of wind data just in front of Serra de Montsià does not allow us to know the exact behaviour of NW (mistral) winds and sea breezes in the mouth of the bay, but the use of numerical models would allow us to approximate the corresponding theoretical wind patterns. The wind module reveals good agreement between both stations (not shown). Thus indicating most of the variability related to direction. All the observations are based on 1-year-long data, so no climatic conclusion is expected from our analysis.

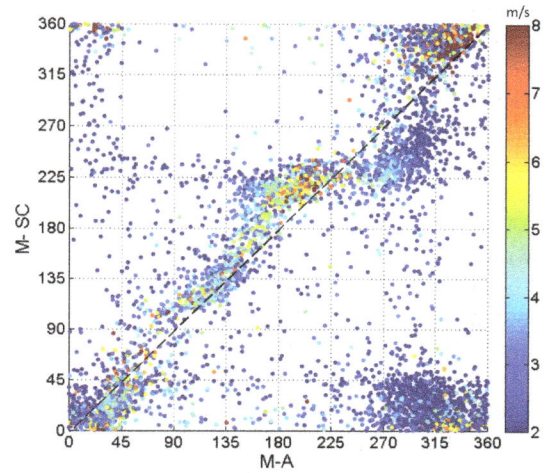

Figure 3. Wind direction comparison between M-A and M-SC for 1 year. The coloured data indicate mean wind velocity values (m s^{-1}) from both stations.

4   Numerical modelling

4.1   Description

Two different atmospheric models, provided by Meteocat, are used in this study to assess the role of spatial heterogeneity in the bay. For this purpose, the Weather Research and Forecasting (WRF) model has been selected (see http://www.wrf-model.org/), in particular the dynamical solver called Advanced Research WRF (WRF-ARW). The WRF-ARW is a model that has a worldwide community of users. It is a fully compressible, non-hydrostatic mesoscale model (Skamarock et al., 2008) and is based on solving the primitive equations of the atmosphere with different physical parameterizations available. Version 3.1.1 is used, with some minor code changes that consider an increase of the surface stress parameter (u-star) by a factor of 20 % in order to address the surface wind overestimation in the WRF model that has been reported by several authors (Mass and Ovens, 2011). By introducing this modification, a better agreement with meteorological observations is obtained. In later versions of the WRF model, more sophisticated corrections have been incorporated in the same direction, such as a surface drag parameterization scheme which allows enhancing the u-star depending on the sub-grid terrain variance (Mass and Ovens, 2011). The main parameterizations used are the Thompson scheme (Thompson et al., 2004) for microphysics, the YSU scheme (Hong et al., 2006) for planetary boundary layer and the unified Noah land surface model (Tewari et al., 2004) for surface physics. More configuration details are available at http://www.meteo.cat. This model is run using Global Forecast System (GFS) data as initial and boundary conditions (with 0.5° and 6 h spatial and temporal resolution). Two different configurations depending on grid resolution have been analysed: 9 and 3 km (named hereafter WRF9 and WRF3).

The main model parameters are summarized in Table 1. Not only spatial resolution is different between them; temporal resolution is also different. These configurations correspond to the available products of the Meteocat meteorological operational forecast system.

On the other hand, an additional simulation has been considered to derive atmospheric data at a very high resolution (400 m). In particular, the WRF-ARW outputs at 3 km are downscaled by a diagnostic meteorological model called CALMET. CALMET, a component of the CALPUFF Modeling System designed for the simulation of atmospheric pollution dispersion (Scire et al., 1999), is a diagnostic three-dimensional meteorological model which includes parameterized treatments of slope flows, kinematic terrain effects and terrain blocking effects, among others. These particular aspects help to better represent regional flows with an efficient computational cost.

The operational numerical model resolutions (spatial and temporal) were selected considering both reliability and computational costs. The computational cost for a 3-day forecast for CALMET (at 400 m resolution) is about 2 h using 10 CPUs. On the other hand, a similar domain and forecast horizon using WRF3 (3 km of resolution) take around 1 h (30 CPUs). In this sense, the application of WRF at higher resolution is not currently considered as an operational product due to computational limitations. For all the verification processes the daily first 24 h of prediction from the operational system are used.

4.2 Model verification

In this section, we present the results of verification studies to assess the performance of wind velocity and direction prediction from the models against the observation. The verification of both WRFs (WRF3 and WRF9) and CALMET is shown in Fig. 4 for summer 2013 (Fig. 4a) and winter 2014 (Fig. 4b). Different oceanographic campaigns were developed at Alfacs Bay during the same periods. We chose that period in order to coincide the meteorological observations and model results with oceanographic data. Verification of all the models and resolutions is done against hourly means from observational data. The wind module velocities measured in both M-A and M-SC are graphically compared to modelling results for both systems (model data is interpolated through bi-lineal interpolation to corresponding points) using a Taylor diagram (Taylor, 2001). In this diagram we can observe the comparison between observations and model through the correlation, the centred root-mean-square difference (CRMSD) and the standard deviations (SDs; see Appendix). In order to compare in the same figure different models against different observations, the standard deviations and CRMSD are normalized over the standard deviation of the corresponding observations (Grifoll et al., 2013). The model skill improves as the points get closer to the observation reference point in the diagram. In summer 2013, better

correlation (around 0.6) is observed in M-SC, while in M-A the values decrease to ≈ 0.5. The standard deviation shows that the model presents lower-amplitude variations than the observed values. In spring 2014 only values from M-A were available. In this period, the correlations are higher for all models, and the SDs are more similar to the observations. Differences through different models are only observable in M-SC locations, revealing the highest correlations and lower errors in higher-resolution models (best results corresponding to CALMET). In M-A, the models do not show noticeable differences.

The wind module velocities measured in both M-A and M-SC are compared in Fig. 5a and b, respectively, for summer 2013 with modelling results using a Weibull distribution, which is defined as a two-parameter function commonly used to fit the wind speed frequency distribution. In both stations, the best fit between the model and observational Weibull distributions is observed for CALMET. In M-A, CALMET and the observational distribution show almost equally shaped coefficient; even observational data present stronger winds. WRF3 also has similar shape, with even more energy distributed at medium wind intensities ($= 3 \, \text{m s}^{-1}$). WRF9 seems to overestimate the mean winds. In M-SC, winds from CALMET are the most similar to observations, even overestimating the frequency of mean winds, and then not reproducing the maximum intensities (6–10 m s^{-1}). On the other hand, WRF3 and WRF9 overestimate both mean wind intensities and corresponding frequencies. In the winter season, observations show higher wind intensities, and both CALMET and WRF3 present Weibull shapes (not shown) similar to observations.

Some characteristic events representing the most usual winds in the area have also been analysed in order to understand the behaviour of each model under different conditions (resumed in Table 2). A period of 3 days is considered to have enough data to compare. Results show that winds from CALMET and WRF3 have higher correlations (except in northwest 2) than WRF9. The worst results are observed during northwesterly winds in summer 2013. This is clearly affected by the topography, and the observed wind in M-A is not reproduced by any of the models. Slightly better results are obtained in M-SC, especially by CALMET (but still with poor correlation). This event was characterized by its shortness (less than 6–8 h) and unsteadiness. On the other hand, in the winter period, another NW wind event (lasting for more than 1 day) was reproduced with noticeable agreement in M-A. In this case, the simulation WRF9 seems to reproduce quite well the wind time series in M-A, but no data for M-SC are available to compare. Southern winds – southeast and diurnal regime of sea breeze – are better reproduced by the finest models, being the highest improvement between WRF3 and WRF9. During sea breezes and considering the complete diurnal–nocturnal cycle, the CALMET model seems to reproduce winds better than the coarser ones. Considering the daily duration of the sea breeze – between 6 and 8 h – all the

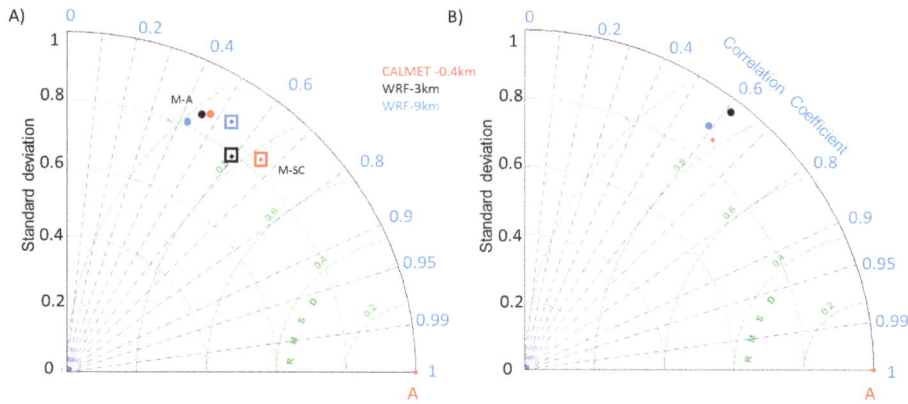

Figure 4. Taylor diagram for summer 2013 (**a**) and winter 2014 (**b**) for both M-A (coloured dot) and M-SC (coloured square filled with dot) model configurations (red for CALMET, black WRF3 and blue for WRF9) compared to corresponding meteorological stations. Both modelled standard deviations and RMSD are normalized over observational standard deviation.

Table 2. Correlation among the three different atmospheric models and observational data for 3-day-long events during summer 2013 and winter 2014. No correlation for winter 2014 (M-SC dismantled on September 2013).

	Day	M-A			M-SC		
		WRF9	WRF3	CALMET	WRF9	WRF3	CALMET
Northwest	8 Aug 2013	.02	.01	.12	.21	.40	.46
Northwest 2	4 Apr 2014	.80	.72	.75	–	–	–
Southeast	13 Aug 2013	.43	.64	.64	.58	.71	.75
Sea breezes	6 Jul 2013	.64	.67	.74	.64	.66	.76
Northeast	28 Mar 2014	.75	.86	.83	–	–	–

models would be able to reproduce it (Table 2). However, the temporal variability of such processes could only be reproduced using high-temporal-resolution models ($\approx$ 1 h).

4.3 Spatial patterns and wind variability

Model wind snapshots for the three different resolutions are used to understand the spatial structures associated with most common winds in the area (Fig. 6). Three events have been chosen, representing a case with higher variability (Fig. 6a, d and g) to one with an almost homogeneous wind field (Fig. 6c, f and i). For northwesterly winds (left column panels in Fig. 6) it is clear that Serra de Montsià exerts a physical barrier on wind fields, thus revealing areas in the inner bay with high wind intensities and areas down the mountain with almost calm winds or with different direction – shadow effect, described in other environments such as the Hawaiian Islands in Chavanne et al. (2002). These effects were also observed for winds coming from the north (not shown). Atmospheric pressure at surface on 4 April 2014 shows low pressure over the North Atlantic and a high-pressure area over north Africa. This synoptic situation promotes winds from the north–northwest (triggered by the Ebro River valley) in the study area. The modelled winds corresponding to observations in M-A and M-SC locations are similar, not showing

all the direction variability measured in observations (Figs. 2 and 3). However, the wind patterns in both WRF3 and CAL-MET are similar and show spatial wind variability inside the bay, thus indicating that the medium-resolution model is able to reproduce topographic constraints under these circumstances. On the other hand, the coarser model (Fig. 6a) does not reveal such a variability – expected for the dimensions of the bay and model resolution, with pixels almost half of the bay size. In summary, CALMET reproduces with higher accuracy these kind of winds in M-SC, while in M-A the errors are similar to coarser models. Both stations are located near the maximum transition zone between high- and lower-intensity winds (Fig. 6g), corresponding to the areas where modelling errors would be more sensitive to topographic effects.

An intensification of sea breezes (Fig. 6b, e and h) at midday in inner areas of the bay is clearly represented as well as a clockwise gyre of wind in M-SC related to M-A. The modelled highest intensities in the inner bay are not able to be validated, due to the lack of more observational data in this area (M-Met has been defined as a bad indicator of wind field). On the other hand, differences from coarser to the finest model configurations are noticeable. Both WRF3 and CALMET show some spatial structures in daily regimes

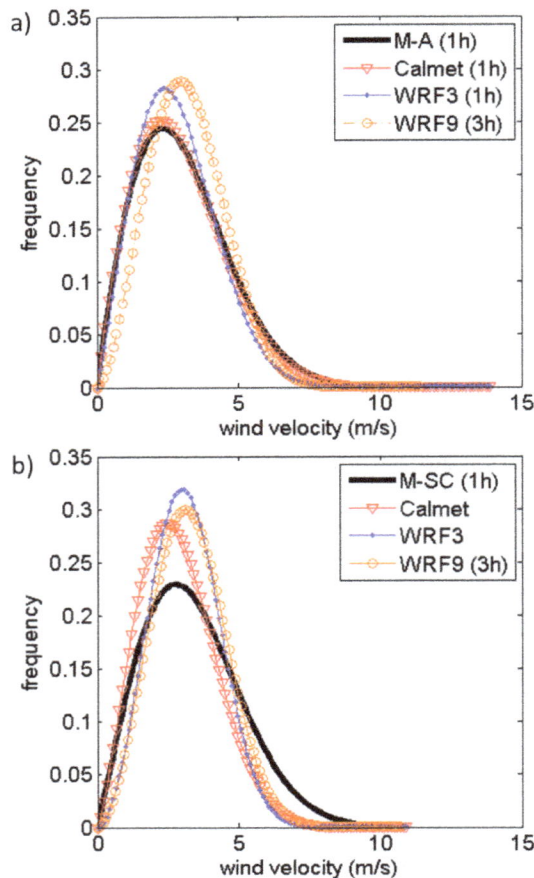

Figure 5. Weibull distributions for summer 2013 in M-A (**a**) and M-SC (**b**). Black line for observational data (hourly) and coloured lines for each model configurations (red for CALMET, blue for WRF3 and orange for WRF9).

not solved by WRF9. In the time series, sea breezes time-lag between M-A and M-SC observations is not reproduced by any of the models.

On the other hand, spatially homogeneous wind fields have also been observed during several events. In this case, winds from the northeast are shown in Fig. 6c, f and i. The wind fields reproduced by observations and atmospherical models indicate homogeneous spatial winds, not affected by topography in the Ebro Delta (winds coming along the coast). For these winds, the coarser model does reproduce the wind pattern similar to the finest model. In Fig. 3a there is an area where directions around 0–45° shows high correlation between M-SC and M-A. However, in Table 2 the NE case shows that CALMET and WRF3 have better correlation than WRF9, indicating that in some cases the temporal resolution would also play an important role in wind prediction.

5 Final remarks

This contribution presents an example of high wind variability in a coastal area (Alfacs Bay, NW Mediterranean Sea) during the period 2012–2013. Observational data demonstrate that wind direction seems to be affected by the surrounding mountains. These effects are maximized during northwesterly winds, in which the local mountains exert noticeable shadowing effects over the mouth of the bay. These results are in agreement with Camp (1994), showing high wind spatial variability at Alfacs Bay, as well as other similar studies which show the wind-channelling effects in some rias of Galicia (Herrera et al., 2005). Other winds, like sea breezes, also show noticeable variability, not only in space but also in time, and probably related to orography, land uses and different sea-water temperature in the bay and in the open sea. Winds from the S–SE not related to sea breezes are likely affected by local orography, but not enough events were recorded to confirm the observed pattern. Due to the short length of observational period (around 1 year), the results have no climatic significance.

The spatial heterogeneity also plays an important role in wind modelling results in this coastal area. The coarse model (9 km) does not reproduce the spatial variability associated with most topographically influenced winds (northwesterlies and sea breezes). The medium-resolution model (3 km) has proven to represent the wind spatial fields with enough accuracy according to the observations. This indicates that the effects of the main topographic structures on the area are recognized by this resolution model, contrary to other places where similar model resolution was not able to reproduce all the wind variability due to orography (Cerralbo et al., 2012). The predicted wind from the CALMET model at the highest resolution (400 m) also improves the spatial variability and shows the highest correlation with observational data under some circumstances. In this sense, the CALMET model could be an interesting and useful product for ocean and wave modelling, minimizing the information losses due to the downscaling processes. In summary, all the systems analysed reproduce with enough accuracy some of the characteristic winds observed at Alfacs Bay. The highest-resolution model shows better responses, since it reproduces more realistically wind fields and discriminates topographic structures such as mountains and gaps between them. However, correlation in some cases is higher in coarse models (WRF9), agreeing with Signell et al. (2005), who demonstrate that sometimes the higher-resolution models would present lower correlation due to higher "noise" (more variability) compared to the coarser models. Other authors (e.g. Miglietta et al., 2012; De Biasio et al., 2014) also argue that local models could show more details (more detailed flow patterns) although worse statistics due to errors in timing and location, whilst global models would produce smoother results and probably much skillful forecasts. At this point, the computational cost

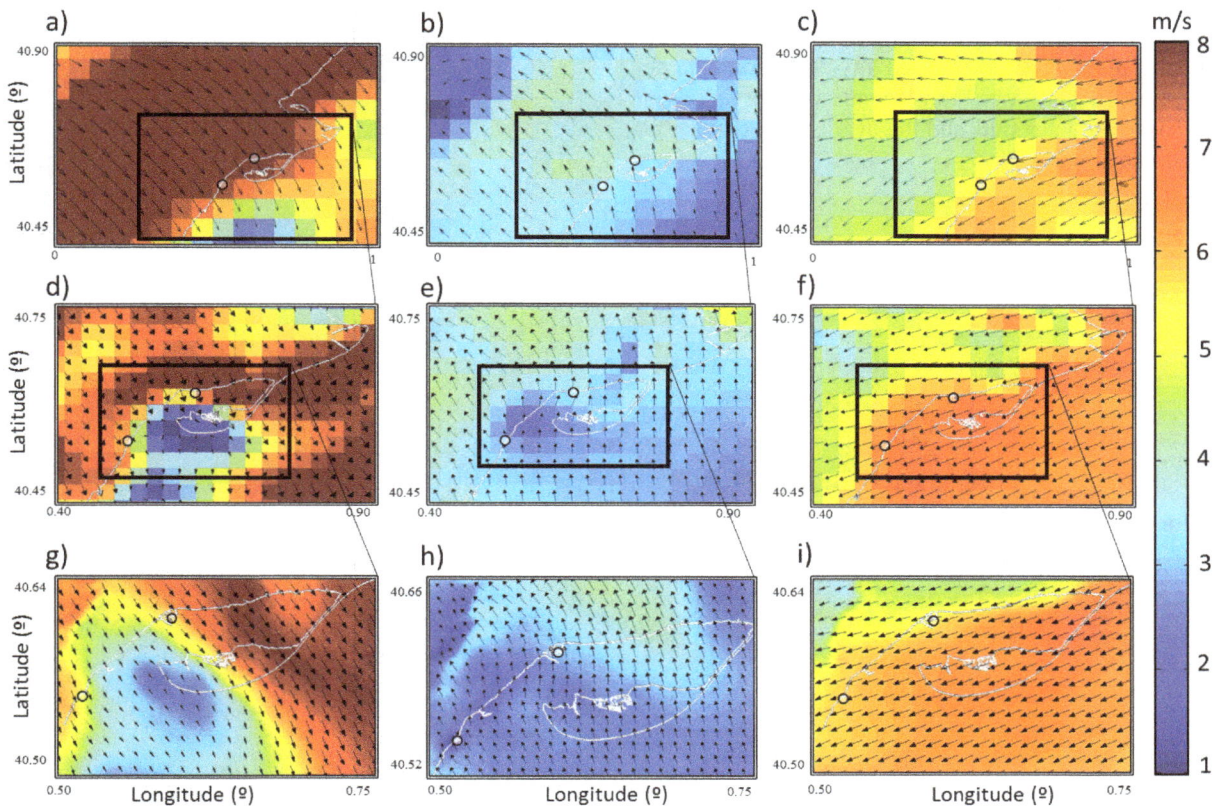

Figure 6. The three different models configurations are plotted for three snapshots of typical wind events at Alfacs Bay. **(a–c)** for WRF9 km, **(d–f)** for WRF3 and **(g–i)** for CALMET 400 m. Events represent winds from the northwest (left column panels) on 4 April 2014, sea breezes (central panels) on 6 July 2013 and northeasterly winds (right panels) on 12 March 2014. White circles indicate meteorological station locations (M-A and M-SC).

would indicate which should be the atmospheric model to be considered depending on the skill assessment requirements.

The effects of wind spatial variability on relatively short length scales would be an important factor to be considered in studies dealing with biology and ecology hazards, e.g. harmful algal blooms as described in Quijano-Scheggia et al. (2008), hydrodynamics (Cucco and Umgiesser, 2006) and water quality parameters (Grifoll et al., 2011) in coastal waters.

Appendix A: Statistics

The statistics used in the normalized Taylor diagram are defined as follows, where "obs" corresponds to observations, m corresponds to model results and the over bar $(-)$ denotes all data length mean values:

$$SD = \frac{\left(\sqrt{\frac{\sum_{i=1}^{n}(m_i-\overline{m})}{n}}\right)}{SD_{obs}}, \tag{A1}$$

$$CRMSE(obs, m) = \frac{\sqrt{\frac{\sum_{i=1}^{n}\left[(obs_i-\overline{obs})-(m_i-\overline{m})\right]^2}{n}}}{SD_{obs}}, \tag{A2}$$

$$Correlation(obs, m) = \frac{\sum_{i=1}^{n}(m_i-\overline{m})\cdot(obs_i-\overline{obs})}{n\cdot SD_{obs}\cdot SD_m}. \tag{A3}$$

Acknowledgements. This work was supported by a FPI-UPC pre-doctoral fellowship from the European project FIELD_AC (FP7-SPACE-2009-1-242284 FIELD_AC), the Spanish project PLAN_WAVE (CTM2013-45141-R) and the Secretariat d'Universitats i Recerca del Dpt. d'Economia i Coneixement de la Generalitat de Catalunya (Ref 2014SGR1253). The campaigns were carried out thanks to the MESTRAL project (CTM2011-30489-C02-01). We would like to thank Joan Puigdefàbregas, Jordi Cateura, Joaquim Sospedra and Elena Pallarés from Laboratori d'Enginyeria Marítima for all their help with campaigns and data analysis, as well as the Ebro Irrigation Community (Comunitat de Regants de la dreta de l'Ebre, http://www.comunitatregants.org) and the XIOM network (Xarxa d'Instruments Oceanogràfics de Catalunya; http://www.xiom.cat) for the information and their commitment to the study. Finally, we thank the two anonymous reviewers for their in-depth criticisms and suggestions.

References

Agterberg, R. and Wieringa, J.: Mesoscale terrain roughness mapping of the Netherlands, Technical Report TR-115, Royal Netherlands Meteorological Institute, Ministerie van Verkeer en Waterstaat, 1989.

Bignami, F., Sciarra, R., Carniel, S., and Santoleri, R.: Variability of Adriatic Sea coastal turbid waters from SeaWiFS imagery, J. Geophys. Res., 112, C03S10, doi:10.1029/2006JC003518, 2007

Bolaños, R., Jorda, G., Cateura, J., Lopez, J., Puigdefabregas, J., Gomez, J., and Espino, M. The XIOM: 20 years of a regional coastal observatory in the Spanish Catalan coast, J. Mar. Syst., 77, 237–260, doi:10.1016/j.jmarsys.2007.12.018, 2009.

Boldrin, A., Carniel, S., Giani, M., Marini, M., Bernardi Aubry, F., Campanelli, A., Grilli, F., and Russo, A.: Effects of bora wind on physical and biogeochemical properties of stratified waters in the northern Adriatic, J. Geophys. Res., 114, C08S92, doi:10.1029/2008JC004837, 2009.

Brecht, B. and Frank, H.: High resolution modelling of wind fields for optimization of empirical storm flood predictions, Adv. Sci. Res., 11, 1–6, doi:10.5194/asr-11-1-2014, 2014.

Camp, J.: Aproximaciones a la dinamica estuarica de una bahia micromareal Mediterranea, Thesis, Universitat de Barcelona, Barcelona, 1994.

Cerralbo, P., Grifoll, M., Espino, M., and López, J.: Predictability of currents on a mesotidal estuary (Ria de Vigo, NW Iberia), Ocean Dynam., 63, 131–141, doi:10.1007/s10236-012-0586-9, 2012.

Chavanne, C., Flament, P., Lumpkin, R., Dousset, B., and Bentamy, A.: Scatterometer observations of wind variations induced by oceanic islands: Implications for wind-driven ocean circulation, Can. J. Remote Sens., 28, 466–474, doi:10.5589/m02-047, 2002.

Cucco, A. and Umgiesser, G.: Modeling the Venice Lagoon residence time, Ecol. Model., 193, 34–51, doi:10.1016/j.ecolmodel.2005.07.043, 2006.

De Biasio, F., Miglietta, M. M., Zecchetto, S., and della Valle, A.: Numerical models sea surface wind compared to scatterometer observations for a single Bora event in the Adriatic Sea, Adv. Sci. Res., 11, 41–48, doi:10.5194/asr-11-41-2014, 2014.

Gracia, V., García, M., Grifoll, M., and Sánchez Arcilla, A.: Breaching of a barrier under extreme events, The role of morphodynamic simulations, J. Coast. Res., 951–956, doi:10.2112/SI65-161.1, 2013.

Grifoll, M., Del Campo, A., Espino, M., Mader, J., González, M., and Borja, A.: Water renewal and risk assessment of water pollution in semi-enclosed domains: Application to Bilbao Harbour (Bay of Biscay), J. Mar. Syst., 109–110, S241–S251, doi:10.1016/j.jmarsys.2011.07.010, 2011.

Grifoll, M., Aretxabaleta, A. L., Pelegrí, J. L., Espino, M., Warner, J. C., and Sánchez-Arcilla, A.: Seasonal circulation over the Catalan inner-shelf (northwest Mediterranean Sea), J. Geophys. Res.-Oceans, 118, 5844–5857, doi:10.1002/jgrc.20403, 2013.

Herrera, J. L., Piedracoba, S., Varela, R. A., and Rosón, G.: Spatial analysis of the wind field on the western coast of Galicia (NW Spain) from in situ measurements, Cont. Shelf Res., 25, 1728–1748, doi:10.1016/j.csr.2005.06.001, 2005.

Hong, S.-Y., Noh, Y., and Dudhia, J.: A new vertical diffusion package with an explicit treatment of entrainment processes, Mon. Weather Rev., 134, 2318–2341, doi:10.1175/MWR3199.1, 2006.

Jiang, H., Farrar, J. T., Beardsley, R. C., Chen, R., and Chen, C.: Zonal surface wind jets across the Red Sea due to mountain gap forcing along both sides of the Red Sea, Geophys. Res. Lett., 36, L19605, doi:10.1029/2009GL040008, 2009.

Martín Vide, J.: Los mapas del tiempo, Volumen 1 de Colección Geoambiente XXI. Davinci Continental, Editorial Davinci, Mataró, Barcelona, 219 pp., 2005.

Mass, C. F. and Ovens, D.: Fixing WRF's high speed wind bias: a new subgrid scale drag parameterization and the role of detailed verification, in: 24th Conference on Weather and Forecasting and 20th Conference on Numerical Weather Prediction, Preprints, 91st American Meteorological Society Annual Meeting, 23–27 January 2011, Seattle, WA, 2011.

Miglietta, M. M., Thunis, P., Pederzoli, A., Georgieva, E., Bessagnet, B., Terrenoire, E., and Colette, A.: Evaluation of WRF model performances in different European regions with the DELTA-FAIRMODE evaluation tool, Int. J. Environ. Pollut., 50, 83–97, 2012.

Oke, T. R.: Boundary Layer Climates, 2nd Edn., Routledge, London, 1987.

Quijano-Scheggia, S., Garcés, E., Flo, E., Fernandez-Tejedor, M., Diogene, J., and Camp, J.: Bloom dynamics of the genus Pseudonitzschia (Bacillariophyceae) in two coastal bays (NW Mediterranean Sea), Scientia Marina, 72, 577–590, 2008.

Sánchez-Arcilla, A., González-Marco, D., and Bolaños, R.: A review of wave climate and prediction along the Spanish Mediterranean coast, Nat. Hazards Earth Syst. Sci., 8, 1217–1228, doi:10.5194/nhess-8-1217-2008, 2008.

Schaeffer, A., Garreau, P., Molcard, A., Fraunié, P., and Seity, Y.: Influence of high-resolution wind forcing on hydrodynamic modeling of the Gulf of Lions, Ocean Dynam., 61, 1823–1844, doi:10.1007/s10236-011-0442-3, 2008.

Scire, J. S., Robe, F. R., Fernau, M. E., and Yamartino, J.: A User's Guide for the CALMET Meteorological Model (Version 5.0), Earth Tech, Concord, MA, 1999.

Serra, P., More, G., and Pons, X.: Monitoring winter flooding of rice fields on the coastal wetland of Ebre delta with multitemporal remote sensing images, IEEE International Geoscience and Remote Sensing Symposium, Barcelona, 2495–2498, doi:10.1109/IGARSS.2007.4423350, 2007.

Signell, R. P., Carniel, S., Cavaleri, L., Chiggiato, J., Doyle, J. D., Pullen, J., and Sclavo, M.: Assessment of wind quality for oceanographic modelling in semi-enclosed basins, J. Mar. Syst., 53, 217–233, doi:10.1016/j.jmarsys.2004.03.006, 2005.

Skamarock, W. C., Klemp, J. B., Dudhia, J., Gill, D. O., Barker, D. M., Duda, M., Huang, X.-Y., Wang, W., and Powers, J. G.: A description of the advanced research WRF version 3, NCAR Tech. Note NCAR/TN-475+STR, Nat. Cent. for Atmos. Res., Boulder, Colorado, 2008.

Taylor, K. E.: Summarizing multiple aspects of model performance in a single diagram, J. Geophys. Res., 106, 7183–7192, 2001.

Tewari, M., Chen, F., Wang, W., Dudhia, J., LeMone, M. A., Mitchell, K., Ek, M., Gayno, G., Wegiel, J., and Cuenca, R. H.: Implementation and verification of the unified NOAH land surface model in the WRF model, in: 20th conference on weather analysis and forecasting/16th conference on numerical weather prediction, American Meteorological Society, Seattle, WA, 11–15, 2004.

Thompson, G., Rasmussen, R. M., and Manning, K.: Explicit forecasts of winter precipitation using an improved bulk microphysics scheme, Part I: Description and sensitivity analysis, Mon. Weather Rev., 132, 519–542, 2004.

Venäläinen, A., Sahlgren, V., Podsechin, V., and Huttula, T.: Small-scale variability of the wind field over a typical Scandinavian lake, Boreal Environ. Res., 8, 71–81, 2003.

12

MATISSE: an ArcGIS tool for monitoring and nowcasting meteorological hazards

V. Rillo[1], A. L. Zollo[1,2], and P. Mercogliano[1,2]

[1]Regional Models and geo-Hydrological Impacts Division, CMCC – EuroMediterranean Centre on Climate Change, Capua (CE), Italy
[2]Meteo Systems and Instrumentation Laboratory, CIRA – Italian Aerospace Research Centre, Capua (CE), Italy

Correspondence to: A. L. Zollo (a.zollo@cira.it)

Abstract. Adverse meteorological conditions are one of the major causes of accidents in aviation, resulting in substantial human and economic losses. For this reason it is crucial to monitor and early forecast high impact weather events. In this context, CIRA (Italian Aerospace Research Center) has implemented MATISSE (Meteorological AviaTIon Supporting SystEm), an ArcGIS Desktop Plug-in able to detect and forecast meteorological aviation hazards over European airports, using different sources of meteorological data (synoptic information, satellite data, numerical weather prediction models data). MATISSE presents a graphical interface allowing the user to select and visualize such meteorological conditions over an area or an airport of interest. The system also implements different tools for nowcasting of meteorological hazards and for the statistical characterization of typical adverse weather conditions for the airport selected.

1 Introduction

Monitoring and nowcasting adverse meteorological conditions are crucial for the safety and optimization of all flight phases, especially for small aircraft or unmanned vehicles that are not equipped with adequate instrumentation. Indeed, adverse weather conditions constitute a major factor causing aviation accidents (Kulesa, 2003; Krozel et al., 2008). For these reasons numerous projects, such as FLY-SAFE (Tafferner et al., 2008), SPADE (Van Eenige and Muehlhausen, 2006), EWENT (Juga and Vajda, 2012) and WxFusion (Gerz et al., 2012), as well as systems such as XMWX (Snyder and Patsiokas, 2004), AWARE (Ruokangas et al., 2006) and AWDSS (Barrere Jr. et al., 2008), have been developed to increase pilots' awareness of in-flight meteorological conditions (so-called weather awareness). In this context the Italian Aerospace Research Center (CIRA) has developed MATISSE (Meteorological AviaTIon Supporting SystEm), an ArcGIS tool for monitoring and nowcasting meteorological hazards, developed using the ArcGIS development environment ArcObjects. The system differs from existing ones in its ability to gather data from different sources (e.g. remote sensing platforms, meteorological center archives, outputs of numerical weather prediction models), handling data of different formats and providing information on monitored and forecasted meteorological conditions in formats tailored to pilots' needs. In addition, it includes innovative nowcasting tools and algorithms able to retrieve information on meteorological parameters and hazards from website images.

A system which presents numerous analogies with MATISSE is Synergie, a weather expertise and forecasting system developed in collaboration with Meteo France experts. Like MATISSE, it integrates meteorological data in an automatic and near real time way from various sources and ensures easy access to data, providing maps and charts. The main difference between the two systems lies in the fact that MATISSE was conceived for aviation purposes, focusing on aviation weather hazards and reporting on board only information relevant to the flight.

The present work is the expanded version of Rillo et al. (2014), including additional functionalities that have been implemented in the system, focusing on nowcasting tools and outlining the context in which the system is placed. The pa-

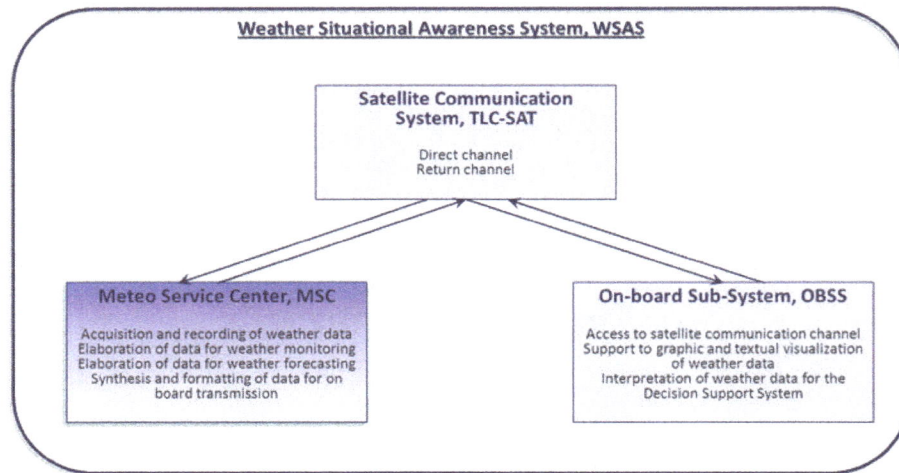

Figure 1. Weather Situational Awareness System block diagram.

per is organized as follows: an overview of the system architecture is proposed, followed by a description of MATISSE input data and of the outputs and functionalities provided to users.

2 Overall system overview

In the framework of the CIRA project TECVOL II (TECnologie per il VOLo autonomo – technologies for autonomous flight) founded by MIUR (Italian Ministry of Education, University and Research), aiming to develop technologies to support pilots of Personal Air Vehicles (PAVs) or automated systems of Unmanned Air Vehicles (UAVs), CIRA is developing the Weather Situational Awareness System (WSAS), a system able to provide real time data for updating mission management and trajectory generation functions (for UAVs) and for enhancing pilots' awareness of the meteorological conditions occurring during the flight (for PAVs). As shown in Fig. 1, the WSAS consists of three blocks: an on-board segment (On-Board Sub-System, OBSS), a ground segment (Meteo Service Center, MSC) and a satellite link between them (Satellite Communication System, TLC-SAT).

The ground segment MSC represents the core of the entire system because it gathers, processes and integrates observational data and forecasts from meteorological center archives and CIRA weather stations. Such data are formatted in order to be transmitted to the On-Board Sub-System using the Satellite Communication System. The on-board segment OBSS accesses the satellite communication channel and supports the graphical and textual visualization of the weather data provided by the ground segment and data provided by sensors installed on board. Visualization will be carried out using tablets, laptops or multi-functional displays integrated into the cockpit. The Satellite Communication System pro-

vides a bidirectional link between the ground segment and the on-board segment using a direct and a return channel.

Within this architecture, MATISSE is the system that accomplishes most of the MSC functionalities by processing and storing raw data collected by the MSC and providing user-friendly information for the detection and forecast of meteorological hazards. Indeed, the system processes meteorological data from different sources (in situ and remote sensing measurements, outputs of numerical weather prediction models, websites), available in different formats (e.g. BUFR, NetCDF, GRIB2, HDF) and stores them in a geodatabase. After storage the meteorological variables can be further elaborated in order to obtain more complex information and provide user-friendly maps, graphs and statistics.

Users can retrieve the information of interest using the MATISSE graphical interface shown in Fig. 2. Different selections on the graphical elements produce SQL (Structured Query Languages) queries to geodatabase able to extract the desired information and to visualize maps, statistics and files of the desired information. MATISSE mainly aims to support flights of PAVs and UAVs. That said, these two kinds of vehicles have different characteristics and of course different "needs". Specifically, in the case of PAVs, graphical visualization of detected hazards could be useful for pilots, while for UAVs (unmanned), graphical information is unnecessary. However, in this case a text file reporting aviation hazard locations and corresponding hazard levels could be of great benefit since such weather information could be fed into on-board Decision Support Systems (DSS), the tool employed on UAVs to improve on-board situational awareness autonomously. In order to take such different needs into account, MATISSE is a very versatile tool since it offers several functionalities that, depending on the case, adapt to different user requirements.

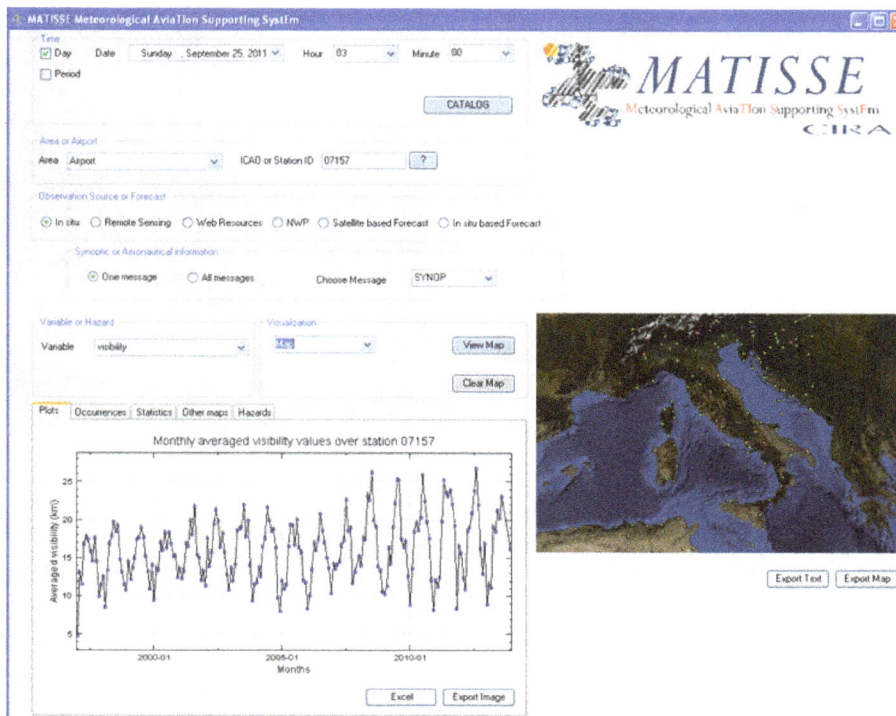

Figure 2. MATISSE graphical interface.

3 Sources of meteorological data

As described above, MATISSE accesses the database containing observed and forecasted meteorological data. In the following section the main features of the imported data are reported.

3.1 Observed data

Thanks to the agreement of CIRA with the Italian Air Force meteorological service, the observed input data are mainly provided by the ECMWF MARS Archive (Hennessy, 1986), an archive making weather data available from stations located all over the world, comprising AIREP data, vertical soundings data (TEMP and PILOT), land surface data (SYNOP and METAR) and sea surface data (SHIP). Such data are available in BUFR (Binary Universal Form for the Representation of meteorological data) format and converted into text files in order to be imported into the database. The data can be used to extract numerous meteorological variables (e.g. visibility, wind speeds and directions, present and past weathers, temperatures, cloud cover, three-hour pressure change), to be visualized or further elaborated. Similarly, MATISSE downloads the latest METARs and TAFs from the Italian Air Force website and converts and imports them into the database.

Additional observed input data are provided by the Meteosat Second Generation (MSG) satellite (Schmetz et al.,

2002). Indeed, MSG data (EUMETSAT, 2007), available in HDF (Hierarchical Data Format) format, have been processed in Matlab for the detection of convective clouds (clouds responsible for the presence of thunderstorms, turbulence, wind shear and icing phenomena) and EUMETSAT products have been elaborated to detect various meteorological hazards for aviation (such as turbulence, icing, snow, cumulonimbus clouds, intense rainfall and volcanic ash) (EUMETSAT, 2011). In particular, three EUMETSAT products have been used: Multisensor Precipitation Estimate (MPE) (Heinemann et al., 2002), Cloud Analysis (CLA) (Lutz et al., 2003) and Atmospheric Motion Vectors (AMV) (Holmlund, 2002).

MPE provides an estimate of precipitation rates observed from satellite and occurring over Europe; CLA provides a classification of types, phases and depths of clouds observed from satellite. Indeed, cloud phases are used in order to detect areas affected by icing conditions (hazardous due to the formation of ice on aircraft walls and instruments). Finally, AMV provides wind speeds and directions, at different pressures, observed from satellite. After the conversion of these products from raw format into text files, they are imported into the database accessed by MATISSE.

In addition, MATISSE recovers information on missing meteorological parameters from free images available on the internet. For example, due to the unavailability of radar data at CIRA, it accesses the Italian Civil Protection website and downloads and visualizes the latest radar image over Italy.

In order to overcome the current lack of raw data, MATISSE implements tools able to retrieve missing meteorological information from free images available on the internet. Current developments concern a tool able to obtain timing and location of lightning flashes over Europe from Blitzortung.org website images and report such information in a text file that can be imported into the geodatabase.

3.2   Forecast data

As previously described, MATISSE is able to handle not only observed data but also outputs of numerical models (forecasting and nowcasting). Forecast data from the COSMO LM model (Doms and Schättler, 2002) at 7 and 2.8 km resolutions are available in MATISSE thanks to the implemented function able to report NetCDF files in a Comma-Separated Value (CSV) format and to store them in the geodatabase. At present, this function has been implemented only for such a limited area model. However, it will be extended in the future to include further forecast models.

In addition MATISSE implements two tools for nowcasting precipitation and the evolution of convective cloud nuclei, i.e. the convective cores inside cloud systems. The former is carried out by applying the uniform advection technique (Austin and Bellon, 1974) to the MPE product. This technique assumes that the precipitation field moves at a uniform and constant velocity and that there are no changes in the rainfall rates during the forecast. As a consequence, a preliminary estimate of the velocity of the precipitation field is computed by evaluating the cross correlations between two consecutive observations. Indeed, the algorithm computes the cross correlations between them for all the possible displacements and chooses the spatial shift that maximizes the correlation. The forecast is obtained by moving the observed precipitation field according to the estimated shift, assuming that the field velocity is constant during the lead time (time interval occurring between the observation time and the forecast one). This technique is used to provide forecasts in text format that can be imported into the database.

The second nowcasting tool is based on MSG-2 data and is able to provide 15 min forecasts of the evolution of convective cloud nuclei. First of all, the algorithm detects the presence of convective clouds using the infrared brightness temperature and identifies cloud nuclei. It then compares the brightness temperatures in the infrared and water vapor channels of two consecutive satellite images forecasting the developing or dissolving of convective cloud nuclei in the following 15 min (Puca et al., 2005).

4   MATISSE outputs

Starting from the meteorological variables stored in the database, MATISSE provides various outputs that will be described below.

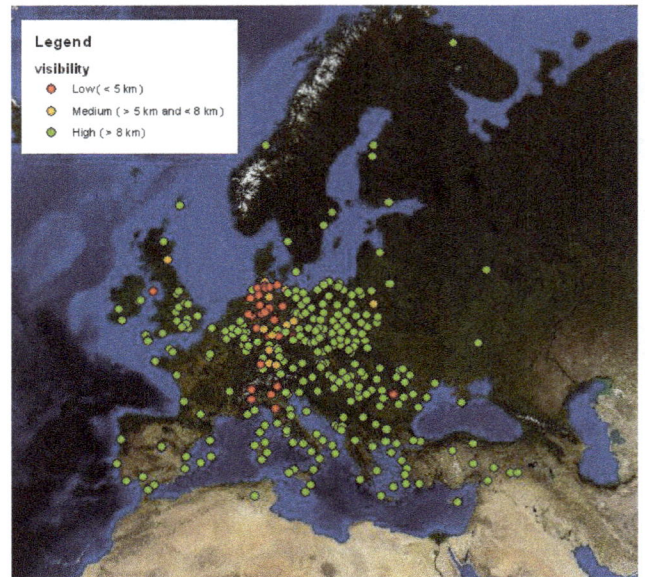

Figure 3. Visibility conditions (km) occurring 05 September 2011, 02:30 UTC.

4.1   Observed and forecasted maps

The main goal of the system is to provide an immediate and concise representation of weather conditions observed or forecasted, with special emphasis on hazards. Such concise representation is provided with maps or text files reporting the risk areas in order to supply only the relevant information during the route or combining different hazards on the same map. For this reason, MATISSE provides maps in a GIS environment displaying the meteorological variables and, in all cases, their corresponding hazard levels with different colors. For instance, in Fig. 3 an observed visibility map is shown: airports with visibility below 5 km are displayed with red dots, airports with visibility between 5 and 8 km are reported with orange dots and airports whose visibility exceeds 8 km are reported with green dots. Another example is shown in Fig. 4 in which areas with the presence of icing conditions are marked in blue.

An additional function of the system allows overlapping of meteorological variables in order to display multiple hazards on a single map. An example is reported in Fig. 5 in which lightning flashes, strong winds and convective clouds are visualized together.

In order to provide a synthetic representation of the meteorological conditions, MATISSE allows potentially hazardous areas to be reported as rectangles in order to reduce the information content to be transmitted on board, representing only the four vertexes of the areas with the presence of meteorological hazards. Indeed, the system implements algorithms able to identify so-called *blobs* (areas that are homogeneous in terms of specific characteristics, Carson et al., 1999) and include them in rectangles. For instance, in Fig. 6 the con-

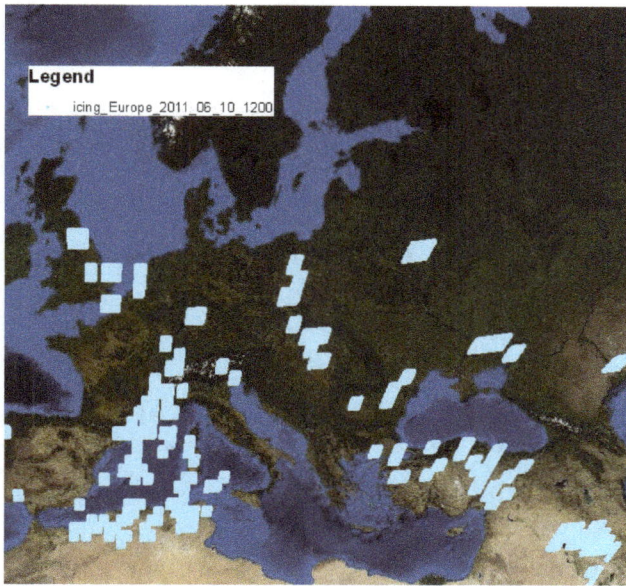

Figure 4. Icing conditions occurring 10 June 2011, 12:00 UTC.

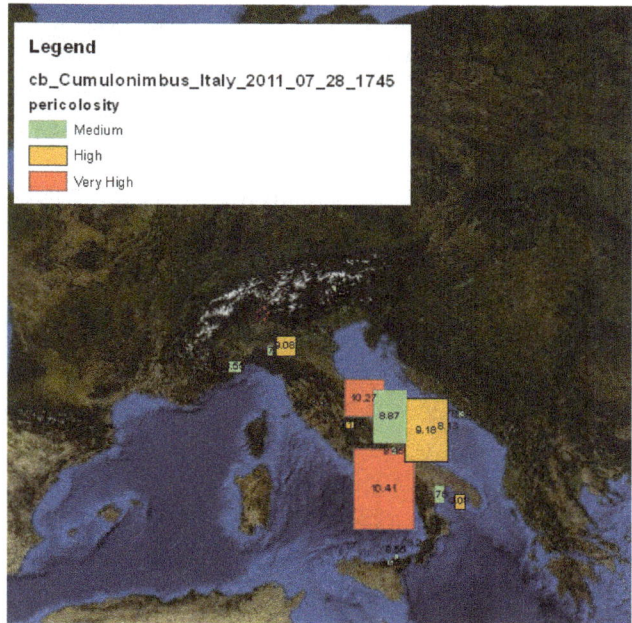

Figure 5. Overlapping of strong winds, convective clouds and lightning occurring 14 June 2014, 17:00 UTC.

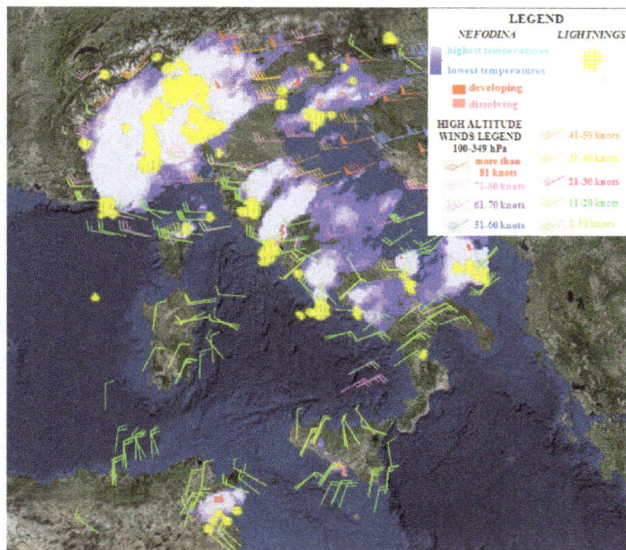

Figure 6. Rectangles representing convective clouds (and their corresponding hazard levels and heights) occurring 28 July 2011, 17:45 UTC.

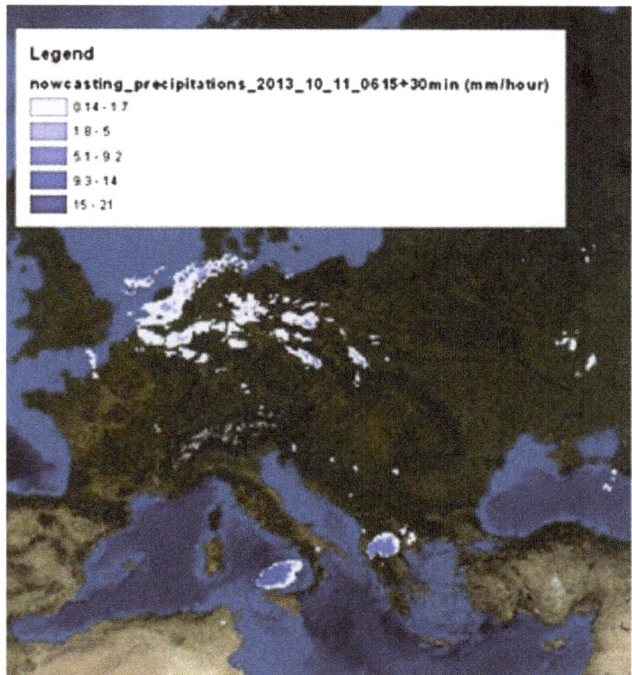

Figure 7. 30 min precipitation forecast based on MPE of 11 October 2013 06:00 UTC.

vective clouds observed during 28 July 2011 at 17:45 UTC are reported as rectangles.

On the other hand, MATISSE is able to display maps of outputs of the convective cloud nowcasting tool (see Rillo et al., 2014), precipitation nowcasting tool (Fig. 7) and the COSMO LM model (Fig. 8). All these maps can be exported in Excel files in order to be further elaborated by users.

4.2 Graphs and statistics

The availability of an historical dataset of meteorological data allows MATISSE to provide graphs and statistics de-

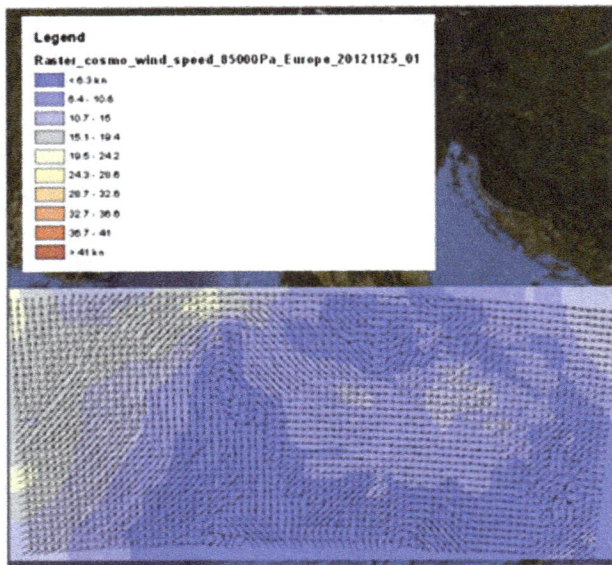

Figure 8. Forecast of wind speeds and directions of 25 November 2012 at 01:00 UTC.

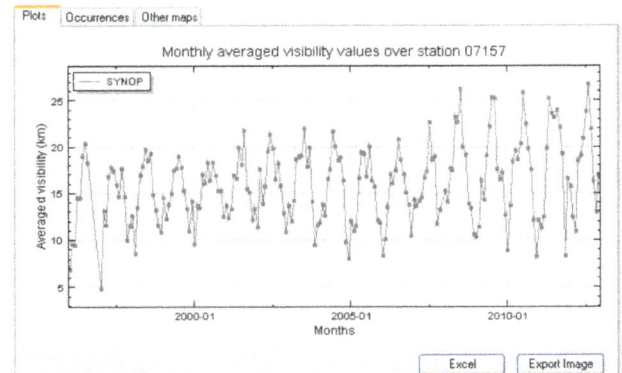

Figure 9. Monthly mean visibility values (km) over Charles de Gaulle airport for period varying from January 1996 to August 2012.

scribing the historical behavior of different meteorological parameters over the main European airports. Users can choose an airport and a period of interest and plot the graph reporting monthly mean, maximum and minimum values of the desired meteorological variable (Fig. 9) or visualize and export the number of occurrences of a parameter below or above a threshold fixed by the user. Another function allows users to compute statistics (mean, variance, maximum, minimum values) of a meteorological parameter over an area of interest by selecting maximum and minimum latitudes and longitudes and a specific time.

Graphs and statistics are all provided to users in text files.

4.3 Hazard representation for supporting flight management

In order to support flight management and flight planning tools, MATISSE is equipped with an additional function able to produce text files reporting aviation hazard locations and corresponding hazard levels. In this way the information content to be transmitted on board is drastically reduced, indicating only the latitudes and longitudes of the potentially dangerous areas due to the presence of one or more meteorological hazard. Such information can be fed into on-board Decision Support Systems (DSS): the tool, usually adopted in UAV, to improve on-board situational awareness autonomously.

5 Conclusions

The MATISSE GIS System described in the present work represents a technology developed to support pilots and automated optimization systems, providing maps, graphs, text files and statistics of observed and forecasted meteorological variables or hazards, in different formats according to different user needs. Meteorological data are displayed in a user-friendly way, indicating the hazard level in order to improve pilots' weather awareness and allowing users to easily identify potentially dangerous areas. Meteorological hazards are reported concisely, providing the relevant information on the hazards present in all the different flight phases. The system is still under development and hence is not currently in use. It will be extended in the future to incorporate additional functions including the forecasting and monitoring of further meteorological hazards and implementing other forecast models (such as ECMWF IFS).

Edited by: T. Hewson
Reviewed by: M. M. Miglietta and another anonymous referee

References

Austin, G. and Bellon, A.: The use of digital weather radar records for short-term precipitation forecasting, Q. J. Roy. Meteorol. Soc., 100, 658–664, 1974.

Barrere Jr., C. A., Eilts, M., Johnson, J., Fritchie, R., Spencer, P., Shaw, B., Li, Y., Ladwig, W., Schudalla, R., and Mitchell, D.: An Aviation Weather Decision Support System (AWDSS) for the Dubai International Airport, in: 13th Conference on Aviation, Range and Aerospace Meteorology, Amsterdam, the Netherlands, 20–24, 2008.

Carson, C., Thomas, M., Belongie, S., Hellerstein, J. M., and Malik, J.: Blobworld: A system for region-based image indexing and retrieval, in: Visual Information and Information Systems, Springer, 509–517, 1999.

Doms, G. and Schättler, U.: A description of the nonhydrostatic regional model LM, Deutscher Wetterdienst, Offenbach, Germany, 2002.

EUMETSAT: MSG Level 1.5 Image Data Format Description, https://www.eumetsat.int/website/wcm/idc/idcplg?IdcService=GET_FILE&dDocName=PDF_TEN_05105_MSG_IMG_DATA&RevisionSelectionMethod=LatestReleased&Rendition=Web (last access date: 6 July 2015), 2007.

EUMETSAT: MSG Meteorological product extraction facility algorithm specification document, Tech. rep., https://www.eumetsat.int/website/wcm/idc/idcplg?IdcService=GET_FILE&dDocName=PDF_TEN_SPE_04022_MSG_MPEF&RevisionSelectionMethod=LatestReleased&Rendition=Web (last access date: 6 July 2015), 2011.

Gerz, T., Forster, C., and Tafferner, A.: Mitigating the Impact of Adverse Weather on Aviation, in: Atmospheric Physics, Springer, Berlin, Heidelberg, Germany, 645–659, 2012.

Heinemann, T., Latanzio, A., and Roveda, F.: The Eumetsat multisensor precipitation estimate (MPE), in: Second International Precipitation Working group (IPWG) Meeting, Madrid, Spain, 2002.

Hennessy, J.: MARS-The ECMWF Meteorological Archive and Retrieval System, in: Proceedings, 2nd International Conference on Interactive Information and Processing Systems for Meteorology, Oceanography and Hydrology (IIPS), AMS, Miami, Florida, United States, 1986.

Holmlund, K.: Current status of the Eumetsat operational and future AMV extraction facilities, in: Proc. of the 6th Int, Winds Workshop, Madison, 45–52, 2002.

Juga, I. and Vajda, A.: The effect of weather on transportation: assessing the impact thresholds for adverse weather phenomena, in: Proceedings of SIRWEC 16th international road weather conference, Helsinki, Finland, 23–25, 2012.

Krozel, J., McNichols, W., Prete, J., and Lindholm, T.: Causality analysis for aviation weather hazards, in: AIAA Aviation Technology, Integration, and Operations Conf., Anchorage, Alaska, 2008.

Kulesa, G.: Weather and aviation: How does weather affect the safety and operations of airports and aviation, and how does FAA work to manage weather-related effects?, in: The Potential Impacts of Climate Change on Transportation, Transportation Research Board (TRB) and the place is Washington DC, USA, 2003.

Lutz, H.-J., Gustafsson, J. B., and Valenzuela-Leyenda, R.: Scenes and cloud analysis from Meteosat Second generation (MSG) observations, EUMETSAT, EUM, Citeseer, in: Proceedings of the 2003 EUMETSAT Meteorological Satellite Conference, Weimar, Germany, p. P39, 2003.

Puca, S., De Leonibus, L., Zauli, F., Rosci, P., and Biron, D.: Improvements on numerical "object" detection and nowcasting of convective cell with the use of SEVIRI data (IR and WV channels) and neural techniques, in: The World Weather Research Programme's Symposium on Nowcasting and Very Short Range Forecasting, Tolouse, France, 5–9, 2005.

Rillo, V., Manzi, M. P., Mercogliano, P., and Galdi, C.: An advanced platform providing multi-sources aviation-critical weather information, in: IEEE 2014 Metrology for Aerospace (MetroAeroSpace), Benevento, Italy, 484–487, 2014.

Ruokangas, C. C., Mengshoel, O. J., Uckun, S., Rand, T. W., Donohue, P., and Tuvi, S.: Aviation weather awareness and reporting enhancements (AWARE) system using a temporal-spatial weather database and a Bayesian network model, US Patent No. 7081834, Washington, DC, US Patent and Trademark Office, 2006.

Schmetz, J., Pili, P., Tjemkes, S., Just, D., Kerkmann, J., Rota, S., and Ratier, A.: An introduction to Meteosat second generation (MSG), B. Am. Meteorol. Soc., 83, 977–992, 2002.

Snyder, J. and Patsiokas, S.: XM satellite radio-satellite technology meets a real market, in: Proceedings of the 22nd AIAA International Communications Satellite Systems Conference and Exhibit, Monterey, CA, USA, 2004.

Tafferner, A., Forster, C., Sénési, S., Guillou, Y., Tabary, P., Laroche, P., Delannoy, A., Lunnon, B., Turp, D., Hauf, T., and Markovic, D.: Nowcasting thunderstorm hazards for flight operations: the CB WIMS approach in FLYSAFE, in: ICAS2008 Conference, International Council of the Aeronautical Sciences Conf. Proc. (8.6.2), Optimage Ltd., Edinburgh, UK, 1–10, 2008.

Van Eenige, M. J. and Muehlhausen, T.: SPADE: Supporting platform for airport decision-making and efficiency analysis, Proceedings of ICAS 2006, Hamburg, Germany, 2006.

13

The FORBIO Climate data set for climate analyses

C. Delvaux, M. Journée, and C. Bertrand

Royal Meteorological Institute of Belgium, Brussels, Belgium

Correspondence to: C. Delvaux (charles.delvaux@meteo.be)

Abstract. In the framework of the interdisciplinary FORBIO Climate research project, the Royal Meteorological Institute of Belgium is in charge of providing high resolution gridded past climate data (i.e. temperature and precipitation). This climate data set will be linked to the measurements on seedlings, saplings and mature trees to assess the effects of climate variation on tree performance. This paper explains how the gridded daily temperature (minimum and maximum) data set was generated from a consistent station network between 1980 and 2013. After station selection, data quality control procedures were developed and applied to the station records to ensure that only valid measurements will be involved in the gridding process. Thereafter, the set of unevenly distributed validated temperature data was interpolated on a 4 km × 4 km regular grid over Belgium. The performance of different interpolation methods has been assessed. The method of kriging with external drift using correlation between temperature and altitude gave the most relevant results.

1 Introduction

The interdisciplinary FORBIO Climate research project wants to scrutinize the adaptive capacity of tree species and predict the future performance of tree species in Belgium under different scenarios of climate change. Towards this objective, the Royal Meteorological Institute of Belgium is in charge of providing high resolution gridded past climate data (i.e. temperature and precipitation) between 1980 and 2013. The gridded daily temperature (minimum and maximum) data set was generated from the network of climatological stations administrated by the Royal Meteorological Institute of Belgium (RMI) since the end of the 19th century. In order to provide the high resolution gridded data set from the data of the station network, data quality control procedures were developed and the performance of different interpolation methods has been assessed.

The RMI network relies on voluntary observers, professional observers on civil and military aerodromes, federal agents, regional agents and employees of private companies. RMI supplies any voluntary observers with a manual rain gauge and a meteorological shelter in about 3/5 of the stations. Air temperature is measured in a shaded enclosure (i.e., Stevenson screen) at a height of approximately 1.5 m above the ground. Maximum and minimum temperatures, for the previous 24 h, are recorded at 08:00 LCT (local clock time). Minimum temperature is recorded against the day of the observation, and the maximum temperature against the previous day. Temperature records from both voluntary and synoptic stations were considered to generate the 34-year-long (1980–2013) high resolution daily gridded temperature data set over Belgium. Within this period, 278 time series from the RMI database of climate observations contain at least one temperature record. However, a large number of time series contain missing observations. The total number of available observations per day varies between 109 and 175 over the considered time period (see Fig. 1, left panel). The spatial distribution of the corresponding stations within the Belgian territory is provided on the right panel in Fig. 1. Stations for which less than 5 % of the temperature records are missing over the considered time period are referred to as reference stations (i.e. 65 stations, see black dots in Fig. 1). The mean temperature over the 34 years obtained using the daily temperature observations from the 65 reference stations are provided in Fig. 2 for *TN* and *TX*, respectively. Because no measuring technique is perfect and errors can run in meteorological observations for a wide variety of reasons (e.g. Aguilar et al., 2003), data were first quality controlled.

To ensure that only valid measurements will be involved in the gridding process, Sect. 2 describes the quality con-

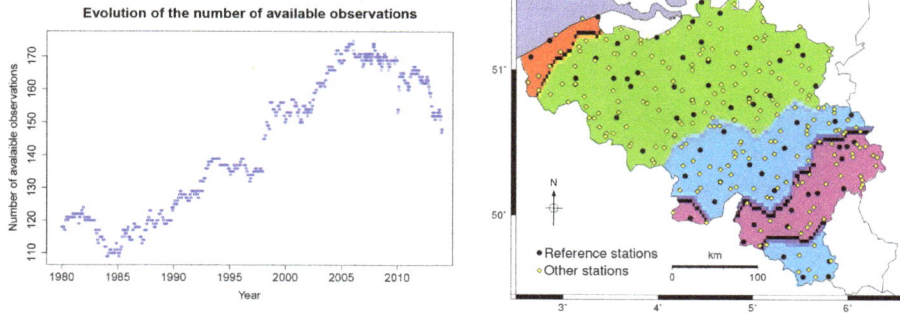

Figure 1. Evolution of the number of available observations per day for the period 1980 to 2013 (left panel) and location of the stations in operation between 1980 and 2013 with the division into 4 climate zones (right panel). Reference stations (i.e. less than 5 % of missing temperature records) are indicated by a black dot while the other stations are represented by a yellow diamond.

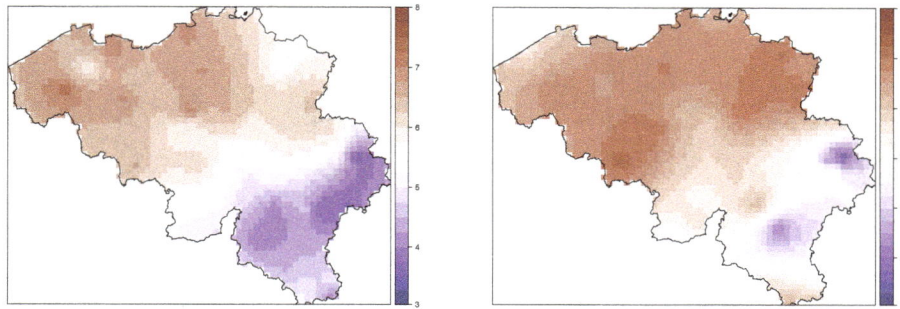

Figure 2. Averaged temperature over the 34 years [°C] for *TN* (left panel) and *TX* (right panels) using reference station records only. The interpolation was made using ordinary kriging method (see Sect. 3). The missing data are estimated by spatial interpolation.

trol (QC) procedures developed. Section 3 presents the considered interpolation methods. Results are presented in Sect. 4 and additional discussions are provided in Sect. 5. Final conclusions are given in Sect. 6.

2 Quality control procedures

2.1 Definition of quality control zones

For quality control purposes, geographical zones of similar temperature characteristics were defined based on the reference stations temperature records. These zones will be used to compare stations of the same zone to each other and to define quality control thresholds specific to each zone. The different zones were identified by k-means clustering approach (Hair, 2009) based on mean and variability of the reference stations *TX* and *TN* time series.

A division into 4 zones was selected (see the right panel in Fig. 1). These zones broadly correspond to the Belgian coast (red), Flanders (green), Ardenne (purple) and the rest of Wallonia (blue).

2.2 Preliminary tests (time series evaluation)

In the first instance, two preliminary tests are applied on 1-year long time series to ensure the validity of the studied station. The first test (variability test) compares the data's standard deviation σ of the given station for a time period of one year to the expected limits. The minimum (maximum) limit corresponds to the half (double) of the reference stations mean standard deviation. If σ lies outside the minimum and maximum thresholds, the entire time series is considered as not usable. The second test (lag test) verifies that no one-day time lag is affecting the data. Each 1-year long temperature record with and without lag (forwards or backwards) were compared to neighboring stations. If the data fit is better with a lag, the entire time series is considered as not usable.

2.3 Daily data quality control

All time series that succeeded the preliminary tests are further checked with the following tests for daily data. After an existence test, a first QC module checks for physical limits and flags the data violating these limits (note that values flagged as erroneous fail immediately and do not require further testing). Second, similarly to Feng et al. (2004) and Boulanger et al. (2010), automated QC procedures check the

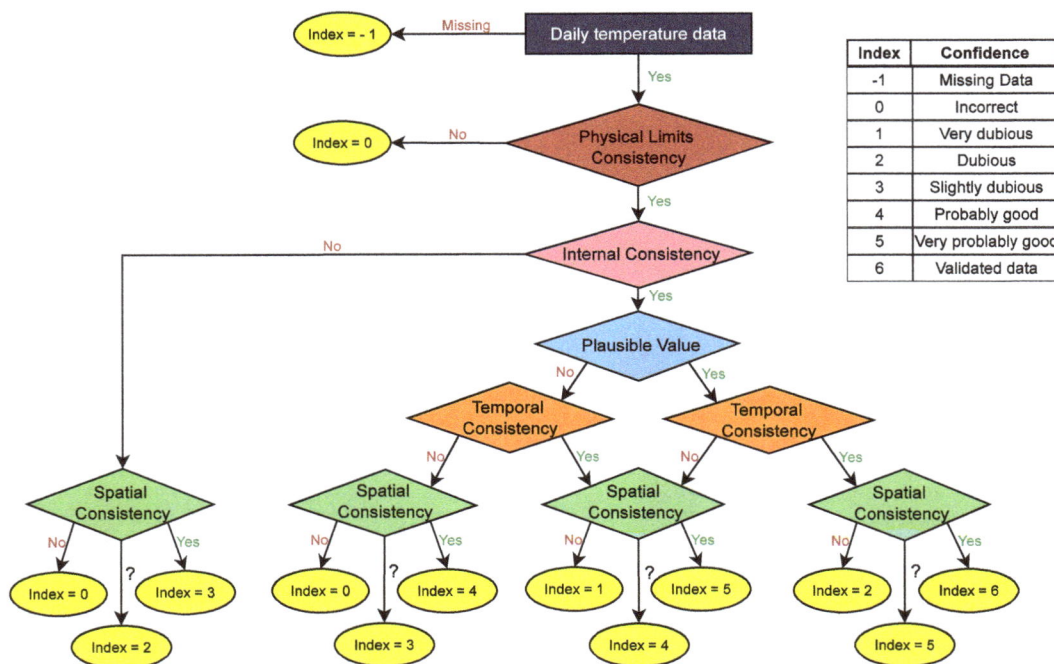

Figure 3. Automated QC applied to temperature records and associated confidence index.

temperature values for more subtle errors. Implementation of the automated QC is diagrammed in Fig. 3. Data are checked for internal consistency, plausible value test (using adapted limits to reflect climatic conditions more precisely than in the first range test), temporal consistency and spatial consistency. At the end of the process, a confidence index ranging from −1 to 6 (i.e. missing data to validated data) is attributed to each datum.

2.3.1 Physical limit consistency test

The aim of the physical limit consistency test is to ensure that temperature records stays within acceptable range limits using extreme climatological boundaries. These boundaries are based on the extreme values of the reference stations for each season and climate zone. To account for extreme values within the different zones, the boundaries are adjusted with a tolerance of 3 °C. Lower and upper limits for the four seasons are given for the Flanders and Ardenne zones in Table 1 for illustration.

2.3.2 Internal consistency test

The internal consistency test ensures that for a given day TN is not higher than TX. Such an incoherence is technically possible because TN is recorded one day later than TX but is attached to the same day as TX.

Table 1. Extreme climatological boundaries for the Flanders/Ardenne zones per seasons used in the physical limit consistency test.

	Lower limit $TN[°C]$	Upper limit $TN[°C]$	Lower limit $TX[°C]$	Upper limit $TX[°C]$
winter	−26.5/−27.6	16.5/15.8	−15.2/−17.6	26.8/25.7
spring	−12.8/−15.4	25.7/24	−2.8/−5.9	39.4/36
summer	−1/−5.4	26/25.3	6.8/6	41.2/40
autumn	−18.2/−23.2	21/18.2	−13.9/−14.2	31/29.9

2.3.3 Plausible value test

For the plausible value test, daily TN and TX values are compared to daily lower and upper bounds. For each of the 4 climate zones, lower and upper bounds for a given temperature data series (i.e. TN and TX) are constructed by retrieving the highest and lowest daily values on each calendar day d of the year from 34 years of data. Assuming that the annual temperature variations follow a sinusoidal wave, a fit of the annual variation of these extreme temperatures is defined using wave function $T_{L/U}(d)$. An observation succeeds the test if it stays within the lower and upper bounds of the corresponding day (Sciuto et al., 2013).

2.3.4 Temporal consistency test

The temporal consistency test analyzes the rate of daily change in order to detect possible anomalies. A spike or step test (Δmax) checks for a plausible rate of change from a pre-

Table 2. Percentage of the different indexes obtained after the QC procedures. Note that to maintain a good comparison between both types of stations, the "missing data" are not included in the index percentage of index 0 to 6.

Index	Index percentage	
	Reference stations [%]	Other stations [%]
−1	3.32	61.22
0	0.001	0.002
1	0.003	0.005
2	0.008	0.008
3	0.029	0.036
4	0.008	0.014
5	0.593	0.640
6	99.358	99.301

ceding acceptable level. The maximum probable change is based on the 99th percentile change for the 34 years of data. The test is applied by season and climate zone with a one-day time step. Similarly a persistence test (Δmin) flags the measurements that fail to change by more than a minimum amount.

2.3.5 Spatial consistency test

The spatial consistency test compares the observations at a given station with the observations of the same day made at neighboring stations. The result is suspicious (represented by a question mark in Fig. 3) if the station record differs for more than 2.5 °C from each of its 3 closest neighboring stations or by more than 4 °C of the averaged values of the 3 closest neighboring stations. The observation fails the spatial consistency tests if the two values above become respectively 5 and 7 °C (these values are based on the expertise of the well qualified RMI's data quality agents). A nearby station is considered only if the difference in elevation with the studied station is below 150 m.

2.4 Confidence Index

More than 5 millions of temperature records (*TN* and *TX*) were faced to the data quality control protocol described above. More than 99.3 % of the analyzed records have succeeded the different tests (confidence index of 6). The proportion of data classified as at least "very probably good" (confidence index greater than 5) was of 99.95 %. Table 2 compares the results in terms of attributed confidence index obtained at the reference stations and at the other stations.

Table 3. Correspondance between Corine Land Cover database and the five land cover types used in this study.

Index	Corine index	Corine description
1	1–11	Artificial surfaces
2	12–15	Arable land
3	16–22	Permanent crops, pastures and heterogenous agricultural areas
4	23–34	Forest and semi-natural areas
5	35–44	Wetlands and water bodies

3 Interpolation methods

Long-term climate patterns observed across the globe are a result of a combination of many different processes that manifest themselves at many spatial scales. It is assumed that most of those patterns occur at scales large enough to be adequately reflected in the station data, and thus are not explicitly accounted for by the major interpolation methods. The main physiographic features affecting spatial patterns of climate are terrain and water bodies (e.g., Daly, 2006). Several additional spatial climate foreigns factors are most important at scales of less than 1 km but may also have effects at larger scales. These factors include slope and aspect, riparian zones and land use/land cover (Lookingbill and Urban, 2003; McCutchan and Fox, 1986; Bolstad et al., 1998; Dong et al., 1998).

For each day of the considered 34-year-long time period (i.e. 1 January 1980 to 31 December 2013), all the validated (i.e. confidence index of 6) stations temperature records (i.e. *TN* and *TX*) were interpolated on a regular 4 km × 4 km grid over Belgium. To predict the unknown values from the unevenly distributed temperature records observed at known locations, the performance of different interpolation methods has been assessed: inverse distance weighting (IDW), ordinary kriging (OK) and kriging with external drift (KED; Wackernagel, 1995) that is able to handle densely sampled auxiliary variables highly correlated with the parameter of interest. In this study, KED was used with either the orography (KED_1) or land cover types (KED_2) as a drift, as well as simultaneously with the two drifts (KED_{12}). For the kriging methods, the parameters used to estimate the semivariogram were fixed in order to have an exact interpolation at the station location (i.e. the estimation satisfies the observation at the station location) with a fixed range of 50 km and an exponential semivariogram model.

Left panel in Fig. 4 displays the Belgian orography at the 4x4 km spatial resolution. Similarly, right panel in Fig. 4 presents the spatial distribution of the 5 land cover classes considered here. These classes are derived from the 100 m spatial resolution CORINE land cover database (Bossard and Feranec, 2000). Basically the 44 classes of the CORINE database are merged in 5 main classes as detailed in Table 3 and reprojected in a 4x4 km regular grid over Belgium (the

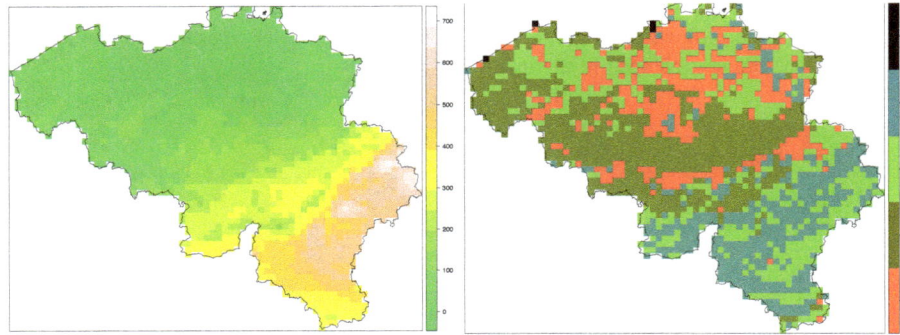

Figure 4. 4 km resolution orography of Belgium [m] (left panel) and 4 km resolution CORINE Land Cover in Belgium (right panel).

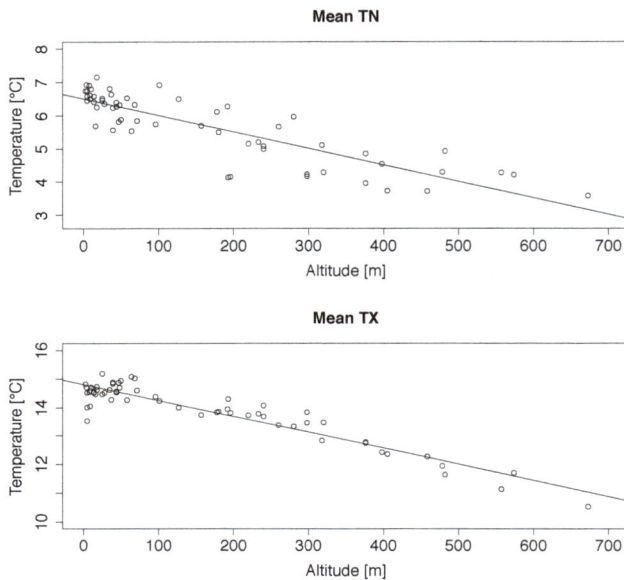

Figure 5. Correlation between station elevation and daily mean temperature (for the reference stations). The correlation coefficients are of -0.85 and -0.93 for *TN* and *TX*, respectively.

dominant land cover class within a given grid point is attributed to the entire grid point).

The various interpolation approaches are validated by leave-one-out cross-validation. For each day of the considered period, the cross-validation root mean square error ($\mathrm{RMSE_{cv}}$) is computed from the differences between the prediction P and the actual measurements M at the n stations:

$$\mathrm{RMSE_{cv}} = \sqrt{\frac{1}{n} \sum_{i=1}^{n} (P(x_i) - M(x_i))^2} \quad \text{where} \quad \{x_i\}_{i=1,\ldots,n} \text{ is}$$

the location of the climate stations.

To determine the best performing method, two indices are derived from the daily $\mathrm{RMSE_{cv}}$. First, the averaged over the 34 years (i.e. $\mathrm{RMSE_{cv}^{AVG}}$) and second, the frequency of the method providing the best daily $\mathrm{RMSE_{cv}}$ over the 34 years (i.e. $\mathrm{RMSE_{cv}^{FREQ}}$).

4 Results

Table 4 summarizes the performance of the investigated interpolation methods in terms of $\mathrm{RMSE_{cv}^{AVG}}$ and $\mathrm{RMSE_{cv}^{FREQ}}$ index for both *TN* and *TX*, respectively. Results indicate that for *TN* all methods perform quite similarly: the averaged accuracy ($\mathrm{RMSE_{cv}^{AVG}}$) ranges from 0.867 for the "worst" method (IDW) to 0.818 for the best one ($\mathrm{KED_1}$). By contrast the interpolated *TX* fields appear more sensitive to the considered methods. As for *TN*, the best performing method is $\mathrm{KED_1}$ and the worst one is IDW but the magnitude of the difference between the two methods in terms of $\mathrm{RMSE_{cv}^{AVG}}$ is larger (i.e. 0.14 °C vs. 0.05 °C for *TN*). This difference between *TN* and *TX* can be explained by the larger correlation found over our domain between the elevation and *TX* than for *TN* (see Fig. 5).

Accounting for the land cover type in the interpolation scheme does not improve the results for both *TN* and *TX*. The difference in terms of $\mathrm{RMSE_{cv}^{AVG}}$ between OK and $\mathrm{KED_2}$ is clearly negligible (i.e. 0.852 °C vs. 0.854 °C for *TN* and 0.721 °C vs. 0.723 °C for *TX*). Similarly using the land cover type in addition to the orography in the kriging with external drift approach does not significantly modify the performance obtained when only considering the orography as drift (e.g. $\mathrm{RMSE_{cv}^{AVG}}$ of 0.818 °C for $\mathrm{KED_1}$ vs. 0.820 °C for $\mathrm{KED_{12}}$ in the case of *TN* and 0.601 °C for $\mathrm{KED_1}$ vs. 0.605 °C for $\mathrm{KED_{12}}$ in the case of *TX*, respectively). This is not so surprising in view of the spatial resolution adopted for the gridding (e.g. 4 km $\times$ 4 km) as the land use/land cover variations are expected to be most important below 1 km. Analysis of the $\mathrm{RMSE_{cv}^{FREQ}}$ provided in Table 4 confirm that the best performing interpolation method in terms of $\mathrm{RMSE_{cv}}$ is the kriging using the orography as external drift ($\mathrm{KED_1}$) for both *TN* and *TX* (e.g. $\mathrm{RMSE_{cv}^{FREQ}}$ of 43 % for *TN* and $\mathrm{RMSE_{cv}^{FREQ}}$ of 77.1 % for *TX*). What is interesting to note regarding the $\mathrm{RMSE_{cv}^{FREQ}}$ in Table 4 is the score obtained by the kriging using the land cover type as external drift ($\mathrm{KED_2}$). Indeed, for both *TN* and *TX*, the value of this index for the $\mathrm{KED_2}$ is lower than for IDW (i.e. 7.9 % vs. 14.5 % for *TN* and 0.3 % vs. 0.6 % for *TX*).

Table 4. Overall performance of the different interpolation methods over the entire time period expressed in terms of RMSE_{cv}^{AVG} and RMSE_{cv}^{FREQ} for both *TX* and *TN*.

	RMSE_{cv}^{AVG} (*TN*) [°C]	RMSE_{cv}^{AVG} (*TX*) [°C]	RMSE_{cv}^{FREQ} (*TN*) [%]	RMSE_{cv}^{FREQ} (*TX*) [%]
IDW	0.867	0.738	14.5	0.6
OK	0.852	0.721	15.1	1.1
KED_1	0.818	0.601	43	77.1
KED_2	0.854	0.723	7.9	0.3
KED_{12}	0.820	0.605	19.5	20.9

Finally, Fig. 6 displays the time evolution of the annual mean daily RMSE_{cv} over the considered 34-year-long time period obtained by using the OK and the KED_1 daily interpolation methods, respectively. Both methods present a clear reduction of the RMSE_{cv} throughout the time for *TX* with the largest reduction obtained by the KED_1 approach (see left panel in Fig. 6). Similar behavior is also observed for *TN* while for this parameter the dispersion is a bit more larger than for *TX* (see right panel in Fig. 6). The reduction of the annual mean daily RMSE_{cv} has to be put in connection with the increasing number of stations involved in the interpolation process with time (see Fig. 1, left panel).

5 Discussion

As an additional validation exercise, the newly developed FORBIO climate data set was compared to existing data sets.

Towards this objective, the RMSE results presented in Sect. 4 are compared to these obtained from the HYRAS data set which cover Germany (Frick et al., 2014). This data set has a similar resolution (5 km) and use the Optimal Interpolation method. The fivefold cross-validation of HYRAS gives an averaged RMSE of 1.39 °C versus values from 0.601 °C to 0.867 °C for the FORBIO climate data set. The better station density and the smaller elevation differences between stations can explain the better results obtained in the FORBIO data set.

The new FORBIO data set can also be compared to the existing E-OBS data set (version 10.0) (Haylock et al., 2008), which provides daily temperature data for the period 1950–2013 on a regular 0.25° grid (approximately 25 km). 16 Belgian synoptic stations are used in the E-OBS data set. To allow a comparison at the same resolution, the FORBIO data set was first degraded at the E-OBS resolution. Figure 7 shows the differences between the two data sets using the mean maximum temperature over the 34 years. The benefit of the original FORBIO data set spatial resolution is evident for the user. At the degraded resolution, a major difference between the two data sets (> 1 °C) appears in the Hainaut Province. Only one station located in the Hainaut Province is considered with the E-OBS data set (i.e. the Chièvre synoptic station, black dot in Fig. 7) and this station appears

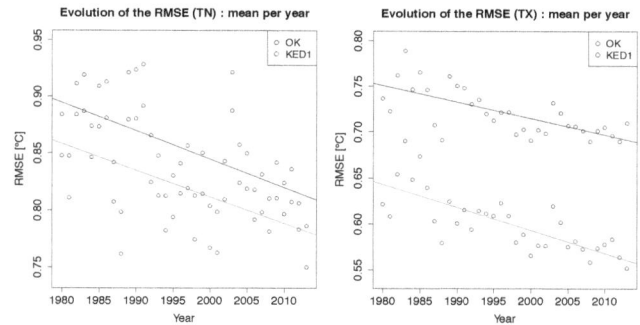

Figure 6. Evolution of the averaged annual RMSE_{cv} and regression line using ordinary kriging (OK) and kriging with external drift using orography (KED_1) for *TN* (left panel) and *TX* (right panel).

too cold compared to the neighboring climatological stations used in the FORBIO data set in addition to the Chièvre station records. This comparison is also true for minimum temperature.

6 Conclusions

To meet the needs of the FORBIO Climate research project, a 34-year-long (1980–2014) daily gridded temperature (minimum and maximum) has been produced over Belgium at a 4 km × 4 km spatial resolution. Because no measuring technique is perfect and errors can run in meteorological observations for a wide variety of reasons, data quality control procedures were developed and applied to temperature records performed within the Belgian climatological station network operated by RMI prior to undergo the daily gridding process. More than 5 millions of daily temperature records (*TN* and *TX*) were analyzed in depth and about 0.7 % of these were discarded. The performance of 5 different interpolation methods was assessed over the Belgian domain ranging from the simple inverse-distance weighting approach to the kriging with external drift methods. Two auxiliary drifts have been considered (i.e. the orography and the land cover type) either individually or in combination. Because of the spatial resolution of 4 km × 4 km adopted for the data set, it has been found that accounting for the land cover type in the interpolation process of temperature over Belgium is not relevant. By contrast using the orography as external drift in the kriging interpolation scheme provides the best results in term of RMSE_{cv} and RMSE_{cv}^{FREQ} for both *TX* and *TN*. It is however worth pointing out that the influence of the selected methods in the accuracy of the interpolated temperature field is only well apparent for *TX*. For *TN* the performance of the different interpolation methods were found rather similar. Based on these results the daily gridded temperature (*TN* and *TX*) were carried out with the kriging with external drift method using the orography as auxiliary highly correlated parameter. The results obtained are similar to the E-OBS data set excepted in

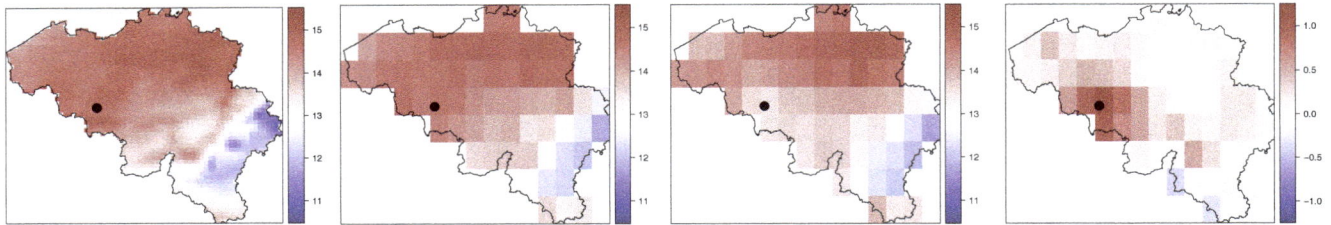

Figure 7. Averaged value for *TX* over the 34 years [°C]: FORBIO (4 km resolution), FORBIO (25 km resolution), E-OBS (25 km resolution), differences between FORBIO and E-OBS. The black dot corresponds to the Chièvre station.

the Hainaut province where the E-OBS data set is colder. A minimum of 80 stations per day were always used in the gridding process for both *TN* and *TX*. Note that in average over the 34 years a bit more stations were involved in the spatial interpolation of *TX* than for *TN* (i.e. 137 vs. 132 stations, respectively). Finally, because the number of stations involved in the daily gridding process has increased with time (even if the total number of available stations has started to decrease in the last decade covered by the dataset), the daily $RMSE_{cv}$ present a clear reduction as a function of time for both *TX* and *TN*. As an example, the annual mean daily $RMSE_{cv}$ has decreased by 0.085 °C from 1980 to 2013 for *TX* and by 0.068 °C for *TN*.

Acknowledgements. This work was supported by the Belgian Science Policy Office (BELSPO) through the BRAIN-be program contract NR BR|132|A1|FORBIO CLIMATE "Adaptation potential of biodiverse forests in the face of climate change". We acknowledge the E-OBS dataset from the EU-FP6 project ENSEMBLES (http://ensembles-eu.metoffice.com) and the data providers in the ECA & D project (http://www.ecad.eu). The "gstat" Package (version 1.0-22) in R (Pebesma, 2004) was used for geostatistical interpolation.

References

Aguilar, E., Auer, I., Brunet, M., Peterson, T., and Wieringa, J.: Guidelines on climate metadata and homogenization, World Climate Programme Data and Monitoring WCDMP-No. 53, WMO-TD No. 1186, World Meteorological Organization, Geneva, 2003.

Bolstad, P. V., Swift, L., Collins, F., and Régnière, J.: Measured and predicted air temperatures at basin to regional scales in the southern Appalachian mountains, Agr. Forest Meteorol., 91, 161–176, 1998.

Bossard, M. and Feranec, J.: Otahel, Jaffrain, Gabriel, CORINE Land Cover technical guide – Addendum 2000, Tech. rep., Technical Report 40, EEA, Copenhagen, http://www.eea.eu.int (last access: 8 January 2015), 2000.

Boulanger, J.-P., Aizpuru, J., Leggieri, L., and Marino, M.: A procedure for automated quality control and homogenization of historical daily temperature and precipitation data (APACH): part 1: quality control and application to the Argentine weather service stations, Climatic Change, 98, 471–491, 2010.

Daly, C.: Guidelines for assessing the suitability of spatial climate data sets, Int. J. Climatol., 26, 707–721, 2006.

Dong, J., Chen, J., Brosofske, K., and Naiman, R.: Modelling air temperature gradients across managed small streams in western Washington, J. Environ. Manage., 53, 309–321, 1998.

Feng, S., Hu, Q., and Qian, W.: Quality control of daily meteorological data in China, 1951–2000: a new dataset, Int. J. Climatol., 24, 853–870, 2004.

Frick, C., Steiner, H., Mazurkiewicz, A., Riediger, U., Rauthe, M., Reich, T., and Gratzki, A.: Central European high-resolution gridded daily data sets (HYRAS): Mean temperature and relative humidity, Meteorol. Z., 23, 15–32, 2014.

Hair, J. F.: Multivariate data analysis, Prentice Hall, Upper Saddle River, 2009.

Haylock, M., Hofstra, N., Klein Tank, A., Klok, E., Jones, P., and New, M.: A European daily high-resolution gridded data set of surface temperature and precipitation for 1950–2006, J. Geophys. Res.-Atmos., 113, D20119, doi:10.1029/2008JD10201, 2008.

Lookingbill, T. R. and Urban, D. L.: Spatial estimation of air temperature differences for landscape-scale studies in montane environments, Agr. Forest Meteorol., 114, 141–151, 2003.

McCutchan, M. H. and Fox, D. G.: Effect of elevation and aspect on wind, temperature and humidity, J. Clim. Appl. Meteorol., 25, 1996–2013, 1986.

Pebesma, E. J.: Multivariable geostatistics in S: the gstat package, Comput. Geosci., 30, 683–691, 2004.

Sciuto, G., Bonaccorso, B., Cancelliere, A., and Rossi, G.: Probabilistic quality control of daily temperature data, Int. J. Climatol., 33, 1211–1227, 2013.

Wackernagel, H.: Multivariable geostatistics: an introduction with applications, Springer-Verlag, Berlin, 1995.

14

The verification of seasonal precipitation forecasts for early warning in Zambia and Malawi

O. Hyvärinen[1], L. Mtilatila[2], K. Pilli-Sihvola[1], A. Venäläinen[1], and H. Gregow[1]

[1] Finnish Meteorological Institute, Helsinki, Finland
[2] Department of Climate Change and Meteorological Services, Lilongwe, Malawi

Correspondence to: O. Hyvärinen (otto.hyvarinen@fmi.fi)

Abstract. We assess the probabilistic seasonal precipitation forecasts issued by Regional Climate Outlook Forum (RCOF) for the area of two southern African countries, Malawi and Zambia from 2002 to 2013. The forecasts, issued in August, are of rainy season rainfall accumulations in three categories (above normal, normal, and below normal), for early season (October–December) and late season (January–March). As observations we used in-situ observations and interpolated precipitation products from Global Precipitation Climatology Project (GPCP), Global Precipitation Climatology Centre (GPCC), and Climate Prediction Centre (CPC) Merged Analysis of Precipitation (CMAP). Differences between results from different data products are smaller than confidence intervals calculated by bootstrap.

We focus on below normal forecasts as they were deemed to be the most important for society. The well-known decomposition of Brier score into three terms (Reliability, Resolution, and Uncertainty) shows that the forecasts are rather reliable or well-calibrated, but have a very low resolution; that is, they are not able to discriminate different events. The forecasts also lack sharpness as forecasts for one category are rarely higher than 40 % or less than 25 %. However, these results might be unnecessarily pessimistic, because seasonal forecasts have gone through much development during the period when the forecasts verified in this paper were issued, and forecasts using current methodology might have performed better.

1 Introduction

Probabilistic seasonal precipitation forecasts in several parts of the world are issued by the Regional Climate Outlook Forums (RCOF, Ogallo et al., 2008). In these forums, national, regional and international climate experts meet to produce real-time regional climate outlooks based on input from National Meteorological Services, regional institutions, Regional Climate Centres, and global producers of climate predictions. The outlooks are consensus-based, implying that the forecast is made for the whole region and downscaling to national level is done afterwards. For the Southern African region, the seasonal outlooks are produced by the Southern Africa Regional Climate Outlook Forum (SARCOF, http://www.sadc.int/news-events/newsletters/climate-outlook/) that had the first meeting in 1997.

Malawi and Zambia are two southern African countries which face multiple challenges related to weather and climate, mainly due to their exposure and vulnerability to weather and climate shocks, particularly prolonged dry spells and floods. For instance, high reliance on rain-fed agriculture, poor disaster preparedness levels, and general lack of capacity in the communities expose the countries and people to a pertinent threat of food insecurity, malnutrition, and loss of lives. This hampers the general development efforts. The seasonal outlook issued by SARCOF and downscaled to Malawi and Zambia is disseminated widely to the disaster preparedness and response authorities, both on the government and major United Nations (UN) organisations, and it is used as one component behind the national contingency plans. To communities and farmers the forecast is issued by some non-governmental organisations (NGO) who work on the community level; the potential of this, however, has not been fully harnessed.

The value of meteorological information to the end user is tightly connected to the accuracy and skill of the information. As seasonal forecasting is a relatively new endeavor in the field of meteorological forecasting, we aim to assess the performance of the seasonal forecasts issued in Malawi and Zambia. The verification results lay a basis for the discussion on the usability and value of seasonal forecasts for the early warning process in the region, as the forecast is widely used in the two countries and the potential benefits of the forecast are high. This study focuses on forecasts of "below normal" precipitation, as they are the most important forecasts for drought and if they prove to be skilful, they could be beneficial for farming practices and as an early warning sign.

The study forms a part of a two-year (2013–2014) research project "Study on risk management of extreme weather related disasters and climate change adaptation in Malawi and Zambia (SAFE-MET)", funded by the Academy of Finland and the Ministry for Foreign Affairs of Finland as a part of the Finnish Research Programme on Climate Change (FICCA). The goals of SAFE-MET were to examine, propose, and test ways to strengthen societies' resilience to climate and weather related hazards and to enhance multidisciplinary climate change research in Zambia and Malawi.

2 Data

Data used consisted of RCOF forecasts and gridded and in-situ observations. The use of more than one observational data set can help in determining the uncertainty of the results.

2.1 Forecasts

The SARCOF forecasts, issued in August, are of summer rainfall accumulations in three categories (above normal, normal, and below normal), for early season (October–December, OND) and late season (January–March, JFM). The forecasts from 2002 to 2013 were available for verification. The forecasts are disseminated only as pictures with forecast probabilities shown as filled contour lines. Therefore, these pictures had to be reverse-engineered into data before they can be compared with the gridded observations.

We divided an area from −8.5 to −18° latitude and from 20 to 37° longitude into $1.0° \times 1.0°$ and $2.5° \times 2.5°$ grids of forecasts (Fig. 1). These grids correspond to the two grids of gridded observations below. The grids were then filled with values from the pictures. In most cases, it was unambiguous what value should be used but when the contour line between two forecast categories crossed a grid square in the picture, the value of forecast category covering the largest area of the square was used. This subjective part of digitization might add some noise to the results. However, the number of ambiguous grid values was small, they amounted to less than

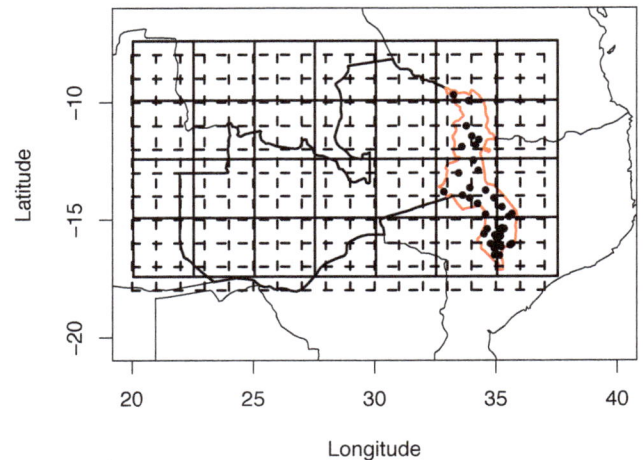

Figure 1. The map of the study area, where Malawi (red lines) and Zambia (blue lines) are emphasized. The grids used for subjective digitalization of forecasts are also shown, the $1.0° \times 1.0°$ grid in dashed line and the $2.5° \times 2.5°$ grid in solid line. Black dots show the stations with in-situ observations in Malawi.

2 % of grid points of the $1.0° \times 1.0°$ grid and 3 % of grid points of the $2.5° \times 2.5°$ grid.

The distribution of forecasts (Fig. 2) shows that forecasts lack sharpness in a verification sense (e.g., Wilks, 2011) as forecasts for one category do not differ much from the climatological values (33 %). Probabilities of forecasts for above normal are usually somewhat smaller than the climatological values, and probabilities of forecasts for normal are usually somewhat greater than the climatological values. Moreover, distributions of forecasts of both grids are very similar, even though grids have different resolutions and grid points are in somewhat different locations. This gives us confidence in our digitalization of forecasts, and noise added by our subjective digitization is probably not significant.

2.2 Gridded observations

As gridded observations, we used interpolated precipitation products from Global Precipitation Climatology Project (GPCP) (Adler et al., 2003), Global Precipitation Climatology Centre (GPCC) (Schneider et al., 2013), and Climate Prediction Centre (CPC) Merged Analysis of Precipitation (CMAP) (Xie and Arkin, 1997). All interpolated products used measurements from rain gauge stations and information from different satellite instruments. For GPCP and CMAP, the resolution was $2.5° \times 2.5°$ and for GPCC, the resolution was $1.0° \times 1.0°$. Years from 1980 to 2001 were used for climatology, so quantiles of 33 and 66% were calculated using those years for each grid point and grid points for years with forecasts (2002–2013) were then classified using those quantiles. Different data sets can have differences in precipitation levels but for our purposes, absolute values need not be exact, if the relative values are consistent.

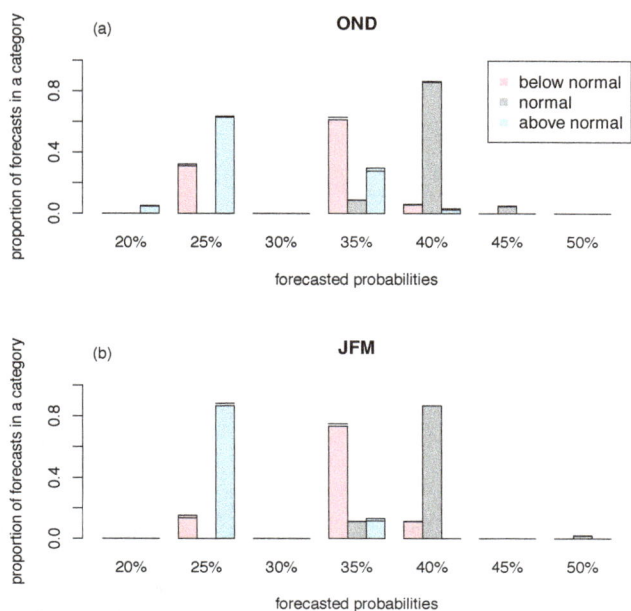

Figure 2. The distribution of forecasted probabilities of early (OND) and late (JFM) season forecasts. Proportions of different forecast probabilities for each category are shown, so bars of the same color add to one. Bars are calculated using the $1.0° \times 1.0°$ grid, results using the $2.5° \times 2.5°$ grid are shown as vertical lines over the bars.

In addition to the interpolated precipitation products, precipitation from the ERA-Interim reanalysis was tested. Unfortunately, there was a clear trend in the precipitation in the area under study, not present in the other products, and the results from ERA-Interim were not used. This non-optimal quality of African precipitation in ERA-Interim is a known problem (Dee et al., 2011).

2.3 In-situ observations

In Malawi, 39 stations were used for the period from 2007 to 2012. Stations covered the whole Malawi for both early season and late season forecasts. The stations used in this analysis are those that are also used to generate the operational forecasts. In Malawi during the study period, below-normal forecasts of 25 and 35 % were issued for OND, while only 25 % forecasts were issued for JFM.

3 Methods

As we focused on below normal forecasts, the methodology for binary forecasts can be used (e.g., Wilks, 2011). Results were assessed graphically using the attributes diagram (which is a refinement of the reliability diagram) and the Receiver Operating Characteristic (ROC) curve, and more quantitatively using the Brier Skill Score (BSS). BSS can be

written as

$$BSS = (\text{Resolution} - \text{Reliability})/\text{Uncertainty}, \qquad (1)$$

where the three terms come from the decomposition of Brier Score. Skill score values larger than zero indicate skillful forecasts compared with climatology.

For gridded data sets, one verification measure was calculated from the whole data set, as point-wise calculations would be based on only 11 data points. The climatology of BSS is also for the whole data set. The block-bootstrapping, as in Hamill (1999), was used to calculate confidence intervals (CIs) in order to take into account the high spatial correlation of grid points. In the standard bootstrap, all data points are sampled with replacement and the correlation of data points is therefore lost, resulting in too narrow CIs. Here we sampled the whole grids (that is, years) so the spatial correlation of the data is not lost, resulting in more realistic CIs. The number of bootstrap samples was 15 000.

For in-situ measurements, block-bootstrapping was not pursued due to limited resources, and only standard bootstrap ($n = 1000$) was used. Because of the proximity of the stations, there is strong autocorrelation, and therefore the standard bootstrap intervals are probably too narrow.

4 Results

4.1 Gridded observations

In Figs. 3a, c, e and 4a, c, e, points of the attribute diagrams should be as near the diagonal as possible, but, in reality, they are rather near the horizontal "No resolution" line, while some data points are in the grey area of skillful forecasts. Similarly, in ROC curves (Figs. 3b, d, f and 4b, d, f), points should be as far as possible from the diagonal line towards the top left corner, but they are very near the diagonal and even below it. Taken at face value, results suggest not very skillful forecasts. However, CIs are very large and cover both skillful and not skillful areas of diagrams and curves. Also, values of BSS (Table 1) indicate that forecasts do not have much skill, as all values are around zero, and no CIs cover only positive values. From BSS decomposition terms, we can conclude that forecasts are rather reliable or well-calibrated (Reliability terms are rather small), but have very low resolution (Resolution terms are almost zero); that is, they are not able to discriminate different events. All in all, differences between data products are smaller than confidence intervals calculated by block-bootstrap. Furthermore, our subjective digitization of forecasts from pictures might add some noise to the results, but small sensitivity tests, by moving the position of forecasts slightly, did not change the results substantially.

From the attributes diagrams (Figs. 3a, c, e and 4a, c, e) the base rate of below normal in different data sets can be estimated, which shows how consistent the different products

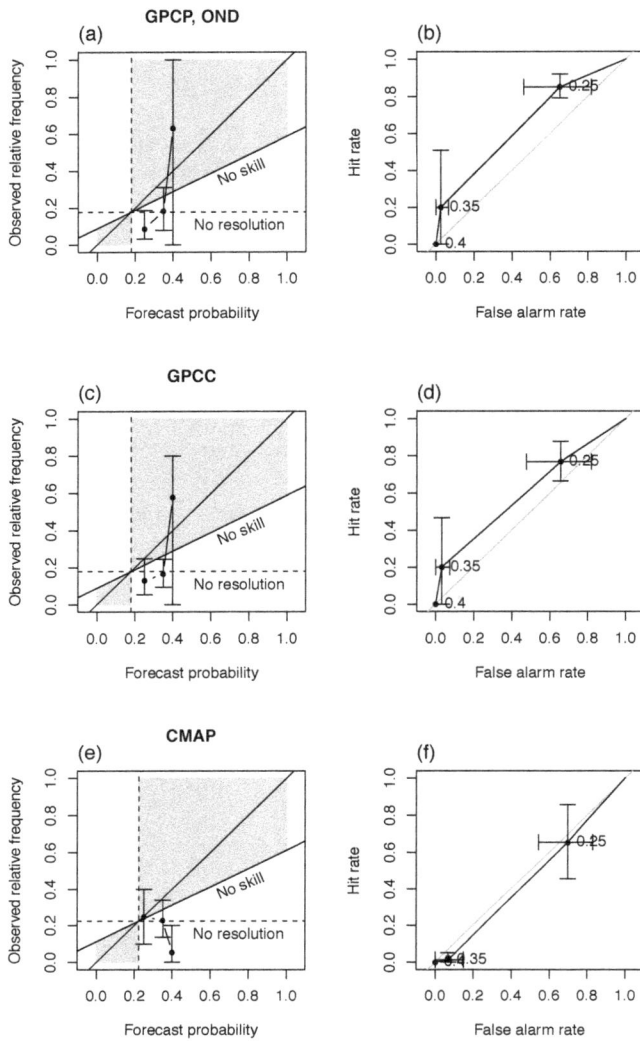

Figure 3. The attributes diagrams and the ROC curves for early season forecasts, based on different data sets for verification. The confidence intervals are calculated using the block bootstrap ($n = 15\,000$).

Figure 4. The attributes diagrams and the ROC curves for late season forecasts, based on different data sets for verification. The confidence intervals are calculated using the block bootstrap ($n = 15\,000$).

are with each other. For early season, the below-normal conditions occur about 20 % in all three products, showing good consistency. In late season, GPCC has a somewhat smaller base rate than others, which then produces a slightly worse BSS, as observations diverge more from forecasts.

4.2 In-situ observations

For in-situ observations, the results are very similar to the gridded results. In the attribute diagrams, all CIs cover the vertical line of "no skill", while points in the ROC curve are very near the diagonal line (Fig. 5). For more quantitative results, BSS is, for practical purposes, zero (Table 2) for both early and late season, and therefore Resolution terms, as well as Reliability terms, are also almost zero. The Resolution of

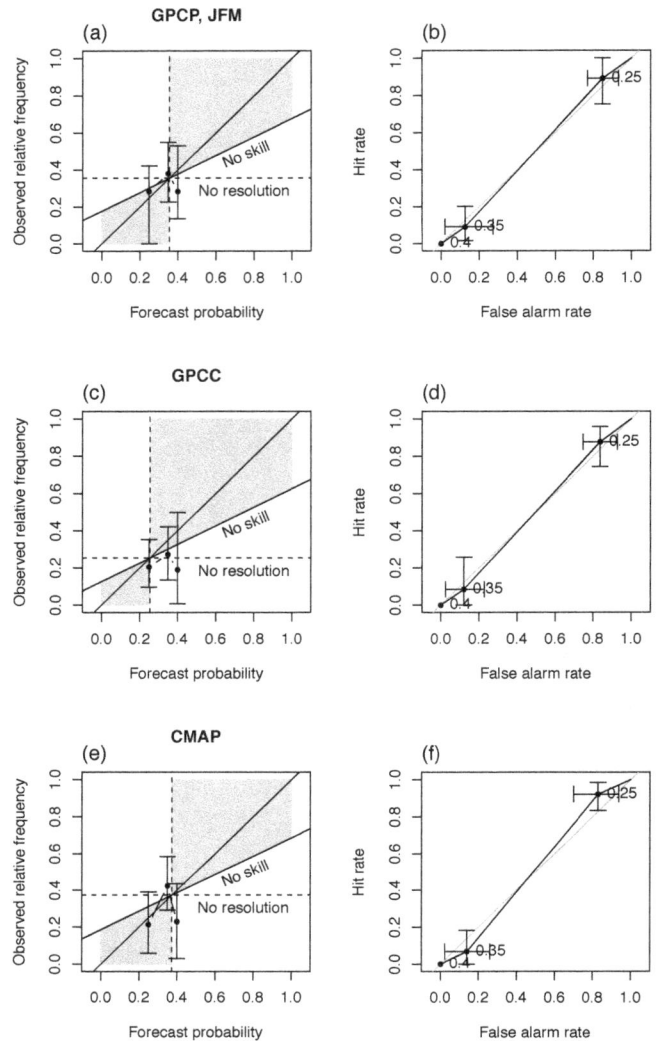

later season is exactly zero, as only forecasts of 25 % were issued for Malawi. So, the forecast skill seems to be very limited.

5 Discussion

Different data sources give slightly different results, but all results indicate that the forecasts have limited skill in the selected area. This is consistent with the results of Mason and Chidzambwa (2008) where they remark that predictability in the vicinity of Malawi is weak because of a transition between distinct ENSO teleconnection signals. It might be possible to increase the skill by recalibrating the forecasts, but that is beyond the scope of this paper.

Table 1. For the early and late season, BSS, its decomposition terms, and its CI for different data sets. The CIs are calculated using the block bootstrap ($n = 15\,000$).

Data name	$\dfrac{\text{Resolution}}{\text{Uncertainty}}$	$\dfrac{\text{Reliability}}{\text{Uncertainty}}$	BSS	BSS $\text{CI}_{0.025}$	BSS $\text{CI}_{0.975}$
OND					
GPCP	0.098	0.192	−0.094	−0.625	0.054
GPCC	0.074	0.184	−0.110	−0.448	0.026
CMAP	0.011	0.092	−0.082	−0.350	−0.014
JFM					
GPCP	0.007	0.011	−0.003	−0.117	0.012
GPCC	0.006	0.048	−0.042	−0.331	0.006
CMAP	0.033	0.033	−0.000	−0.106	0.021

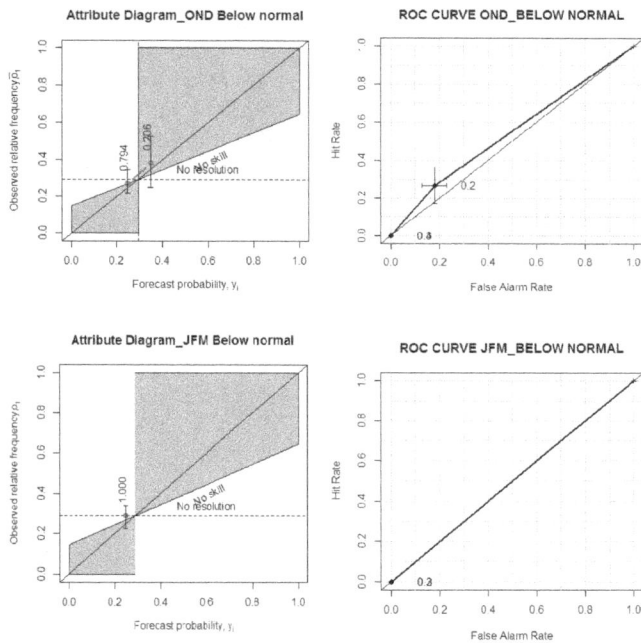

Figure 5. The attributes diagrams and the ROC curves for early and late season forecasts, based on in-situ observations for verification. The confidence intervals are calculated using the standard bootstrap ($n = 1000$).

Table 2. For the early and late season, BSS, and its decomposition terms for in-situ observations in Malawi.

	$\dfrac{\text{Resolution}}{\text{Uncertainty}}$	$\dfrac{\text{Reliability}}{\text{Uncertainty}}$	BSS
OND	0.009	0.003	0.006
JFM	0.000	0.008	−0.008

However, the question which arises is if verification should focus on validating the past performance or be more future-oriented and estimate how forecasts will perform in the future. Our verification is based on historical data (what was really forecasted), not on reprocessed data (what would have been forecasted, if the present system had been available then). Moreover, because of the unreproducible way RCOF forecasts are made, getting large enough data sets will take years, and even then they will not be of constant quality as the underlying forecast systems continue to evolve. Therefore the results can be too pessimistic, and for a more optimistic outlook, we should compare our results with results using hindcasts. This is a natural way of presenting verification results by the model developers (e.g., Landman, 2014). Some data is available also for the larger verification community from, for example, the Climate-system Historical Forecast Project of the World Climate Research Program (Kirtman and Pirani, 2009), but then the temporal range of data might not be as comprehensive.

Of course, seasonal forecasts can be very beneficial for agriculture. For example, a review (Hansen et al., 2011) shows how farmers can use and benefit from the information, given enabling environment and improved communication of the information and its uncertainties, and bundled with historic observations. Furthermore, as long as food insecurity remains a threat in Malawi and Zambia, the government sector, supported by major UN organisations and NGOs, benefits from the information, for instance, in vulnerability assessments and contingency plans (personal communication with the end users). However, the uncertainty of the information needs to be clearly communicated and understood by the end users to provide benefits and avoid losses from unskilful forecasts compared to efficient, science-based use of climatological information. Furthermore, the forecast, currently issued once during the rainy season, could be updated in the light of new information closer to the rainy season. And finally, the seasonal information must be complemented with accurate and efficiently communicated short-range weather forecasts to reap the full benefits of weather and climatological information in southern Africa.

6 Conclusions

Based on our dataset, the SARCOF forecasts seem to have limited skill, which is partly explained by the climatological conditions in the area under study. However, it might not be prudent to make drastic conclusions about the usability of the current SARCOF forecasts based on this dataset alone, because ten years of seasonal forecasts is a rather small data set, as shown by large confidence intervals for all measures. This small number of forecasts, that are also constantly evolving, presents a challenge for verification in an operational setting. Especially, due to the partly subjective nature of RCOF, getting ten data points takes ten years of time. Therefore, it can take decades before a reasonable number of forecasts is available. For more automatic forecasts, hindcasts offer more forecasts in a shorter time frame, and new forecast systems can be assessed as new versions becomes available.

Acknowledgements. The research leading to these results has received financial support from the Academy of Finland through the FICCA SAFE-MET research project (no. 264058).

References

Adler, R. F., Huffman, G. J., Chang, A., Ferraro, R., Xie, P.-P., Janowiak, J., Rudolf, B., Schneider, U., Curtis, S., Bolvin, D., Gruber, A., Susskind, J., Arkin, P., and Nelkin, E.: The Version-2 Global Precipitation Climatology Project (GPCP) Monthly Precipitation Analysis (1979–Present), J. Hydrometeorol., 4, 1147–1167, doi:10.1175/1525-7541(2003)004<1147:TVGPCP>2.0.CO;2, 2003.

Dee, D. P., Uppala, S. M., Simmons, A. J., Berrisford, P., Poli, P., Kobayashi, S., Andrae, U., Balmaseda, M. A., Balsamo, G., Bauer, P., Bechtold, P., Beljaars, A. C. M., van de Berg, L., Bidlot, J., Bormann, N., Delsol, C., Dragani, R., Fuentes, M., Geer, A. J., Haimberger, L., Healy, S. B., Hersbach, H., Hólm, E. V., Isaksen, L., Kållberg, P., Köhler, M., Matricardi, M., McNally, A. P., Monge-Sanz, B. M., Morcrette, J.-J., Park, B.-K., Peubey, C., de Rosnay, P., Tavolato, C., Thépaut, J.-N., and Vitart, F.: The ERA-Interim reanalysis: configuration and performance of the data assimilation system, Q. J. Roy. Meteor. Soc., 137, 553–597, doi:10.1002/qj.828, 2011.

Hamill, T.: Hypothesis Tests for Evaluating Numerical Precipitation Forecasts, Weather Forecast., 14, 155–167, 1999.

Hansen, J. W., Mason, S. J., Sun, L., and Tall, A.: Review of Seasonal Climate Forecasting for Agriculture in sub-Saharan Africa, Exp. Agr., 47, 205–240, doi:10.1017/S0014479710000876, 2011.

Kirtman, B. and Pirani, A.: The State of the Art of Seasonal Prediction: Outcomes and Recommendations from the First World Climate Research Program Workshop on Seasonal Prediction, B. Am. Meteorol. Soc., 90, 455–458, doi:10.1175/2008BAMS2707.1, 2009.

Landman, W. A.: How the International Research Institute for Climate and Society has contributed towards seasonal climate forecast modelling and operations in South Africa, Earth Perspectives, 1, 22, doi:10.1186/2194-6434-1-22, 2014.

Mason, S. and Chidzambwa, S.: Verification of African RCOF Forecasts, in: World Meteorological Organization RCOF Review 2008, IRI Technical Report 09-02, 2008.

Ogallo, L., Bessemoulin, P., Ceron, J.-P., Mason, S., and Connor, S. J.: Adapting to climate variability and change: the Climate Outlook Forum process, WMO Bulletin, 57, 93–101, 2008.

Schneider, U., Becker, A., Finger, P., Meyer-Christoffer, A., Ziese, M., and Rudolf, B.: GPCC's new land surface precipitation climatology based on quality-controlled in situ data and its role in quantifying the global water cycle, Theor. Appl. Climatol., 115, 15–40, doi:10.1007/s00704-013-0860-x, 2013.

Wilks, D. S.: Statistical Methods in the Atmospheric Sciences, Academic Press, 3 edn., 2011.

Xie, P. and Arkin, P. A.: Global Precipitation: A 17-Year Monthly Analysis Based on Gauge Observations, Satellite Estimates, and Numerical Model Outputs, B. Am. Meteorol. Soc., 78, 2539–2558, doi:10.1175/1520-0477(1997)078<2539:GPAYMA>2.0.CO;2, 1997.

15

Diurnal temperature cycle deduced from extreme daily temperatures and impact over a surface reanalysis system

F. Besson[1], E. Bazile[2], C. Soci[2], J.-M. Soubeyroux[1], G. Ouzeau[1], and M. Perrin[3]

[1]Meteo-France, DCSC/AVH, Toulouse, France
[2]Meteo-France, CNRM/GAME, Toulouse, France
[3]University of Toulouse III – Paul Sabatier, Toulouse, France

Correspondence to: F. Besson (francois.besson@meteo.fr)

Abstract. Due to the evolution of the observation network, hourly 2 m temperature analysis performed by re-analysis systems shows temporal inhomogeneities. The observation network gap is less present for extreme daily temperature observations. In order to reduce inhomogeneities and enable a climatological use of temperature analysis, information from extreme temperatures could be useful. In this study, the diurnal temperature cycle has been reconstructed for stations which only record extreme temperatures. These new "pseudo" hourly temperature observations are then provided to the analysis system. Two methods have been used to deduce hourly temperatures from extremes and compared to real observations. The results have shown that using those new pseudo-observations as an input for two different reanalysis systems enables reducing the bias in temperature analysis.

1 Introduction

Observation network heterogeneity (both in time and space) has to be addressed in order to produce unbiased climatological indexes (Peterson and Vose, 1997). Focusing on France, a big gap in the hourly 2 m temperature observation network occurred in the early 1990s related to the automatic station deployment. From the end of the 1950s to the end of the 1980s, the number of hourly temperature observations available ranges from about 200 to about 500, and then it increases and reaches 2000 in the present day; the number has increased by a factor of 10 in 55 years.

As a consequence the hourly 2 m temperature analysis, performed by SAFRAN (Durand et al., 1993, 1999), and using those hourly temperatures, shows inhomogeneities (Vidal et al., 2009; Soubeyroux et al., 2011). The change in observation network density is not so steep for the extreme daily temperatures (6-fold increase in 55 years, evolution quite linear). So we have tried to extract the information coming from extreme temperatures to deduce hourly data and reduce the gap.

Thus, the diurnal temperature cycle has been reconstructed and used as pseudo-observations for stations recording only extreme temperatures employing two methods. The first one only uses extreme temperatures and geographical characteristics of the station, whereas the second one needs extreme temperatures and hourly observation from the neighbourhood. Those two methods will be briefly described in Sect. 2 and compared in Sect. 3.1. Comparison of reanalyses performed with pseudo-observations using respectively SAFRAN and MESCAN (Soci et al., 2013) analysis systems are shown in Sect. 3.

2 Methods

2.1 Temperature temporal downscaling methods

2.1.1 Astro method

This method is described in Reicosky et al. (1989) as WCALC. We have called it the astro (astronomical) method because we deduce the hourly temperature at the station from extreme temperatures and sunrise/sunset hours. The temperature diurnal evolution is subdivided into three segments: from midnight to sunrise $+2$ h, for daylight hours and from sunset to midnight. For each of these segments coefficients are computed depending on extreme temperatures and sunset/sunrise hours (of the day and/or previous/following days).

The advantage of this method is that very few data are needed and it can be applied whatever the density of the network. However, it is not suitable for days with atypical diurnal cycle (Perrin et al., 2013).

2.1.2   Alpha method

$$T(h) = \text{Tmin} + \alpha(h)(\text{Tmax} - \text{Tmin}) \qquad (1)$$

We assume that hourly temperature can be defined as in Eq. (1) with $T(h)$ being an hourly temperature, Tmin the minimum temperature, Tmax the maximum temperature and $\alpha(h)$ an hourly coefficient varying from 0 to 1. This equation can be applied for all station with hourly data.

The aim of this method is to determine the $\alpha(h)$ coefficient for stations without hourly data. This can be done by using average values of $\alpha(h)$ available for stations with hourly data in the neighbourhood. Thus with this $\alpha(h)$ value we can apply Eq. (1) and get a pseudo hourly observation of temperature for a station without hourly observation.

A previous study (Perrin et al., 2013) of the spatial correlation of the $\alpha(h)$ showed that a criterion is to be used in order to select stations in the neighbourhood. Thus we only keep in the neighbourhood stations with a difference of elevation lower than 200 m.

Furthermore we can vary the number of stations selected in the neighbourhood. In our study several experiments are performed:

- one experiment (called Alpha_All) with all the stations filling the elevation criterion;

- experiments with the N closest stations filling the elevation criterion (called Alpha_N), with N varying from 1 to 10.

Thanks to this method we have quite similar quality even if the diurnal cycle is atypical (Perrin et al., 2013). The inconvenience is that, unlike the astro method, we need a network with hourly observations to apply this method.

2.1.3   Reanalysis systems

Two different reanalysis systems have been used: SAFRAN and MESCAN. Both use a guess and observations to perform an optimal interpolation. MESCAN does it on a regular grid of 5.5 km, whereas SAFRAN does it on 615 climatically homogeneous area with a 300 m vertical gradient. In our study SAFRAN uses as a guess ECMWF's operational archives projected on its analysis points (Quintana-Seguí et al., 2008), and MESCAN uses a HIRLAM guess downscaled from 22 to 5.5 km.

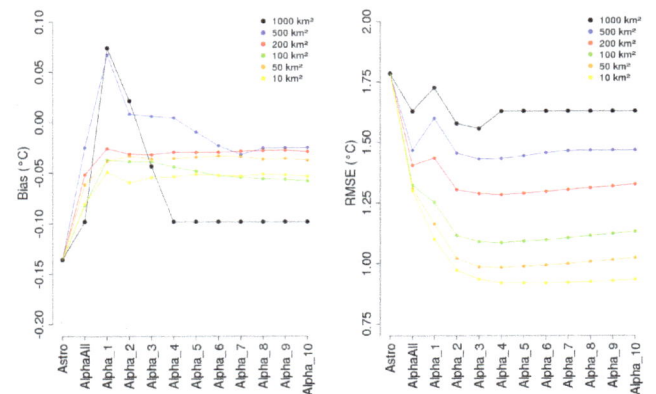

Figure 1. Bias (left) and RMSE (right) of pseudo-observations for hourly temperatures using different methods (x axis) and different network densities (colours). If we only have N hourly data usable for alpha method, the pseudo-observation for Alpha_M (with $M >$ N) is set to the Alpha_N 's value.

3   Results

3.1   Comparison of temporal downscaling methods

The comparison of the astro and alpha methods is performed over a 12-month period from August 2009 to July 2010 every 6 h, at 318 independent hourly observations stations over France. We performed six experiments with different densities (keeping one observation per 10, 50, 100, 200, 500, 1000 km^2) for hourly observations available to deduce pseudo-observations at the 318 stations. Thus we can evaluate the astro and alpha methods over different network densities.

In Fig. 1 we notice that, whatever the network density used, alpha methods perform better than the astro method (lowest bias and root-mean-square error (RMSE) values for alpha methods).

Focusing on alpha methods in Fig. 1, we observe that

- all of them improve when the network density increases (fairly constant low bias and increasing low RMSE);

- using only one neighbour degrades the pseudo-observations compared to the use of two or more neighbours;

- with a high-density network (< 200 km^2) the Alpha_all experiment gets the worst results because we use information coming from the whole area, whereas when reducing the number of neighbours we keep local features.

Then we did four extreme experiments using only one hourly observation to deduce pseudo-observation with alpha method. Evaluation is done over the 318 independent stations. For all experiments, bias and RMSE values over the whole domain are better for astro (or quite similar to alpha

Table 1. Bias and RMSE of pseudo-observation for the four extreme experiments for astro and alpha methods.

	Exp. 1		Exp. 2		Exp. 3		Exp. 4	
	Bias	RMSE	Bias	RMSE	Bias	RMSE	Bias	RMSE
Astro	−0.18	1.76	−0.18	1.76	−0.18	1.76	−0.18	1.76
Alpha	0.27	1.78	−0.15	1.87	−0.36	2.8	−0.19	2.5

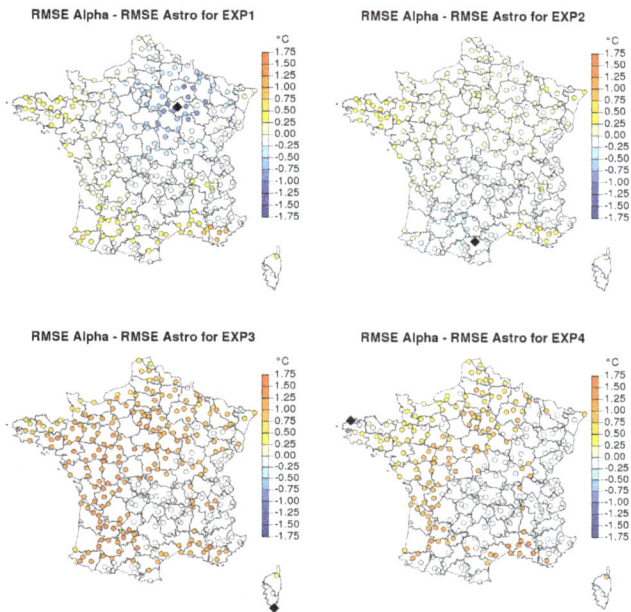

Figure 2. Comparison of the RMSE between alpha and astro for the four extreme experiments over 318 independent stations (circle). RMSE alpha–RMSE astro is represented by colours of the circles' background: cold (warm) colours show that alpha is better (worse) than astro. White circle for station where the alpha method cannot be applied (because of the elevation criterion). The black diamond is the location of the hourly observation used to deduce pseudo-observation.

method); see Table 1. Looking at the geographical repartition of the differences of RMSE between the two methods (see Fig. 2), we notice that for Exp. 1 and Exp. 2 where hourly observations are located on a plain far from the sea, close to the station, alpha is better than astro, but the further away the better is astro. Whereas for Exp. 3 and Exp. 4 with observation location along the coast (with specific weather type), astro is almost always the best.

3.2 Impact over reanalysis systems

We performed the reanalysis over France and in the period from October 2009 to June 2010. We did it using real observations and then using pseudo-observations from three selected methods: Alpha_All, Alpha_5 and astro. We added a criterion for alpha methods regarding the selection in the

Table 2. Bias and RMSE of pseudo-observation at 1555 observation points.

	Pseudo-obs.		
	Astro	Alpha_All	Alpha_5
Bias	−0.09	0	0
RMSE	1.73	0.9	0.85

neighbourhood: to be selected, an observation has to be less than 100 km from the pseudo-observation location.

Results are finally interpolated over an irregular grid built with 1555 observations' location. For SAFRAN, as the temperature analysis is done in a homogeneous area in increments of 300 m of elevation, we can do a linear interpolation between those levels, whereas for MESCAN we determine the gradient using the 4 closest model grid points.

Table 2 shows characteristics of statistic errors for pseudo-observations used as input for analysis systems; results are fairly similar to those obtained in Sect. 3.1. We just notice that the new criterion on the maximal distance improves the results of Alpha_All experiments. In Table 3, statistics for SAFRAN and MESCAN using those pseudo-observations and the real observations, and additional statistics for MESCAN's guess, are presented. Both reanalysis systems have low bias values; they slightly increase RMSE values, but experiment performance is sorted in the same way before and after the reanalysis systems: alpha methods give better results compared with astro.

Using Alpha_5 or Alpha_All pseudo-observations gives results very close to the experiment with real observations. Furthermore bias and RMSE coming from the guess are reduced whatever the pseudo-observations used.

Looking at error statistics month by month (see Fig. 3), we notice that bias and RMSE for the astro method (for both input pseudo-observation and output of MESCAN) is monthly dependent. This is not the case for alpha methods, which show rather stable statistics. Results using SAFRAN are similar (not shown here).

4 Conclusion and perspectives

In order to deal with variations in the network's density of hourly 2 m temperature observations during the reanalysis period, two methods have been tested to produce pseudo-

Table 3. Bias and RMSE for MESCAN and SAFRAN (using different data) and MESCAN's guess at 1555 observation points.

	MESCAN					SAFRAN			
	Astro	Alpha_All	Alpha_5	Real obs.	Guess	Astro	AlphaAll	Alpha_5	Real obs.
Bias	0.01	0.09	0.09	0.09	−0.72	−0.1	−0.06	−0.05	0
RMSE	1.96	1.30	1.26	1.12	2.09	1.85	1.15	1.06	0.93

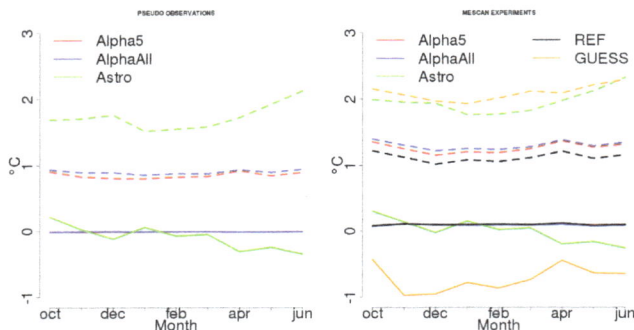

Figure 3. Bias (plain lines) and RMSE (dashed lines) for pseudo-observations (left) and MESCAN experiments (REF only uses real observations) (right).

observations. The first method (astro) is the simplest to use and only needs extreme temperatures and geographical data. The second one (alpha) needs hourly data from stations in the neighbourhood. Even if both methods have low bias, alpha method has the best results. The impact of the hourly observation network density available to use the alpha method has been tested. Even in a sparse network (one observation per 1000 km^2 over France), results with alpha remain slightly better. However in extreme experiments with only one hourly observation over the whole country, best results are obtained with the astro method (especially when the weather type of the local observation is particular). Furthermore with a high-density network, using two neighbours for the alpha method leads to satisfying results.

Using the pseudo-observations as an input for reanalysis systems gives results in accordance; i.e. results for alpha methods are better than astro. In addition performances of reanalysis systems with pseudo-observations with respect to performances with real observations are close; even if there is a slight deterioration, the bias coming from the guess is still reduced (even if for astro the bias is monthly dependent). The results of this work show that pseudo-observations can be useful in a reanalysis system. But as we introduce observation errors using pseudo-observation in a reanalysis system, we should use specific error statistics for pseudo-observations in the optimal interpolation.

Finally we will work on the extension over a longer period to be able to compute climatological indexes and compare them to the ones deduced from long-term series.

Acknowledgements. The authors would like to thank the reviewers for their comments and suggestions, which improved the quality of this manuscript. The research leading to these results has received funding from the European Union's Seventh Framework Program (FP/2007-2013) under grant agreement no. 242093 and agreement no. 607193.

References

Durand, Y., Brun, E., Merindol, L., Guyomarc'h, G., Lesaffre, B., and Martin, E.: A meteorological estimation of relevant parameters for snow models, Ann. Glaciol., 18, 65–71, 1993.

Durand, Y., Giraud, G., Brun, E., Merindol, L., and Martin, E.: A computer-based system simulating snowpack structure as a tool for regional avalanche forecasting, J. Glaciol., 45, 469–484, 1999

Perrin, M., Besson, F., Soubeyroux, J. M., and Bazile, E.: Reconstitution du cycle quotidien de température pour le futur système d'analyse atmosphérique MESCAN. Rapport M1/SID – University of Toulouse III – Paul Sabatier, Toulouse, France, 2013

Peterson, T. and Vose, R.: An overview of the global historical climatology network temperature database, B. Am. Meteor. Soc., 78, 2837–2849, doi:10.1175/1520-0477(1997)078<2837:AOOTGH> 2.0.CO;2, 1997.

Quintana-Seguí, P., Le Moigne, P., Durand, Y., Martin, E., Habets, F., Baillon, M., Canellas, C., Franchisteguy, L., and Morel, S.: Analysis of near surface atmospheric variables: validation of the SAFRAN analysis over France, J. Appl. Meteorol. Climatol., 47, 92–107, doi:10.1175/2007JAMC1636.1, 2008.

Reicosky, D. C., Winkelman, L. J., Baker, J. M., and Baker, D. G.: Accuracy of hourly air temperatures calculated from daily minima and maxima, Agr. Forest Meteorol., 46, 193–209, 1989.

Soci, C., Bazile, E., Besson, F., Landelius, T, Mahfouf, J.F., Martin, E., and Durand, Y.: EURO4M Project – REPORT D 2.6 Report describing the new system in D2.5, 2013

Soubeyroux, J. M., Schneider, M., and Besson, F.: Faisabilité de construction d'un indicateur thermique régionalisé sur la France à partir de la réanalyse SAFRAN, DCLIM/AVH, Météo-France: Toulouse, France, 2011

Vidal, J. P., Martin, E., Franchisteguy, L., Baillon, M., and Soubeyroux, J. M.: A 50-year high-resolution atmospheric reanalysis over France with the Safran system, Int. J. Climatol., 30, 1627–1644, doi:10.1002/joc.2003, 2009

16

Forecasting wind power production from a wind farm using the RAMS model

L. Tiriolo[1], R. C. Torcasio[1], S. Montesanti[1], A. M. Sempreviva[2], C. R. Calidonna[1], C. Transerici[3], and S. Federico[3]

[1]Institute of Atmospheric Sciences and Climate of the Italian National Council of Research, ISAC-CNR, UOS of Lamezia Terme, zona Industriale Comparto 15, 88046 Lamezia Terme, Italy
[2]Wind Energy Department, Danish Technical University, Frederiksborgvej 399, 4000-Roskilde, Denmark
[3]ISAC-CNR, UOS of Rome, via del Fosso del Cavaliere 100, 00133-Rome, Italy

Correspondence to: L. Tiriolo (l.tiriolo@isac.cnr.it)

Abstract. The importance of wind power forecast is commonly recognized because it represents a useful tool for grid integration and facilitates the energy trading.

This work considers an example of power forecast for a wind farm in the Apennines in Central Italy. The orography around the site is complex and the horizontal resolution of the wind forecast has an important role.

To explore this point we compared the performance of two 48 h wind power forecasts using the winds predicted by the Regional Atmospheric Modeling System (RAMS) for the year 2011. The two forecasts differ only for the horizontal resolution of the RAMS model, which is 3 km (R3) and 12 km (R12), respectively. Both forecasts use the 12 UTC analysis/forecast cycle issued by the European Centre for Medium range Weather Forecast (ECMWF) as initial and boundary conditions.

As an additional comparison, the results of R3 and R12 are compared with those of the ECMWF Integrated Forecasting System (IFS), whose horizontal resolution over Central Italy is about 25 km at the time considered in this paper.

Because wind observations were not available for the site, the power curve for the whole wind farm was derived from the ECMWF wind operational analyses available at 00:00, 06:00, 12:00 and 18:00 UTC for the years 2010 and 2011. Also, for R3 and R12, the RAMS model was used to refine the horizontal resolution of the ECMWF analyses by a two-years hindcast at 3 and 12 km horizontal resolution, respectively.

The R3 reduces the RMSE of the predicted wind power of the whole 2011 by 5 % compared to R12, showing an impact of the meteorological model horizontal resolution in forecasting the wind power for the specific site.

1 Introduction

Wind farms power prediction is of great importance, since a good forecast allows better integration of the renewable energy in the grid. A suitable use of wind energy needs the setup of methodologies able to reduce the uncertainty of the wind resource. The prediction system is usually based on meteorological models e.g. Limited Area Models (LAM) (Pinson et al., 2007; Alessandrini et al., 2013; Holmgren et al., 2010). These models predict the wind speed and direction in the target region and, by a power curve or other meth-

ods (Giebel et al., 2011), this output is converted to the wind power forecast for the wind farm. Hence, the quality of the power forecast at different forecasting ranges depends on the quality of the wind prediction over the area of the wind power plant.

This paper shows the wind power prediction for a wind farm in Central Italy, starting from wind velocity forecast of the Regional Atmospheric Modeling System (RAMS; Cotton et al., 2003). In Italy, wind farms are usually located in complex terrain, where wind prediction is more difficult than in flat orography (Giebel et al., 2011). In these conditions the

Figure 1. (a) Wind farm in Abruzzo (Central Italy), and (b) turbines layout.

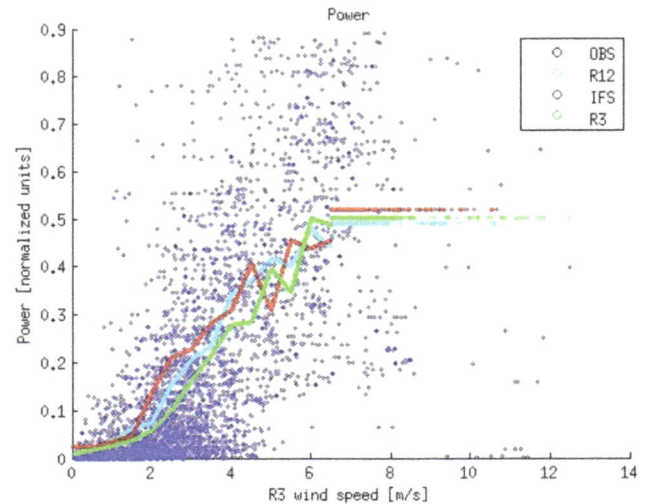

Figure 2. The power curve for R3, R12 and IFS. Blue diamonds show the pairs (wind, power) for the R3 case. R12 and IFS (wind, power) pairs are not shown for clarity.

horizontal resolution of LAMs has an important role and we discuss this point by comparing two 48 h wind power forecasts for the whole year 2011, using two wind forecasts at 3 (R3) and 12 km (R12) horizontal resolutions.

For completeness the wind power forecasts using the winds of R3 and R12 are compared with that issued using the winds of the ECMWF IFS 12:00 UTC analysis/forecast cycle. This cycle also gives the initial and dynamic boundary conditions for the R3 and R12 forecasts. It is important to note that, while the differences between R3 and R12 are only caused by their different horizontal resolution, the differences between IFS and RAMS forecasts are not only due to the horizontal resolution of the models, but also to their different physical and dynamical parameterizations, as well as to their different numerical coding.

Because of the natural variability of the Mediterranean climate, results are shown seasonally. Moreover, a case study is considered to better focus on the differences found for the models, while statistics are considered for 14 cases.

2 Data and methodology

The wind farm considered in this study is located in a complex orographic area in the Abruzzo region, Central Italy. The wind farm has wind turbines in 6 different zones away few kilometers each other. There are different kinds of turbines, with capacity of 0.6 MW (Fig. 1).

The wind power prediction of this study can be divided in two-steps: (a) finding a power curve for the whole wind farm, and; (b) using the wind forecast of R3 and R12 and the corresponding power curve of the step (a) to issue the power forecast. While two years of data (2010–2011) were used for step (a), the power forecast is for 2011 only.

For the step (a) wind measurements were not available for the period considered, so we used meteorological analyses at 00:00, 06:00, 12:00 and 18:00 UTC to derive the wind in correspondence of the wind farm. Moreover, because we want to assess the impact of the horizontal resolution on the wind power forecast, two different dynamical downscaling

(hindcast) were produced by R3 and R12, i.e. at 3 and 12 km horizontal resolution, respectively. Initial and boundary conditions of the R3 and R12 hindcasts are derived from the ECMWF operational analyses.

The R3 and R12 models share the same physical and dynamical parameterizations, which are as in Federico (2011), the only differences being their grids and horizontal resolutions. The R3 model uses two two-ways nested grids: the first grid has a horizontal resolution of 12 km and covers the Central part of the Mediterranean Basin, while the second grid extends over Central Italy with 3 km horizontal resolution. The R12 model uses only the first grid of the R3 model.

The surface wind speed of the RAMS hindcasts, as well as that of IFS, were bi-linearly interpolated to the position of the wind farm (14.500° E, 41.875° N) at 00:00, 06:00, 12:00 and 18:00 UTC for each day and these values were used, with the corresponding values of the observed power produced by the wind farm, to find the power curve of the whole wind farm (Fig. 2).

To fit the data, the wind speed sample has been divided in bins $0.5 \, \mathrm{m \, s^{-1}}$ wide; for each bin, we computed a fitting power by minimizing the variance between the power values inside the bin and the fitting power.

In addition to the large scatter of the data of Fig. 2, which is a common feature of all models considered in this paper and is, at least in part, caused by the lack of wind observations, it is noticed that IFS and R12 power curves lie above that of R3 for wind speeds in the range 2.0–$4.5 \, \mathrm{m \, s^{-1}}$. Stated in other terms, the power predicted by IFS and R12 is larger than that of R3, for the same velocity forecast in the range 2.0–$4.5 \, \mathrm{m \, s^{-1}}$.

It is also noticed that there are three different values of the power plateau for wind speed larger than $6.5 \, \mathrm{m \, s^{-1}}$. This

Table 1. Bias (BIA), mean absolute error (MAE), RMSE, coefficient of determination (r^2) and skill score (SKILL) compared to the 3 h (first number of the cell) and 9 h (second number of the cell) persistence forecast for the wind power. P3 and P9 are the three and nine hours persistence forecast.

	Statistics	R3	R12	IFS	P3	P9
Winter	BIA (W)	4112	4055	3256	−9	4
	MAE (W)	12 236	12 884	13 555	7988	13 981
	RMSE (W)	16 473	17 122	18 057	12 706	20 444
	r^2	0.45	0.39	0.32	0.66	0.26
	SKILL (%)	−30 ; 19	−35; 16	−42; 12	–	–
Spring	BIA (W)	−1253	−1826	−547	17	3
	MAE (W)	10 383	11 606	11 552	7670	12 965
	RMSE (W)	14 103	15 589	15 663	12 624	19 695
	r^2	0.63	0.55	0.54	0.72	0.40
	SKILL (%)	−12; 28	−23; 21	−24; 20	–	–
Summer	BIA (W)	−1113	−41	−2856	−2	20
	MAE (W)	10 616	11 434	11 508	8502	12 986
	RMSE (W)	16 790	17 182	17 705	15 634	22 500
	r^2	0.34	0.30	0.29	0.51	0.16
	SKILL (%)	−7; 25	−10; 24	−11; 23	–	–
Fall	BIA (W)	−2512	−2574	−2856	4	8
	MAE (W)	9691	10 597	11 508	6476	11 379
	RMSE (W)	14 845	15 857	17 705	13 163	21 301
	r^2	0.79	0.75	0.62	0.77	0.46
	SKILL (%)	−13; 30	−20; 25	−34; 17	–	–
Year	BIA (W)	−209	−119	278	2	9
	MAE (W)	10 726	11 627	12 187	7658	12 825
	RMSE (W)	15 578	16 445	17 201	13 581	21 001
	r^2	0.54	0.49	0.44	0.68	0.34
	SKILL (%)	−15; 26	−21; 22	−27; 18	–	–

arises because the plateau is computed as the average of the observed power for wind speeds larger than 6.5 m s^{-1} for each model. The number of speeds larger than 6.5 m s^{-1} varies for R3, R12 and IFS giving the different values of the plateau shown in Fig. 2. It is noted, however, that the plateau difference among the models has a small impact on the results of this paper (see next section).

For the step (b), a one-year forecast (2011) of R3 and R12 for the following 48 h was made. Initial and boundary conditions were derived from the ECMWF-IFS 12 UTC operational analysis/forecast cycle. The surface wind speeds of the R3, R12 and IFS forecasts were interpolated to the position of the wind farm (14.500° E, 41.875° N) and used to produce the power forecast by employing the corresponding power curve of Fig. 2. As stated in the previous section the differences between R3/R12 and IFS are not only caused by their different spatial resolutions.

Before concluding this section it is noted that both steps, (a) and (b), are important for the power forecast issued by each model. Indeed, while the quality of the wind forecast directly impacts the power forecast in the step (b), each model uses its own power curve, whose quality is in turn determined by the ability of each model to simulate the wind for the specific site.

3 Results

Considering the comparison between R3 and R12 (Table 1), it is noticed that the Bias is similar for all season but summer, when R12 performs better than R3. The Bias is positive in winter for both R3 and R12, while it is negative for other seasons as well as for the whole year (−209 W for R3 and −119 W for R12).

The MAE and RMSE show that R3 scores better than R12. Focusing on the RMSE, there is a reduction of 5 % of the RMSE for the whole year for R3 compared to R12. The largest RMSE reduction occurs in spring (10 %), while the lowest in summer (2 %). This shows the importance of the seasonal forcing on the results.

The coefficient of determination (r^2) also shows that R3 performs better than R12 because the correlation between the predicted and observed power is larger for R3. This occurs for all seasons as well as for the whole year. It is also noticed

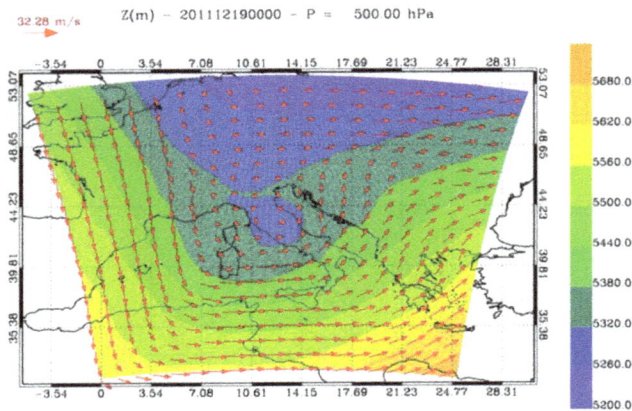

Figure 3. Geopotential height and wind vectors at 500 hPa on 19 December 2011 at 00:00 UTC. A cut-off (5320 m) is apparent over central Italy.

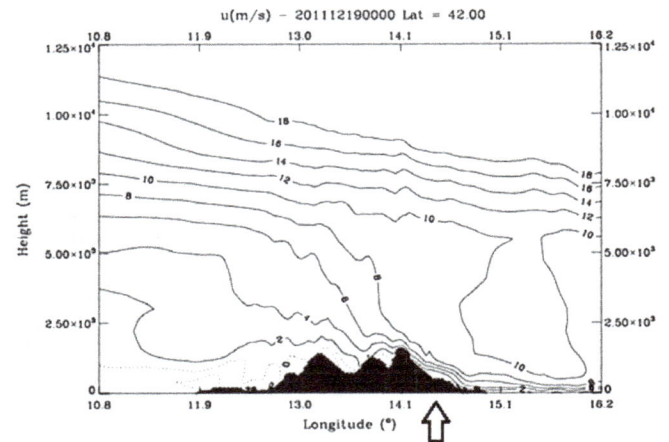

Figure 4. Vertical cross section at 42° N of the zonal wind showing the deep gravity wave generated by the Apennines. The arrow shows the approximate longitude of the wind farm.

the decrease of the coefficient of determination in summer compared to other seasons, showing the difficulty of forecasting the wind power for the site when local circulations play a major role.

Another statistic shown in Table 1 is the skill of the model compared to the 3 and 9 h persistence forecast. The skill is computed for RMSE, i.e.:

$$SKILL = 100 \frac{RMSE_P - RMSE_F}{RMSE_P}. \qquad (1)$$

In Eq. (1) the $RMSE_P$ is the RMSE of the persistence, while $RMSE_F$ is the RMSE of the forecast. A positive value of the skill means an improvement of the forecast compared to persistence.

The results of Table 1 show that both R3 and R12 have skill compared to the 9h persistence, while the 3h persistence performs better than the RAMS forecast. This is true for all seasons and for the whole year, showing a stable result. Comparing the results of R3 and R12 we notice the larger skill for R3, showing again the positive impact of the model horizontal resolution on the power forecast for the specific site.

It is also noticed that the statistics of Table 1 were recomputed assuming a unique power plateau in the power curve (Fig. 2) for all models. This value was, in turn, that of R3, R12, IFS and the average of the three. However, the results were similar to that of Table 1, computed assuming for each model its own value of the plateau. So, the statistics of Table 1 are not sensitive to the choice of the plateau value.

Considering the results for IFS, it is noticed that IFS has the largest RMSE and the lowest coefficient of determination among the models compared in this work. This is true for all seasons and for the whole year. Considering the whole year, the RMSE for R3 is 10 % lower compared to IFS, while the coefficient of determination (r^2) is 0.54 for R3 and 0.44 for IFS. The IFS has skill compared to the 9h persistence, while the 3h persistence performs better, similarly to the re-

sult found for R3 and R12. However, the IFS skill is lower than those of R12 and R3.

We note that the skill of the model compared to the 6 h persistence (not shown) is in between the skills compared to the 3 and 9 h persistence. In particular R3 and R12 have skill compared to the 6 h persistence for all seasons as well as for the whole year, while the IFS doesn't have skill compared to the 6 h persistence in winter. The values of the R3 and R12 skills compared to the 6 h persistence are often below 10 %.

To show a case study for which the forecast resolution has an important role, we focus on the period from 15 to 20 December 2011.

This period was characterized by the passage of cyclones over the Central Mediterranean Basin. On 19 December 2011 the RAMS forecast at 00:00 UTC shows the passage of one of these cyclones over Central Italy (Fig. 3). The cyclone evolved in a cut-off at 500 hPa and air masses crossed the Central Italy from west to east. In these conditions, the orography of the Apennines can generate deep gravity waves as shown by the zonal velocity of Fig. 4.

The linear theory of standing gravity waves predicts an increase of the wind speed on the lee of the orographic barrier, consistently with the results of Fig. 4. However, nonlinear effects become significant in a multilayered atmosphere even at small values of the Froude number ($Fr = Nh/U$, where U is the mean undisturbed flow, h is the height of the orographic barrier and N is the buoyancy frequency of the atmosphere), and the behavior of the interaction between the flow and the orography is more complex than that depicted by the linear theory (Durran, 1986).

The characteristics of the interaction between the atmospheric flow and the orography can be described considering the value of the Froude number. If Fr is larger than 1 the flow is referred to as supercritical, if the Froude number is less than 1 the flow is subcritical. In everywhere (i.e. before

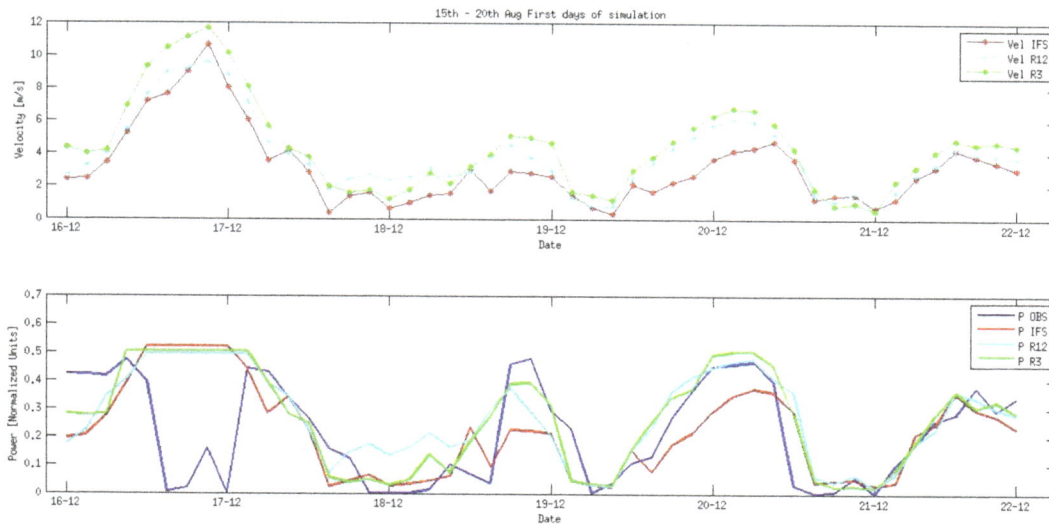

Figure 5. Upper panel: the wind velocity forecasts for the period 15–20 December 2011 (R3 is the green curve, R12 is the cyan curve and IFS is the red curve). Lower panel: comparison between the observed (blue curve), R3 (green curve), R12 (cyan curve) and IFS (red curve) wind powers for the same period of the upper panel. Note that the power forecast reached the plateau of Fig. 2, which differs for each model, for part of 16 and 17 December 2011 (wind speeds larger than $6.5\,\mathrm{m\,s^{-1}}$).

and after the topography peak) supercritical flow, the wind speed has a maximum on the lee side of the orographic barrier, whose position depends on the value of the static stability of the atmosphere, on the value of the undisturbed flow velocity and on the height and shape of the orography. In supercritical flow high winds develop on the lee of the mountain. In everywhere subcritical flow the velocity has a maximum over the mountain crest.

The largest winds on the leeside of the orographic barrier occur when there is a transition from subcritical to supercritical flow at the top of the orography. This situation occurs in windstorms and, roughly speaking, the flow recovers its subcritical state near the bottom of the orographic barrier with an hydraulic jump (Durran, 1986). Wave breaking also plays an important role in windstorms.

Figure 4 shows the development of a deep gravity wave and intense winds on the lee of the orographic barrier, where the wind farm is located, suggesting the passage to supercritical flow on the lee of the mountains. In these conditions, the horizontal resolution of the model has an important role in the forecast of the evolving cyclone and of the wind speeds, specifically at lower atmospheric levels where wind turbines are located. The interaction between air-masses and orography is simulated in more detail for increasing horizontal resolutions and differences arise between R3 and R12. For the specific case study we found that R3 velocities are larger than those of R12, even if it is not always the case, as shown for part of the 17 and 18 December (Fig. 5a). In general, the velocities simulated by IFS are smaller than those of R3 and R12.

Figure 5b shows the comparison among wind power calculated for R3, R12 and IFS and the observed values in the period considered. All forecasts show a similar behavior; nevertheless the R3 follows more closely the observations, especially for 18 December. The IFS underestimates the observed power as a consequence of the lower wind speed simulated. We also note that the velocities forecast by all models are larger than $6.5\,\mathrm{m\,s^{-1}}$ for most of 16 and part of the 17 December. The power forecast reaches the plateau for each model, causing the (small) difference of the power forecast of Fig. 5.

The situation shown in Figs. 3–5 is not uncommon in Central Italy. To better assess the importance of the interaction between the large-scale systems and the local orography, we considered the cases when, for the whole 48 h of forecast, R3 has a 10 % RMSE improvement compared to R12. For these cases we also require that the averaged observed power for the two forecast days is larger than 10 kW to exclude cases when the wind is low, the power production small, and the behavior of the (small) RMSEs becomes erratic.

We found a total of 49 days and, among them, 23 (roughly 50 % of the cases) were associated with synoptic scale disturbances acting over Central Italy. The RMSE for those cases is shown in Table 2, with a short description of the synoptic environment. For some cases the situation is similar to that of Figs. 3–5, with a cyclone crossing Central Italy (for example the 15 May 2011), for other cases the cyclone did not cross Central Italy, nevertheless its action extended over the target area.

The numbers above show that the interaction between the large scale flow and the Apennines orography is a key feature

Table 2. Selected events for which the performance of R3 is better than R12 (plain text). A short synoptic description of the events is shows in the last column. Each cell for R3, R12 and IFS shows the value of the RMSE ([W], first value) and of the coefficient of determination (r^2, second value). In bold two cases when R12 is better than R3 are shown. The number in parenthesis in the first column shows the consecutive days of better/worse R3 performance compared to R12. For cases lasting more than one day, statistics are shown for the first day of the event.

Date	R3	R12	IFS	Synoptic description
14 Jan 2011 (2)	15 879 W, 0.42	18 236 W, 0.19	18 032 W; 0.21	Wave trough over the Balkans
13 Feb 2011 (3)	16 500 W, 0.89	19 032 W, 0.84	22 250 W; 0.78	Cyclone developing on the west Mediterranean (Lyon Gulf)
18 Feb 2011	15 173 W, 0.79	19 092 W, 0.71	20 116 W; 0.68	Influence of the cyclone of 20110215 (above row) while evolving toward the South-East
15 Mar 2011 (3)	29 059 W, 0.60	33 894 W, 0.50	36 726 W; 0.19	Cyclone developed over the western Mediterranean and crossing central Italy
15 May 2011 (3)	16 776 W, 0.83	19 722 W, 0.73	18 890 W; 0.76	Cyclone (developed on the lee of the western Alps) crossing Central Italy
18 Jun 2011	13 646 W, 0.78	16 642 W, 0.63	16 061 W; 0.69	Wave trough crossing central Europe
27 Jul 2011	18 260 W, 0.89	21 379 W, 0.76	23 422 W; 0.51	Wave trough crossing central Europe
20 Sep 2011 (2)	12 487 W, 0.90	15 106 W, 0.82	15 382 W; 0.75	Cyclone (developed on the lee of the Alps) crossing Southern Italy
18 Oct 2011 (2)	24 322 W, 0.83	27 366 W, 0.78	29 811 W; 0.66	Cyclone on the lee of the Eastern Alps
22 Nov 2011 (2)	13 952 W, 0.25	16 357 W, 0.21	15 923 W, 0.01	Cyclone evolving on the Western Mediterranean
17 Dec 2011	10 211 W, 0.66	13 747 W, 0.39	12 635 W; 0.54	Cyclone developing on the lee of the Alps and crossing Central Italy
18 Dec 2011 (2)	12 275 W, 0.67	14 098 W, 0.55	14 265 W; 0.46	Cyclone developing on the lee of the Alps and crossing Central Italy
20 Jul 2011 (3)	**19 680 W, 0.23**	**17 606 W, 0.34**	**21 164 W; 0.03**	**Cyclone over the Central Europe**
24 Dec 2011 (2)	**12 856 W, 0.76**	**9340 W, 0.73**	**12 312 W; 0.56**	**Cyclone developing on the lee of the Alps and crossing Central Italy**

for the successful forecast of the wind power for the case considered, as the higher resolution of the forecast resolves in more detail this interaction.

It is important to highlight that the R3 is not always the best forecast. The impact of the model horizontal resolution on the wind speed forecast has been studied by several authors (for example Möhrlen et al., 2002; Rife and Davies, 2005; Hashimoto et al., 2007; see Giebel et al., 2011, for a review). In several of these studies it is found that the increase of the model horizontal resolution improves the forecast of the intensity and gradient of the wind speed, so that the wind field resembles more closely the reality at higher resolution. Nevertheless, traditional verification metrics, as those used in this paper, often improve slightly or even show worse performance for higher horizontal resolution because they penalize forecasts with small temporal or spatial errors (phase errors). In other terms, because the wind and its gradient are, in gen-

eral, more intense at higher horizontal resolutions, the phase error is amplified, and this penalize the traditional scores of comparatively higher resolution forecasts.

For the forecast considered in this paper, we found 8 days when the RMSE of R12 had a 10 % improvement compared to R3. Also for these cases we required that the averaged observed power for the whole forecast is larger than 10 kW. For five of these days a clear synoptic system was acting over the area (Table 2, cases in bold).

For the two days of July 2011 the R3 model underestimates the observed power (the R3 bias is −8115 W, while the R12 bias is −4444 W). This is caused by the slower winds simulated by R3 compared to R12. This error is amplified by the power curves of R3 and R12 (Fig. 2) for wind speed in the range 2–4.5 m s^{-1}.

For the case of December 2011, R3 simulates too strong winds (the bias of the R3 power forecast is 10 803 W, while

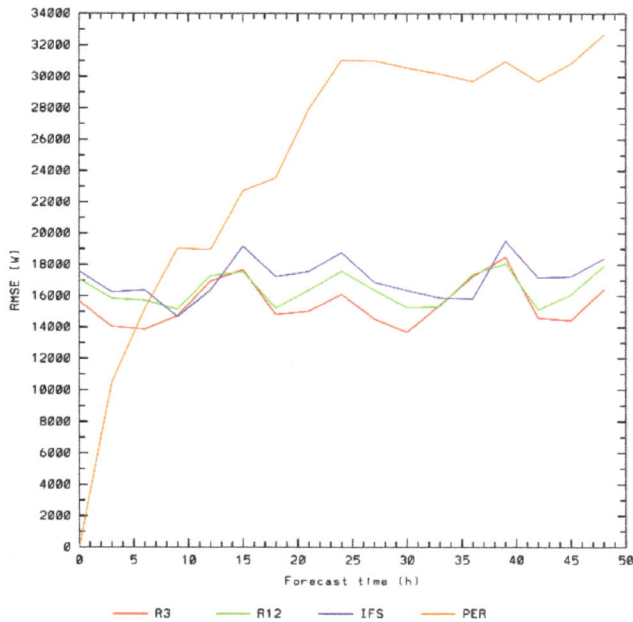

Figure 6. RMSE for R3 (red), R12 (green), IFS (blue) and PER (orange) as a function of the forecasting time for the first and second forecast days.

that of R12 is 8505 W). The too strong wind speeds are simulated as a consequence of a cyclone that was located over central Italy on 24 December 2011. Likely, the position and/or the intensity of the cyclone are not well simulated by RAMS for this case and the error is amplified by R3 compared to R12.

Another point considered in this paper is the performance of the forecast with forecasting time. Figure 6 shows the RMSE for the R3, R12, IFS and persistence (PER) forecasts every three-hours for the two forecast days. It is apparent that the RAMS and IFS RMSEs do not increase considerably with forecasting time. This is confirmed by the RMSEs for the first and second day forecasts. Using the triple (RMSE_R3, RMSE_R12, RMSE_IFS) we have, respectively, for the first and second forecast day (15 490, 16 437, 17 139 W) and (15 842, 16 665, 17 399 W). So, for the second forecast day, we have an increase of the forecast error below 5 % of the RMSE for the first forecast day.

In a recent paper on the performance of the RAMS forecast for Southern Italy, run using a configuration similar to that used in this paper for the year 2013, Tiriolo et al. (2015) show that the RMSE for the wind and for the third forecast day increases by less than 10 % of its value at the initial time, finding a small increase of the error with forecasting time. Even if the power law is not linear with the wind, this paper confirms the small increase of the error with forecasting time for the wind power too.

4 Conclusions

This paper shows the 48 h wind power forecast for a wind farm located in complex terrain in Central Italy and focuses on the impact of the horizontal resolution of the wind forecast on the power prediction.

The power forecast is divided in two steps: (a) finding a power curve for the whole wind farm, and; (b) applying the power curve along with the wind forecast to issue the power forecast. We evaluate the performance of two models, R3 and R12, differing only for their horizontal resolutions (3 and 12 km, respectively) and grids. Moreover, as a further comparison, we show the performance of the IFS model of ECWMF.

Wind observations were not available at the site for the period considered, and ECMWF-IFS operational analyses were used to compute the power law for the whole wind farm. To increase the horizontal resolution of the wind field, in order to better account for the local orography, a two-years hindcast of the RAMS model was made at 3 and 12 km horizontal resolution, and the power law was computed using the surface wind speed of the hindcast as well as the analyses of the ECMWF-IFS. In this way, each model has its own power curve, which is a key feature of the power forecast.

The results show the importance of the horizontal resolution for the power prediction. Considering the whole year, the comparison between the power forecast of R3 and that of R12 shows a RMSE reduction of about 5 % when using the higher resolution. The improvement, however, has a noticeable seasonal variability, reaching the maximum value (10 %) in spring and the lowest value (2 %) in summer.

The coefficient of determination is larger for R3 compared to R12 for all seasons and for the whole year, confirming the importance of the horizontal resolution of the forecast, for the specific site.

The comparison with the persistence forecast shows that R3 and R12 forecasts are better than the 6 and 9 h persistence, while they are worse than the 3h persistence. Moreover, the R3 skill is better than that of R12.

The IFS forecast has the worst performance among all models considered. While the horizontal resolution of the IFS forecast is about 25 km for the period and area considered, it is noted that the differences between the RAMS and IFS forecasts are not only attributable to the different resolutions of the models. Several other factors as the difference in numerical and physical parameterizations contribute.

A closer investigation of the model performance for different cases shows the importance of the interaction between the large-scale flow and the orography. This interaction is resolved in more detail at finer horizontal resolution that give, in general, better results. There are occasions, however, when the position/intensity of meteorological systems are not well represented in space and/or time. For these cases the higher resolution may amplify the phase errors that penalizing traditional scores.

Acknowledgements. This work was carried out during a model benchmark exercise within the COST Action ES1002 WIRE (Weather Intelligence for Renewable Energies). It was also supported by the projects PON04a2_E Sinergreen-ResNovae - "Smart Energy Master for the energetic government of the territory" and PONa3_00363 "High Technology Infrastructure for Climate and Environment Monitoring" founded by Italian Ministry of University and Research (MIUR) PON 2007-2013. The ECMWF and CNMCA – "Centro Nazionale di Meteorologia e Climatologia Aeronautica" are acknowledged for giving the access to the analyses and forecasts of the IFS. We are grateful to Edison Energie Speciali Spa for the wind farm power data.

Both reviewers are acknowledged for their useful and constructive comments, which improved the quality of the paper.

References

Alessandrini, S., Sperati, S., and Pinson, P.: A comparison between the ECMWF and COSMO Ensemble Prediction Systems applied to short-term wind power forecasting on real data, Appl. Energy, 107, 271–280, 2013.

Cotton, W. R., Pielke Sr., R. A., Walko, R. L., Liston, G. E., Tremback, C. J., Jiang, H., McAnelly, R. L., Harrington, J. Y., Nicholls, M. E., Carrio, G. G., and McFadden, J. P. : RAMS 2001: Current satus and future directions, Meteor. Atmos., 82, 5–29, 2003.

Durran, D. R.: Another look at downslope wind storm. Part I: The development of analogs to supercritical flows in an infinitely deep, continuously stratified flow, J. Atmos. Sci., 43, 2527–2543, 1986.

Federico, S.: Verification of surface minimum, mean, and maximum temperature forecasts in Calabria for summer 2008, Nat. Hazards Earth Syst. Sci., 11, 487–500, doi:10.5194/nhess-11-487-2011, 2011.

Giebel, G., Brownsword, R., Kariniotakis, G., Denhard, M., and Draxl, C.: The State of the Art in Short-Term Prediction of Wind Power: A Literature Overview, Deliverable Report D1.l of the Anemos project (ENK5-CT-2002-00665), available at http://www.anemos-plus.eu/images/pubs/deliverables/aplus.deliverable_d1.2.stp_sota_v1.1.pdf, last access: April 2015.

Hashimoto, A., Hattori, Y., Kadokura, S., Wada, K., Sugimoto, S., Hirakuchi, H., and Tanaka, N.: Effects Of Numerical Models On Local-Wind Forecasts Over A Complex Terrain with Wind Farm Prediction Model, Proceedings of the European Wind Energy Conference, Milano (IT), 7–10 May 2007.

Holmgren, E., Siebert, N., and Kariniotakis, G.: Wind Power Prediction Risk Indices Based on Numerical Weather Prediction Ensembles, Oral presentation at EWEC 2010 Conference, Warsaw, Poland, 20–23 April 2010.

Möhrlen, C., Jørgensen, J. U., and McKeogh, E. J.: Power predictions in Complex Terrain with an operational Numerical Weather Prediction Model in Ireland including Ensemble Prediction, Proc. World Wind Energy Conference "Clean Power for the World", Berlin, Germany, 2002

Pinson, P., Nielsen, H. Aa., Møller, J. K., Madsen, H., and Kariniotakis, G.: Nonparametric probabilistic forecasts of wind power: required properties and evaluation, Wind Energy, 10, 497–516, 2007.

Rife, D. L. and Davis, C. A.: Verification of Temporal Variations in Mesoscale Numerical Wind Forecasts, Mon. Weather Rev., 133, 3368–3381, 2005.

Tiriolo, L., Torcasio, R. C., Montesanti, S., and Federico, S.: Verification of a real time weather forecasting system in southern Italy, Adv. Meteorol., 2015, 14 pp., doi:10.1155/2015/758250, 2015

17

The SASSCAL contribution to climate observation, climate data management and data rescue in Southern Africa

F. Kaspar[1], J. Helmschrot[2], A. Mhanda[3], M. Butale[3], W. de Clercq[4], J. K. Kanyanga[5], F. O. S. Neto[6], S. Kruger[7], M. Castro Matsheka[3], G. Muche[2], T. Hillmann[2], K. Josenhans[2], R. Posada[1], J. Riede[1], M. Seely[8], C. Ribeiro[9], P. Kenabatho[10], R. Vogt[11], and N. Jürgens[2]

[1]Deutscher Wetterdienst, National Climate Monitoring, Frankfurter Str. 135, 63067 Offenbach, Germany
[2]University of Hamburg; Biodiversity, Evolution and Ecology of Plants; Hamburg, Germany
[3]Department of Meteorological Services (DMS), Gaborone, Botswana
[4]Stellenbosch University, Stellenbosch, South Africa
[5]Zambia Meteorological Department (ZMD), Lusaka, Zambia
[6]Instituto Nacional de Meteorologia e Geofisica (INAMET), Luanda, Angola
[7]National Botanical Research Institute, Windhoek, Namibia
[8]Gobabeb Research and Training Centre, Walvis Bay, Namibia
[9]Instituto Superior Politécnico Tundavala, Lubango, Angola
[10]University of Botswana, Gaborone, Botswana
[11]University of Basel, Basel, Switzerland

Correspondence to: F. Kaspar (frank.kaspar@dwd.de)

Abstract. A major task of the newly established "Southern African Science Service Centre for Climate Change and Adaptive Land Management" (SASSCAL; www.sasscal.org) and its partners is to provide science-based environmental information and knowledge which includes the provision of consistent and reliable climate data for Southern Africa. Hence, SASSCAL, in close cooperation with the national weather authorities of Angola, Botswana, Germany and Zambia as well as partner institutions in Namibia and South Africa, supports the extension of the regional meteorological observation network and the improvement of the climate archives at national level. With the ongoing rehabilitation of existing weather stations and the new installation of fully automated weather stations (AWS), altogether 105 AWS currently provide a set of climate variables at 15, 30 and 60 min intervals respectively. These records are made available through the SASSCAL WeatherNet, an online platform providing near-real time data as well as various statistics and graphics, all in open access. This effort is complemented by the harmonization and improvement of climate data management concepts at the national weather authorities, capacity building activities and an extension of the data bases with historical climate data which are still available from different sources. These activities are performed through cooperation between regional and German institutions and will provide important information for climate service related activities.

1 Introduction

Africa is considered being the most vulnerable continent regarding climate variability and change (e.g. as stated by the Working Group 2 of the IPCC 5th Assessment Report; IPCC, 2014). The preparation of an adequate response to the expected changes leads to an increasing demand for climate information and associated service capacities. Such activities are a basis for impact assessment research, climate adaptation measures and climate services at global, continental and regional scales. Addressing these issues, the World Meteorological Organization (WMO) established the Global Framework for Climate Services (GFCS) in order to provide a worldwide mechanism for coordinated actions to enhance the quality, quantity and application of climate services. In that context it was noted that the majority of developing countries still lack the resources and expertise they need to provide effective climate services to their citizens (WMO, 2014a). Observations of various parameters are required as a basis for such services. They have to be of adequate quality and quantity and should be measured at high temporal and spatial resolutions (WMO, 2014b).

Global infrastructures for exchange of meteorological observations have been in operation for decades, but their status and performance depend on the region. Figure 1 illustrates the status of the exchange of the so-called "CLIMAT"-reports. The figure reveals that especially for large parts of Africa only a small amount of the expected data is available. It further indicates that existing national networks in sub-Saharan Africa are either not sufficient or service capacities are strongly limited. Improvements of the existing observational networks are urgently needed in order to achieve a better regional and global coverage with high quality and reliable climatological data. Thus, many national and international efforts are being undertaken to improve national station networks (e.g. within the GCOS (Global Climate Observing System) Cooperation Mechanism) and to implement advanced technologies for information management and exchange, often aligned with capacity development (e.g. within the WMO Voluntary Cooperation Programme).

In addition to the required improvements in the data availability of current weather conditions, the availability of long-term climate records is also limited for several Southern African countries. Partly, observations have been performed over long periods and have been documented manually on hard copies and stored in archives, but are not available in digital format, and are therefore not easily accessible for climate applications. Figure 2 shows an example of time series from Angola. The figure illustrates that observations for historical periods are available in international archives, but may vary in their availability over time. Historic documents with weather observations are partly available in the archives of the national meteorological services, but older data might also be available elsewhere, e.g. in archives of European meteorological services. In the archives of Germany's national

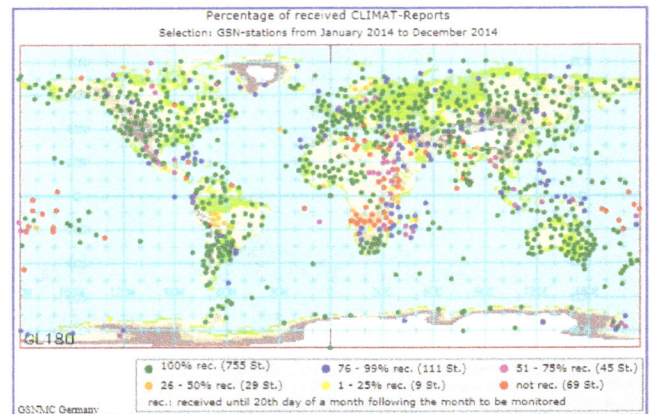

Figure 1. Percentage of received monthly CLIMAT-reports from the "GCOS Surface Network stations (GSN)" for the period January 2014 to December 2014. CLIMAT-reports are monthly summaries for special observing stations on the land surface that are collected and disseminated via the Global Telecommunication System (GTS) of the WMO within the so-called GCOS Surface Network ("Global Climate Observing System – Surface Network": GSN). The performance of this data exchange is monitored by two 'GSN Monitoring Centers' ("GSNMC"), one of them operated by Deutscher Wetterdienst (DWD). Colors indicate the percentage of reports that were received by the GSN Monitoring Center of DWD until the 20th day of a month following the month to be observed (Source: www.gsnmc.dwd.de).

meteorological service (Deutscher Wetterdienst, DWD), historical documents with weather observations from overseas stations and marine weather observations from ships (sailing ships, buoys steamers and light vessels) are available from several years at the beginning of the 20th century (Kaspar et al., 2015).

In summary, several activities are needed to improve the availability of climate data for Southern Africa: an enhancement of the observation network, an improvement of data management concepts and data rescue activities. With this paper, we give an overview on SASSCAL contributions addressing this deficit and specifically refer to the technical implementation which is conducted as a joint effort of various key institutions from Southern Africa and Germany.

2 The Southern African Science Service Centre for Climate Change and Adaptive Land Management

The implementation of such activities requires financial support and the cooperation of relevant institutions, especially the involvement of the national meteorological services. As a regional initiative, SASSCAL is an example of an activity that provides support for cooperation, technical infrastructure enhancement and implementation, and the improvement of capacities of regional institutions and service providers.

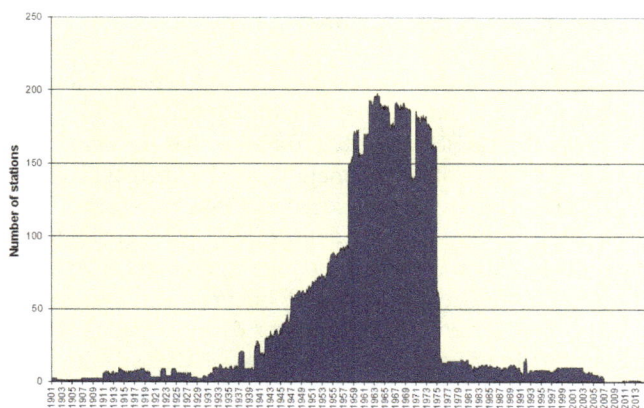

Figure 2. Number of stations with monthly precipitation data from Angola available in the database of Global Precipitation Climatology Center (GPCC) from 1901 to 2014 (status: 2014). Those stations were used for the GPCC Full Data Product (Schneider et al., 2014). GPCC is hosted at DWD.

The overarching aim of SASSCAL is to improve the capacities to provide sound science-based solutions for current problems and future risks in the region, in particular regarding climate change and the associated demands concerning land management practices of local players. As many of the global change challenges go beyond national boundaries, SASSCAL has a regional scope. It is a joint initiative of Angola, Botswana, Namibia, South Africa, Zambia and Germany. The main objectives therefore are:

1. trans-disciplinary, problem-oriented research in the area of adaptation to climate change and sustainable land management in order to improve the livelihoods of people in the region,

2. services and advice for policy, decision makers and stakeholders and

3. capacity development.

A full description of aims, objectives and intended services is given at www.sasscal.org.

One specific intention is to make data and information available for further scientific uses as well as to generate, compile, analyze and process relevant data on various topics, such as climate, the water cycle, soil fertility, forest resources, agriculture, biodiversity and socio-economy, in cooperation with national and, if required, with international authorities.

In the following sections, we only focus on SASSCAL activities related to climatological ground-based observations.

3 SASSCAL WeatherNet

The SASSCAL WeatherNet (www.sasscalweathernet.org) has been established in order to support climate data avail-

ability for the SASSCAL region and is continuously under development. It comprises two major components, the regional station network itself and the online platform linking various tools for data repository, management and online visualization.

The station network covering all SASSCAL countries unifies

1. existing automatic weather stations of the previous initiative BIOTA AFRICA (www.biota-africa.org) and contributions of the Namibian Ministry of Agriculture, Water and Forestry,

2. new automatic weather stations (AWS) installed as a joint effort of SASSCAL, the national authorities and a Namibian NGO to improve national networks,

3. weather stations installed by individual SASSCAL research projects (experimental networks) and

4. rehabilitated stations from former monitoring efforts.

Complementing the former BIOTA network consisting of 40 automatic weather stations and in support of national efforts of Angola, Botswana and Zambia to improve their national weather monitoring systems, SASSCAL installed 10 fully automatic weather stations in each of the three countries in 2013/2014. For detailed studies on climatological phenomena, nine AWSs were installed in the Namib Desert near Gobabeb (Namibia) as well as 10 AWSs in the Namibe Province, south-west Angola in 2014. In early 2015, further 37 AWSs spread over the region and supporting specific research projects are put into operation by SASSCAL respectively.

All new and rehabilitated stations are equipped with WMO-certified sensors measuring rainfall, temperature, solar radiation, wind speed and direction, relative humidity and barometric pressure, all at a temporal resolution of either 15, 30 or 60 min intervals. Depending on research requirements, stations are upgraded with additional equipment measuring sunshine duration, leaf moisture, soil moisture and fog in selected locations. In order to facilitate near-real time data availability, data transmission is primarily based on mobile communication technology (GPRS/GSM). However, 8 stations at remote sites use satellite-based data transmission (supported by EUMETSAT, based on METEOSAT), and additional 7 satellite-based stations are planned for installation in 2015. The observations are transmitted to the SASSCAL WeatherNet server, and additionally to various SASSCAL partner institutions. In Namibia, data are transferred to the National Botanical Research Institute as well as the Gobabeb Research and Training Centre, while in Angola, Botswana and Zambia, the data are also collected by the national meteorological services. Altogether, the current SASSCAL station network comprises of 113 stations in the region and will, in cooperation with other activities, increase to 148 in 2015 (see Fig. 3).

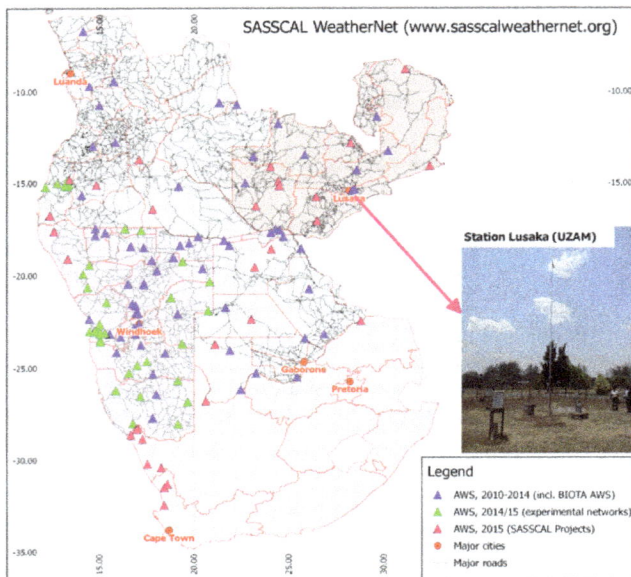

Figure 3. Location of observing stations supported by SASSCAL (including stations installed by BIOTA AFRICA). Data from these stations are available at www.sasscalweathernet.org. Note: Additional five AWS will be installed within a radius of 20 km South East of Gaborone City, i.e. upstream of Gaborone Dam.

According to the SASSCAL mandate the data are made freely available to the public at the SASSCAL WeatherNet website. Currently, data of 105 stations are available online, with time series partly starting in 2010. Technically, station data are provided by six different transmission systems which vary by country and provider. Thus, data flow procedures and data protocols were unified for harmonization purposes. Incoming data are (automatically) checked in terms of homogeneity and consistency and corrected before getting stored in a relational data base system. A user-friendly web-interface (Fig. 4) based on MySQL and PHP technology was developed allowing users to get an overview on the actual condition at each station, to visualize and download data as well as to explore various statistics. All measured values are available for download in 15 or 60 min resolution. In addition, daily and monthly averages, diagrams for individual parameters, near real-time rainfall information and station information sheets are provided on the website. Users can also subscribe for an e-mail-based daily weather report.

With these efforts, SASSCAL WeatherNet provides not only data and information for its research community as for example projects in the water sector (Helmschrot and Jürgens, 2015), but also for collaborating agencies and activities (e.g. FEWS NET, the Famine Early Warning Systems Network), many local and regional stakeholders (e.g. farmers, lodge owners) and the public (e.g. radio stations).

4 Climate data management

For sustainable archiving and efficient utilization of the climate data within the national meteorological and hydrological services (NMHSs), a specialized data base system is required, typically called "Climate Data Management System (CDMS)". During the last decades different systems have been in use in developing countries (see Stuber et al., 2011) and different approaches for their development can be identified (e.g. developed by an NMHS or commercial products).

The improvement of concepts for data management and the completion of the data bases at the NMHSs is an additional aim of SASSCAL. This topic was discussed among the NMHSs of Angola (Instituto Nacional de Meteorologia e Geofisica, INAMET), Botswana (Department of Meteorological Services, DMS), Zambia (Zambia Meteorological Department, ZMD) and Germany (DWD) during a SASSCAL Climate Workshop (http://www.sasscal.org) held in Namibia in April 2014. The group agreed that harmonization of concepts in the region is desirable. A discussion of available options led to the conclusion that the approach of CLIMSOFT ("CLIMatic SOFTware") was a promising option to fulfill the specific requirements for these countries, considering the enhancements that are being planned (see below).

CLIMSOFT was originally developed by an African team of 3 developers located in Zimbabwe, Kenya and Guinea (Stuber et al., 2011). The aim for the development of CLIMSOFT was to provide a free and easy-to-use CDMS for developing countries. It has an intuitive Graphical User Interface, a key-entry module, quality control procedures and allows data import from various sources, including data from automatic weather stations. It allows data export into several formats and generation of products and summaries.

The software is currently based largely on the data base management system (DBMS) Microsoft® Access and Microsoft® Visual Basic 6, but enhancements to the software, and a switch to open-source solutions are currently in preparation. However, the open-source data base system MySQL has already been widely used as a back-end to the MS® Access DBMS to meet the needs of NMHS with larger volumes of climate data and also at regional centers like the African Centre of Meteorological Applications for Development (ACMAD).

5 Capacity development and training activities

A major component of all SASSCAL activities is the development of capacities and associated training in collaborating institutions. As a capacity development effort during the buildup of the SASSCAL WeatherNet, technical staff at all involved institutions was trained on technical requirements regarding the installation and maintenance of the different types of automatic weather stations during the installation campaigns, but also on the technical aspects of data transmission, management and publication.

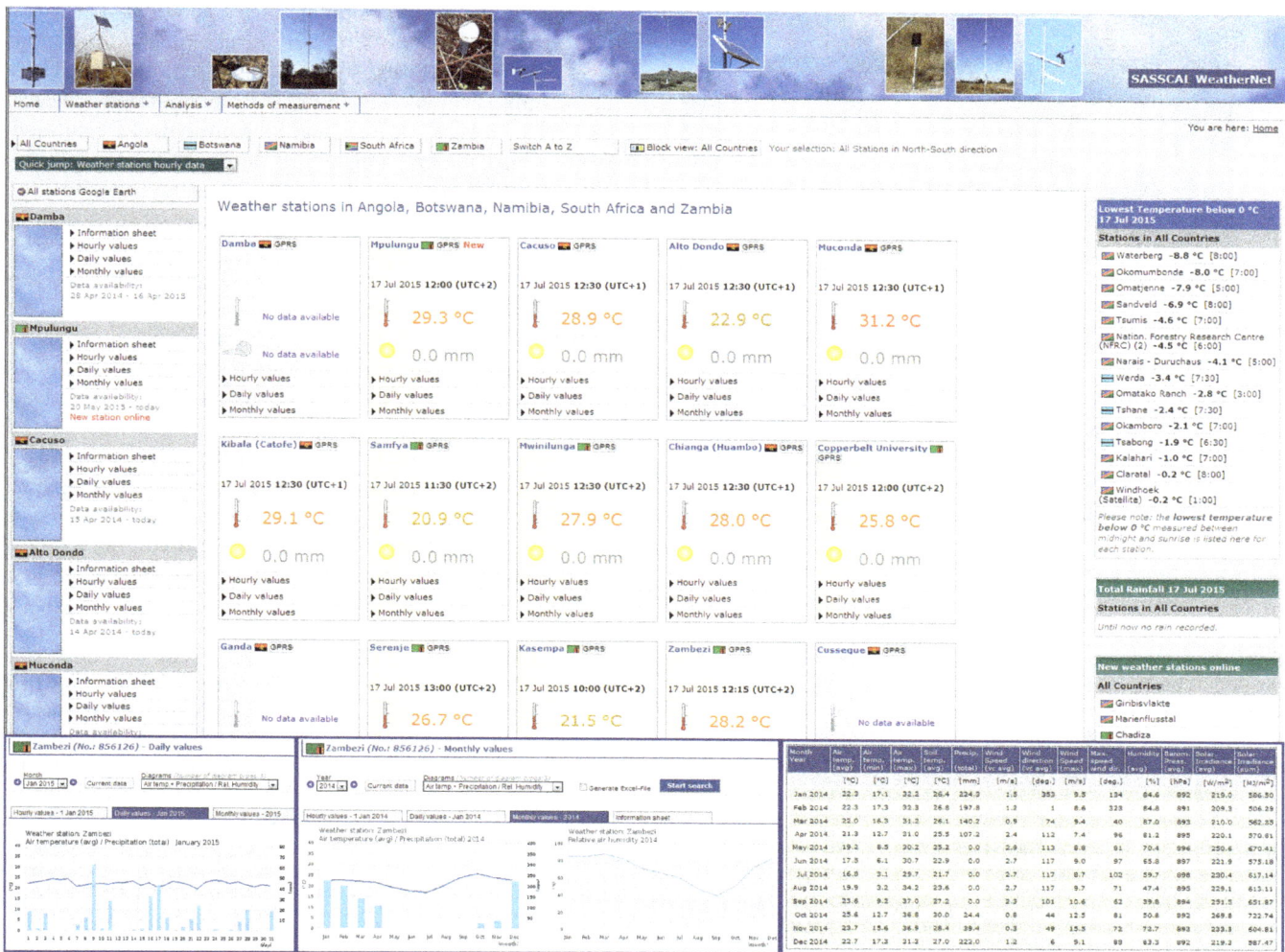

Figure 4. User interface and tools of the SASSCAL WeatherNet (www.sasscalweathernet.org). The upper figure shows the welcome page providing a general overview on actual climate conditions at each station and the actual rainfall status. The lower figures illustrate summaries of station-based daily rainfall and temperature in January 2015 (lower left), monthly summaries for 2014 (lower central) and statistics for 2014 (lower right) at an example for the Zambezi station.

One important conclusion of the SASSCAL Climate Workshop was that training activities related to climate data management are needed for the staff of the meteorological services. DWD now works together with the national meteorological services of Angola, Botswana and Zambia to train the local staff in charge of data management. One major aim of the cooperation is to ensure the proper operation and more holistic operation of CLIMSOFT in the long-term.

When conceiving the priorities of the training, the specific needs of each service are taken into account and therefore, the capacity building activities differ from one country to another. However, two common priorities can be pointed out: to improve the skills of the local staff in dealing with climate data (e.g. the importance of metadata and homogenization); and to raise awareness for the importance of documenting the processes and routines related to data management.

6 Extension of archives and data rescue

During the SASSCAL workshop it was also stated by the NMHSs that there is a need for data rescue activities in order to complete the data bases and to assure that historic data do not get lost. Data Rescue is the ongoing process of (1) preserving all data at risk of being lost due to deterioration of the medium and (2) digitizing current and past data into computer compatible form for easy access (Tan et al., 2004). Gaps in existing time series of the national meteorological services can eventually be filled with data from existing national or international archives. One example are the time series of precipitation that are available in the archive of the Global Precipitation Climatology Center (GPCC; http://gpcc.dwd.de), hosted at DWD. It collects precipitation observations from various sources and provides

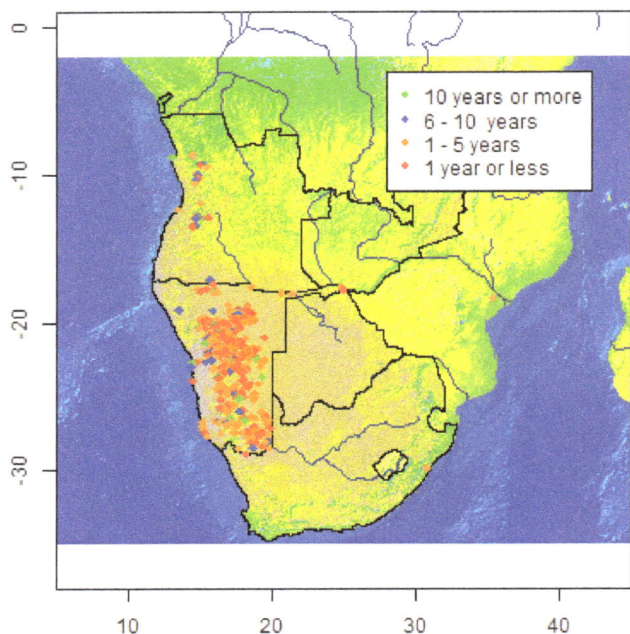

Figure 5. Location and length of the longest observation interval of overseas stations of the German Naval Observatory ("Deutsche Seewarte"; 1830–1943) that are available in the archives of Deutscher Wetterdienst as original paper documents. DWD works on the digitization of these data (Kaspar et al., 2015).

gridded global precipitation data covering the period 1901 to today (Becker et al., 2013). It receives precipitation data from various NMHSs, partly on a regular basis, but according to its mandate, the original data are not distributed to third parties. However, the assessment of national data bases of the SASS-CAL countries revealed that the original data were not available anymore in some of the national archives. First steps have now been undertaken to integrate these time series into the national archives again. DWD currently also works on the digitization of observations that are available in its archive from colonial times (Fig. 5; Kaspar et al., 2015). Time series from several stations in the regions are also available on paper at the Gobabeb Research and Training Center (Namibia).

7 Conclusions

SASSCAL, in close cooperation with the national weather authorities, supports the rehabilitation of existing station networks and the installation of new fully automated weather stations in Angola, Botswana, Namibia, South Africa and Zambia and thereby contributes to national and international climate initiatives. This is achieved by provision of consistent, reliable and up-to-date information, as well as providing capacity development and training activities at the meteorological services. With the integration of these stations into the SASSCAL WeatherNet, the measured time series are made available to a large number of stakeholders in near-real time and open access. With future installations, the observation network will be expanded. In addition, other activities like a United Nations Development Program project on early warning systems at the Zambian Meteorological Department or a new installation campaign in Botswana built on the developed skills and the technological infrastructure and align their efforts regarding the positioning of new weather stations and technological integration.

Technical support and capacity building activities are performed in order to improve the data management at the NMHSs. The NMHSs agreed on a harmonized approach for their internal climate data management systems based on a freely available Climate Data Management Software. Data sources have been identified that will allow the extension of data bases with additional historic data. In total, the ongoing activities will contribute to the strengthening of regional capacity to provide information on regional climate change and thereby providing important information for the science-based decision making for adaptation to climate change.

Acknowledgements. The authors would like to thank the German Ministry for Education and Research (BMBF) and its cooperating African ministries for funding SASSCAL. We are also grateful to the technical staff assisting the operation of the system at all involved institutions. We thank EUMETSAT for supporting the satellite-based data transmission. Two anonymous reviewer gave helpful comments on the manuscript.

References

Becker, A., Finger, P., Meyer-Christoffer, A., Rudolf, B., Schamm, K., Schneider, U., and Ziese, M.: A description of the global land-surface precipitation data products of the Global Precipitation Climatology Centre with sample applications including centennial (trend) analysis from 1901–present, Earth Syst. Sci. Data, 5, 71–99, doi:10.5194/essd-5-71-2013, 2013.

Helmschrot, J. and Jürgens, N.: Integrated SASSCAL research to assess and secure current and future water resources in Southern Africa. Hydrological Sciences and Water Security: Past, Present and Future, Proc. IAHS, 366, 168–169, doi:10.5194/piahs-366-168-2015, 2015.

IPCC: Climate Change 2014: Impacts, Adaptation, and Vulnerability. Contribution of Working Group II to the Fifth Assessment Report of the Intergovernmental Panel on Climate Change, Cambridge University Press, Cambridge, UK and New York, NY, USA, 2014.

Kaspar, F., Tinz, B., Mächel, H., and Gates, L.: Data rescue of national and international meteorological observations at Deutscher Wetterdienst, Adv. Sci. Res., 12, 57–61, doi:10.5194/asr-12-57-2015, 2015.

Schneider, U., Becker, A., Finger, P. Meyer-Christoffer, A., Rudolf, B., and Ziese, M.: GPCC Full Data Reanalysis Version 6.0 at 0.5°: Monthly Land-Surface Precipitation from Rain-Gauges built on GTS-based and Historic Data, Deutscher Wetterdienst, Global Precipitation Climatology Center, doi:10.5676/DWD_GPCC/FD_M_V6_050, 2011.

Stuber, D., Mhanda, A., and Lefebvre, C.: Climate Data Management Systems: status of implementation in developing countries, Clim. Res., 47, 13–20, doi:10.3354/cr00961, 2011.

Tan, L. S., Burto, S., Crouthamel, R., van Engelen, A., Hutchinson, R., Nicodemus, L., Peterson, T. C., and Rahimzadeh, F.: Guidelines on Climate Data Rescue, World Meteorological Organization, WMO-TD No. 1210, 2004.

World Meteorological Organization (WMO): Annex to the Implementation Plan of the Global Framework for Climate Services – Capacity Development, Geneva, Switzerland, 2014a.

World Meteorological Organization (WMO): Annex to the Implementation Plan of the Global Framework for Climate Services – Observing and Monitoring Component, Geneva, Switzerland, 2014b.

18

How seasonal forecast could help a decision maker: an example of climate service for water resource management

Christian Viel, Anne-Lise Beaulant, Jean-Michel Soubeyroux, and Jean-Pierre Céron

DCSC, Météo-France, Toulouse, France

Correspondence to: Christian Viel (christian.viel@meteo.fr)

Abstract. The FP7 project EUPORIAS was a great opportunity for the climate community to co-design with stakeholders some original and innovative climate services at seasonal time scales. In this framework, Météo-France proposed a prototype that aimed to provide to water resource managers some tailored information to better anticipate the coming season. It is based on a forecasting system, built on a refined hydrological suite, forced by a coupled seasonal forecast model. It particularly delivers probabilistic river flow prediction on river basins all over the French territory.

This paper presents the work we have done with "EPTB Seine Grands Lacs" (EPTB SGL), an institutional stakeholder in charge of the management of 4 great reservoirs on the upper Seine Basin. First, we present the co-design phase, which means the translation of classical climate outputs into several indices, relevant to influence the stakeholder's decision making process (DMP). And second, we detail the evaluation of the impact of the forecast on the DMP. This evaluation is based on an experiment realised in collaboration with the stakeholder. Concretely EPTB SGL has replayed some past decisions, in three different contexts: without any forecast, with a forecast A and with a forecast B. One of forecast A and B really contained seasonal forecast, the other only contained random forecasts taken from past climate. This placebo experiment, realised in a blind test, allowed us to calculate promising skill scores of the DMP based on seasonal forecast in comparison to a classical approach based on climatology, and to EPTG SGL current practice.

1 Introduction

Past climate information is very often used by stakeholders who are aware of their climate-sensitivity. It generally consists in using past climate events to force their impact model, to anticipate potential crises in the next months. The logic behind is a "stationary climate" hypothesis, which could be misleading considering the current rapid rhythm of climate change.

Seasonal forecasts are an obvious alternative to take into account the climate evolution. It has effectively partly replaced the former strategy in different economic sectors in some parts of the world where seasonal predictability is reasonably good: for example in Australia, in Brazil and in Africa in the agricultural sector (Hammer et al., 2001; Lemos et al., 2002) and for water resource management (Bader et al.,

2006). Over Europe, such practical applications of seasonal forecasts are very rare, mainly due to the low skill of such forecasts in this region (Doblas-Reyes, 2012).

The FP7 project EUPORIAS (http://www.euporias.eu/) aimed to fill this gap, by involving both providers and potential users of seasonal and decadal climate information. They could engage specific co-design of prototypes of climate services, and continue up to the evaluation of the real added-value of this approach, compared to their current practice. This paper presents an application developed within this project, concerning dam management at seasonal time scale. It particularly focuses on assessing the real usefulness of seasonal forecast in a specific decision making process. The problematic of dam management in France is presented in Sect. 2, illustrated by a practical example. The prototype proposed by Météo-France in the framework of EUPO-

RIAS is described in Sect. 3. Section 4.1 presents the original methodology used to assess the advantage of using seasonal forecast compared to their current practice. In Sect. 4.2, we present the very first results of this method.

2 A typical climate sensitive application: dam management

Water resource management generally requires long range forecasts, especially when it consists in optimizing a dam's water stock. This kind of programming is exactly one of the role of "Etablissement Public Territorial de Bassin Seine – Grands Lacs" (EPTB SGL) in France.

In accordance with EPTB SGL, we have chosen to deal with the Marne basin and its dam-reservoir (Fig. 1). Built in 1974, it could stock $350 \, \text{Mm}^3$ of water. The corresponding lake is the greatest artificial lake in France. It drains a $2900 \, \text{km}^2$ watershed. In this paper, we will focus on the summer management, especially the early programming of water release to sustain downstream river flow during summer and early autumn. The main stakes are to guarantee the provision of drinkable water downstream (especially in the Paris urban area), to allow enough river flow for navigation, for crops, for energy production and to preserve the ecological balance. The reservoir has to be emptied in autumn to ensure its winter mission of flood protection of the downstream urban areas.

Each year in May, EPTB SGL needs to program the dam water release for the entire dry period, i.e. up to November. This decision concretely consists in drawing a water release curve that would be approved in a decision committee and eventually applied by the dam operators. Of course this projection could be adjusted if needed, but the more this program is known in advance by the actors of the basin, the best it is.

To achieve this, EPTB SGL has defined a method based on the knowledge of the initial conditions (actual filling of the dam, previous month's upstream river flow), simple tools to extrapolate river flow evolution based on past years, and a great dose of expertise of its team that has now a long experience in such an exercise.

Their expectation is very clear: to dispose of a service that could strengthen their current practice.

3 The Météo-France EUPORIAS prototype "RIFF"

Our prototype is based on a seasonal hydrological forecasting suite. In particular, it is able to deliver river flow forecasts on more than 880 river gauging stations in France.

The hydrological suite is composed by a SVAT (soil vegetation atmosphere transfer model) called ISBA, coupled to a water routing model called MODCOU (Habets et al., 2008). This suit, forced by a daily atmospheric analysis, has been used operationally at Météo-France since 2004 for real time climate monitoring of soil wetness indices or snow cover for

example. Moreover, a reanalysis from 1958 has been calculated, allowing daily outputs to be compared to climate references.

These models have also been used in a forecast mode, forced by ECMWF medium range forecasts (Rousset-Regimbeau, 2007). An objective evaluation (through classical scores) has been performed for high and low flows, it has concluded that the system is efficient for early flood warning, especially for large river basins such as the Seine.

For the EUPORIAS prototype, we have forced ISBA-MODCOU by seasonal forecasts coming from ARPEGE-System 3 model (see Fig 2), to deliver river flow ensemble seasonal forecasts. This hydrological suite had been run over the long hindcast period of ARPEGE-System 3 (1979–2007), composed of 11 members and delivering 7 months' forecasts. Objective river flow scores (Singla et al., 2012) show that this suit is able to provide significant skilful forecasts up to several months, especially in river basins influenced by snow melting or groundwater.

The construction of the RIFF prototype (river flow forecast over France, http://riff.predictia.es/en) is the result of a close collaboration with EPTB SGL. A specific work of codesigning was necessary to define a set of products, compromise between user's need and limited abilities of such a long range forecast. Indeed, seasonal forecast skill over Europe is known to be low (Kim et al., 2012), this led to recommend time-integrated products.

Figure 3 shows two examples of these tailored probabilistic products, presenting what would be the situation without any river flow sustain from the dam. The left one shows the forecast of monthly river flow downstream from the dam, compared to climatology. The right one is a time-integrated product that represents the forecast of the number of days below a specific river flow threshold, called "vigilance". This threshold is specified by EPTB SGL, it corresponds to a level below which use restrictions could be taken by authorities.

These products were assessed by classical scores (Brier scores, ROC scores, correlations) firstly to evaluate their intrinsic value and secondly to provide some recommendation of using. For example, for the May initial condition forecast, it was specified to the stakeholder that the mean monthly skill downstream from the dam does not extent more than August.

4 Evaluation of the effective impact on the DMP

4.1 Method

The assessment of the usefulness of the climate information was a major ambition of EUPORIAS. To achieve this goal, Météo-France has proposed a specific protocol, so called Placebo Protocol, as a reference to the methodology used to test new medication.

The principle of this protocol is to provide two sets of impact forecasts, one issued with "real" seasonal forecasts (i.e. N scenarios coming from the hydrological suite forced by a

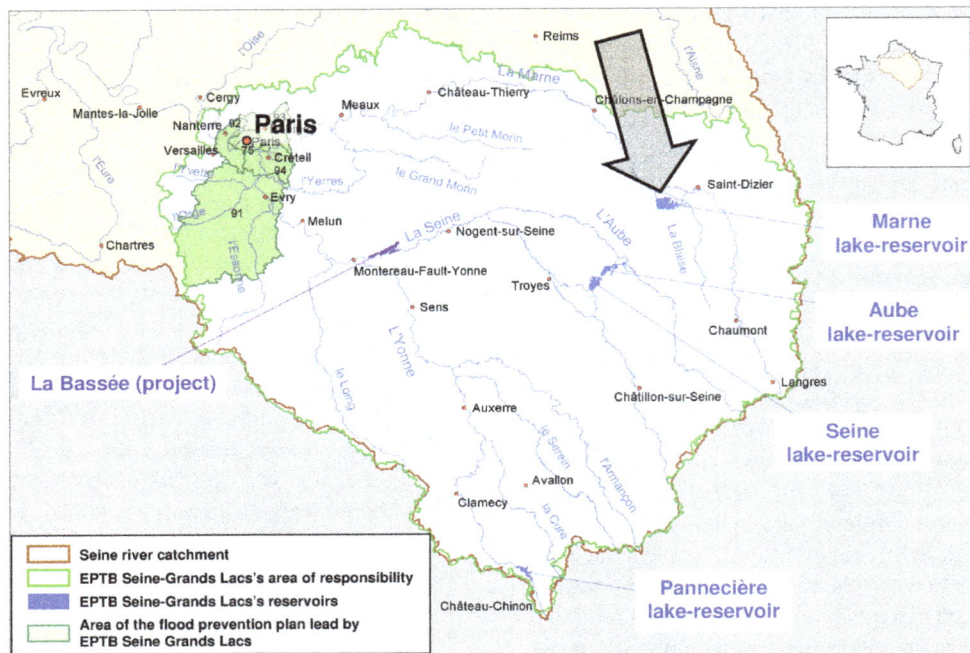

Figure 1. Area managed by EPTB Seine-Grands Lacs and location of the Marne lake-reservoir (grey arrow). Source: EPTB Seine-Grands Lacs.

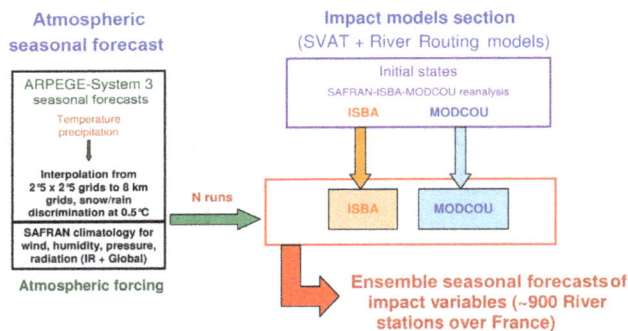

Figure 2. The seasonal hydrological forecast suite used to produce riverflow forecasts.

For a specific situation, the stakeholder has to make its decision with each set of forecast. Precisely, for each situation, in addition to seasonal forecast, it had access to the filling of the dam (for certain years the filling could be incomplete) and to the last months upstream river flow (to have an insight into the recent watershed condition). Its decision consists in drawing the most appropriate release curve considering the context and the forecast.

The quality of each decision is evaluated using a specific metric, defined in agreement with EPTB SGL. It is based on the number of days below its "vigilance" threshold. It means that for a specific decision, we have calculated the downstream river flow as the sum of "natural" flow plus the flow coming from the water release curve. We could then calculate the resulting number of days below the "vigilance" threshold.

Redoing this on a set of situation the added-value of seasonal forecasts compared to climatological forecasts could be assessed.

4.2   First results

This protocol was applied by EPTB SGL for its May programming, for 29 situations corresponding to the 1979–2007 hindcast (also called "re-forecast") period of the seasonal forecast model.

To complement this experiment, EPTB SGL has accepted to make decisions without any forecast information (neither seasonal forecast, nor Placebo), which corresponds more or less to its current practice.

climate forecast model composed of N members); the other one called "climatological forecasts" is issued from the same hydrological suite but forced by random atmospheric forcing (i.e. a selection of N years coming from the atmospheric re-analysis). These two sets are to be used in a blind test, over several past situations, to make decisions.

The main advantage of this protocol is that it ensures a perfect objectivity when comparing the decisions made with the Placebo and with Seasonal Forecasts. But of course, it could not take into account all the context in which the decision is made (specific environmental context, or pressure from the decision committee for instance).

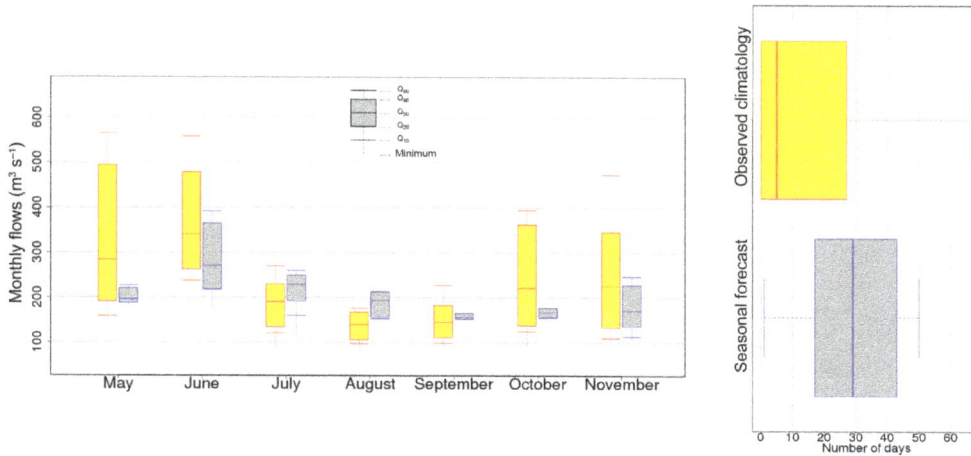

Figure 3. Examples of tailored products. On the left, monthly mean river flow forecast (grey boxplot) compared to climatology (yellow boxplot) for the next seven months. On the right, forecast of the number of days of low flow (grey boxplot) compared to climatology (yellow boxplot), integrated over the whole dry season.

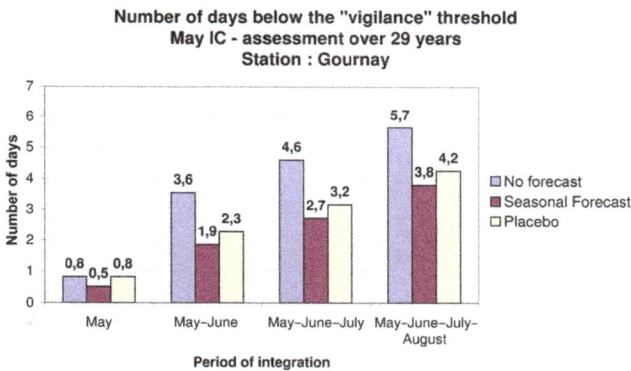

Figure 4. Evaluation of the quality of the May initial conditions forecasts. The metric is applied for four periods of integration, from 1 month (May) to 4 months (May to August). The best decisions would lead to "zero day below the threshold"

In Fig. 4, we can see that for this station, the added-value of seasonal forecast compared to Placebo is anything but obvious: looking carefully at the different situations, it appears that the little advantage of seasonal forecast is very fragile. We could not reasonably conclude on this only result, based on one station.

These good results of Placebo, which may at first sight seem surprising, could be explained by the presence of aquifers, main source of predictability of this basin. Indeed, during summer, the percentage of groundwater contribution to river flow reaches 80 % (Singla, 2012) and Placebo benefits from MODCOU's good representation of the Seine aquifer, as much as seasonal forecasts.

However, the "no forecast" experiment obtains the worse scores than Seasonal Forecast and Placebo. It means that our hydrological system brings relevant information to the stakeholder, comparing to its current practice. Moreover, the difference between "no forecast" in the one hand, and Placebo and Seasonal Forecast in the other hand, is smaller for May only than for longer integrations: this reinforces stakeholder's interest in using our climate information.

5 Conclusions and perspectives

This work in the framework of EUPORIAS presents a promising example of application of seasonal forecast over Europe. Our interaction with EPTB SGL, and other similar French institutions, shows the great expectation they have in seasonal forecast. It also shows the gap between the classical way of presenting seasonal forecasts and the tailoring needs, to be able to connect this climate information to their tools and decision making processes.

Once the interaction between climate experts and potential users has been set up, a crucial phase is to assess its real usefulness. "Low skill" doesn't mean "no skill", and the little added-value (from the climate expert point of view) could sometimes correspond to a significant interest (from the stakeholder point of view). The Placebo protocol we have proposed and experimented in this project intends to objectively measure the impact of climate information on decision made. It implies a strong involvement from the stakeholder, who has to remake past decisions in a kind of theoretical framework. But at the end it allows to have a measure of the added-value (or loss) of seasonal forecast, with a metric adapted to the stakeholder. In the example presented in this paper, we have shown that our forecast system was able to improve dam management at seasonal timescale, compared to current practice.

Our technical perspectives are to consolidate our experimental system by experimenting it in a real-time mode, to update model versions (seasonal forecasts and hydrological suite) and to develop a delivery web platform.

Concerning applications, we have already enlarged our panel of potential users to be able to test other products and other potentialities of our suite: such as soil wetness or snow cover forecast.

Acknowledgements. We acknowledge Stéphane Dermerliac and Claudine Jost from EPTB Seine-Grands Lacs for their strong involvement in this experiment. This work was supported by the EUPORIAS project, funded by the European Commission 7th Framework Programme for Research, grant agreement 308291.

References

Bader, J. C., Piedelièvre, J.-P., and Lamagat, J.-P.: Prévision saisonnière du volume de crue du fleuve Sénégal: utilisation des résultats du modèle Arpège Climat, Hydrolog. Sci. J., 51, 406–417, 2006.

Doblas-Reyes, F. J.: Seasonal prediction over Europe, in: Seminar on Predictability in the European and Atlantic regions from days to years, 6–9 September 2010, 171–185, 2012.

Habets, F., Boone, A., Champeaux, J. L., Etchevers, P., Franchistéguy, L., Leblois, E., Ledoux, E., Le Moigne, P., Martin, E., Morel, S., Noilhan, J., Quintana Segui, P., Rousset-Regimbeau, F., and Viennot, P.: The SAFRAN-ISBA-MODCOU hydrometeorological model applied over France, J. Geophys. Res., 113, D06113, doi:10.1029/2007JD008548, 2008.

Hammer, G. L., Hansen, J. W., Phillips, J. G., Mjelde, J. W., Hill, H., Love, A., and Potgieter, A.: Advances in application of climate prediction in agriculture, Agr. Syst., 70, 515–553. 2001.

Kim, H.-M., Webster, P., and Curry, J.: Seasonal prediction skill of ECMWF System 4 and NCEP CFSv2 retrospective forecast for the Northern Hemisphere winter, Clim. Dynam., 39, 2957–2973, doi:10.1007/s00382-012-1364-6, 2012.

Lemos, M. C., Finan, T. J., Fox, R. W., Nelson, D. R., and Tucker, J.: The use of seasonal climate forecasting in policymaking: lessons from Northeast Brazil, Climatic Change, 55 , 479–507, 2002.

Rousset-Regimbeau, F.: Modélisation des bilans de surface et des débits sur la France, application à la prévision d'ensemble des débits, PhD thesis, Université Paul Sabatier, Toulouse, France, 2007.

Singla, S.: Prévisibilité des ressources en eau à l'échelle saisonnière en France, PhD thesis, Institut National Polytechnique de Toulouse, Toulouse, 2012.

Singla, S., Céron, J.-P., Martin, E., Regimbeau, F., Déqué, M., Habets, F., and Vidal, J.-P.: Predictability of soil moisture and river flows over France for the spring season, Hydrol. Earth Syst. Sci., 16, 201-216, doi:10.5194/hess-16-201-2012, 2012.

19

Variability of atmospheric circulation patterns associated with large volume changes of the Baltic Sea

A. Lehmann[1] **and P. Post**[2]

[1]GEOMAR Helmholtz Centre for Ocean Research, Kiel, Germany
[2]Institute of Physics, University of Tartu, Tartu, Estonia

Correspondence to: A. Lehmann (alehmann@geomar.de)

Abstract. Salinity and stratification in the deep basins of the Baltic Sea are linked to the occurrence of Major Baltic Inflows (MBIs) of higher saline water of North Sea origin, which occur sporadically and transport higher saline and oxygenated water to deeper layers. Since the mid-1970s, the frequency and intensity of MBIs have decreased. They were completely absent between February 1983 and January 1993. However, in spite of the decreasing frequency of MBIs, there was no obvious decrease of larger Baltic Sea volume changes (LVCs). A LVC is defined by a total volume change of at least $100\,km^3$. LVCs can be identified from the sea level changes at Landsort which is known to represent the mean sea level of the Baltic Sea very well. Strong inflows leading to LVCs are associated to a special sequence of atmospheric circulation patterns. Our analysis based on Jenkinson-Collison circulation (JCC) types confirms that most effective inflows occur if about a month before the main inflow period, eastern air flow with anticyclonic vorticity over the western Baltic prevails. These conditions reduce the mean sea level of the Baltic Sea and lead to an increased saline stratification in the Belt Sea area. An immediate period of strong to very strong westerly winds trigger the inflow and force LVCs/MBIs. The lack of MBIs coincide with a negative trend of eastern types and a parallel increase of western type JCCs.

1 Introduction

The salt budget of the Baltic Sea (Fig. 1) is determined by a balance between saline inflow from the Kattegat and brackish water outflow from the Baltic Sea through the Danish Straits. River runoff and precipitation cause dilution while evaporation acts in the opposite direction. Ice formation and melting act as evaporation and precipitation, respectively, but have no influence on an annual timescale. Generally, during dry periods the mean salinity of the Baltic Sea increases while during wet periods a decrease will happen. These long-term changes are overlaid by the atmospheric-driven water exchange between the North Sea and the Baltic Sea. Moderate saline waters flow in permanently, but highly saline inflows that are able to replace the stagnant bottom waters, happen relatively rarely.

The salinity and stratification in the deep basins are linked to the occurrence of major Baltic inflows (MBIs) which occur sporadically and transport highly saline and oxygenated water of North Sea origin to deeper layers of the Baltic Sea. These major inflows are often followed by stagnation periods with no strong saline inflows, during which the permanent halocline weakens, even disappears in some basins, and extended areas of oxygen deficiency develop in those regions where the salinity stratification remains (Leppäranta and Myrberg, 2009, Väli et al., 2013). Since the early eighties the frequency of highly saline inflow events has dropped drastically from 5 to 7 events per decade to only one inflow per decade. Major inflow events occurred in January 1993 and 2003. After more than 10 years without major Baltic inflows, in December 2014 a strong MBI brought large amounts of saline and well oxygenated water into the Baltic Sea (Mohrholz et al., 2015).

In spite of the decreasing frequency of MBIs and increasing periods of stagnation, large volume changes (LVCs) of the Baltic Sea still took place which were reflected in the mean sea level of the Baltic Sea (Franck and Matthäus, 1992; Lehmann et al., 2002). A LVC is typical to occur in the Baltic

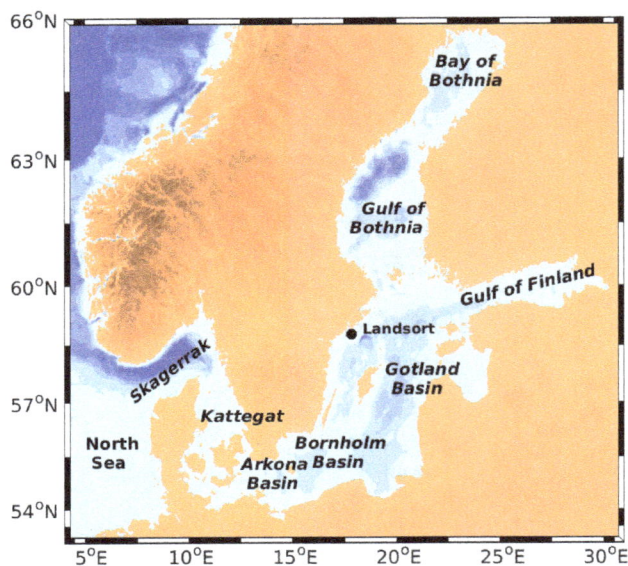

Figure 1. Map of the Baltic Sea and marked position of the Landsort tide gauge.

Sea, being connected to favorable atmospheric forcing, but these do not always lead to MBIs even if the incoming water volumes might be large. The difference between MBI and LVC is that during a LVC only relative low-saline water can be re-imported to the Baltic Sea. A MBI is always related to an effective salt transport which is able to substitute the bottom water in the deep basins of the Baltic Sea. Thus, MBIs can be considered as subset of LVCs transporting additionally to the large water volume a huge amount of salt into the Baltic Sea. Large volume changes have also been observed in previous studies (Jacobsen, 1981; Lass and Schwabe, 1990). The associated volume change is typically about $100\,\text{km}^3$ and the duration is about 40 days (Lehmann et al., 2002).

Atmospheric conditions leading to MBIs have comprehensively been studied (e.g. Schinke and Matthäus, 1998; Lass and Matthäus, 1996; Matthäus and Schinke, 1994). Strong inflows are associated with certain sequences of atmospheric flow patterns over the larger North Atlantic/North European region. The atmospheric forcing of major Baltic inflows has two phases: at first, high atmospheric pressure over the Baltic region with easterly winds followed by several weeks of strong westerly winds over the North Atlantic and Europe. The intensity of single events depends on how well both phases are developed and how closely they are related in time. Previous studies concentrated mostly on MBI periods, describing the average situation during and before the events, but also trying to find some earlier predictors of the events.

Recently, Schimanke et al. (2014) presented an algorithm based on mean sea level pressure (MSLP) fluctuations to identify major Baltic inflow events. The algorithm which used daily SLP-fields as only parameter identified the majority of major inflow events between 1961 and 2010. Fur-

thermore, they found more favorable atmospheric conditions for inflow events than have been identified by observations. They argued that in spite of favorable atmospheric conditions other factors such as runoff could prevent MBIs. It is reasonable to assume that the atmospheric forcing of MBIs and LVCs should be similar. As LVCs and MBIs are very variable in length and intensity we tried to find common characteristics and to somehow generalize the pressure fields and their sequences.

One possible method to aggregate the main variability of atmospheric pressure patterns into a reasonable small set of patterns is using classifications of atmospheric circulation types (Huth et al., 2008). Circulation type classification (CTC) is widely used in order to describe the variability of pressure patterns in a certain domain. During this process exact information about the individual configuration of the pressure field is lost, replaced by a single number per pressure field and by a sequence of circulation type numbers for periods of time. Regardless of losing detailed information, suitably chosen classification methods reduce also noise which might be seen as advantage. The gain is the simplicity of the result, which makes it attractive for a wide range of applications dealing with the linkage between large-scale circulation and weather-related surface conditions. If any relationship between the occurrence of a distinct circulation type and an associated surface condition is detected, this technique easily offers insights into physical reasons for this relationship. In recent years, there has been a real avalanche of papers with applications from many fields beginning from searching trends in temporal variability of atmospheric circulation to climate models output downscaling.

The idea of this paper is to separate the atmospheric conditions necessary to force large volume changes of the Baltic Sea. We studied the atmospheric circulation patterns forcing LVCs by Lamb automated weather types (Jenkinson and Collison, 1977) or synoptic weather types that are easily interpretable. With the knowledge of sequences of main patterns which cause large inflows to the Baltic Sea, it is possible to analyze the variability of these patterns in time to reveal possible changes in the atmospheric circulation which potentially could explain the lack of large inflows in recent decades.

2   Materials and methods

2.1   Sea level data at Landsort

Hourly sea level data at the tide gauge station Landsort in Sweden (Fig. 1) have been downloaded from the SMHI Öppna data bank system (http://opendata-download-ocobs.smhi.se/explore/) for the time period 1887–2013. Sea level data have been detrended from effects of land uplift and climate change. Furthermore, to reduce local effects of sea level changes daily averages have been calculated. Landsort sea level data are known to describe mean sea level changes of

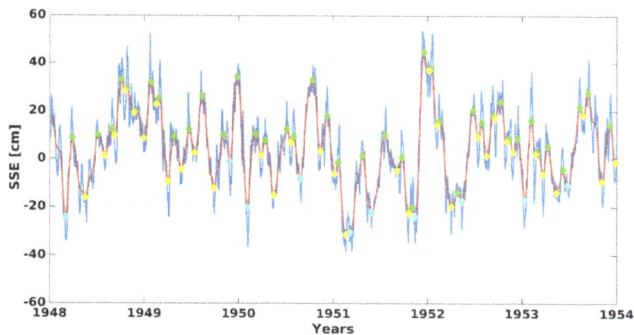

Figure 2. Sea surface elevation daily averages (blue) and filtered curve (red) at Landsort tide gauge for the period 1948–1954, minima (yellow) and maxima (green) as well as detected LVCs (cyan) based on the threshold of 29 cm.

the entire Baltic Sea very well (Franck and Matthäus, 1992). The up- and down movements can be related to corresponding volume transports through the entrance area of the Baltic Sea (Lehmann et al., 2002). The mean sea level of the Baltic Sea is a function of in- and outflow through the Danish Straits and the net fresh water flux. Due to the limited transport capacity of the Danish Straits, sea level changes in the Kattegat result in a delayed response of the mean sea level. The entrance area of the Baltic Sea acts as a low-pass filter for signals entering from the Kattegat, short-term variations are effectively filtered out (e.g. Andersson, 2002).

The main interest of our study are sea level changes occurring on weekly to monthly timescales, thus being evident in the mean sea level variation. So we used the method proposed by Pasanen et al. (2013) to smooth the sea level time series and filter out high frequency fluctuations. Pasanen et al. (2013) proposed a method for extracting time series features in different scales. It produces a multi-resolution analysis of the time series as the sum of scale-dependent components. These components were obtained from differences of smooth. The smoothing levels were determined using derivatives of smooths of the original time series. Details of the multi-resolution method for extracting time series features may be found in Pasanen et al. (2013).

From the smoothed curve local minima and maxima of the sea level have been determined. Furthermore, from the difference between minima and maxima we detected larger inflows resulting in large volume changes. A LVC is defined by the sea level difference of at least 29 cm corresponding to about 100 km^3 of volume change. Figure 2 shows the original and smoothed time-series of sea surface elevation (SSE) for the period 1948–1954 which includes the strongest MBI so far in December 1951. This inflow resulted in a total volume change of about 240 km^3 (Fig. 3). For the threshold of 29 cm we determined all LVCs from the smoothed times series of sea level/volume changes at Landsort occurring over the period from 1948–2010 (Fig. 3). All in all 74 LVCs have been detected. Nearly all MBIs coincide with LVCs, differences

are due to the applied smoothing of the sea level time-series. This gives confidence that the applied method is suitable to detect LVCs.

On average LVCs were associated with about 40 cm of sea level difference, 141 km^3 of volume change (inflow) over a period of about 53 days. It should be noted that these events are very different from the point of view of inflow duration or time between maximum and minimum SSE values, ranging from 31 to 116 days. The main season of LVCs is from October to December (OND, 42 %), followed by January–March (JFM, 22 %), April–June (AMJ, 20 %) and July to September (JAS, 15 %).

2.2 Atmospheric circulation classification

We classified air pressure patterns in order to describe the variability of atmospheric circulation in the Baltic Sea region. The classification of circulation weather types developed by Jenkinson and Collison (1977) as an automatic version of Lamb's classification (Lamb, 1972) was selected because of the interpretation simplicity. The latter offers easy comparison of circulation types frequencies over different domains. It is widely used elsewhere in Europe (e.g. Trigo and DaCamara, 2000; Post et al., 2002; Linderson, 2001; Jones et al., 2013).

MSLP values from 16 points in and around the study area (Fig. 4) are used to calculate 6 different flow indices, which quantify the geostrophic airflow and vorticity (see Post et al., 2002 for details). Jenkinson Collison' circulation types (JCT) and classes (JCC) are defined by comparing the numeric values of the indices. There are altogether 26 circulation types that could be merged to 10 classes. If a straight airflow dominates over the vorticity then, depending on the direction of the flow, 8 directional types are specified (W, NW, N, NE, E, SE, S, SW). If the vorticity Z is at least two times larger than the geostrophic flow F, then the cyclonic (C) or anticyclonic (AC) types are assigned. There are also 16 so-called hybrid types: 8 cyclonic and 8 anticyclonic ones where the geostrophic flow and positive or negative vorticity play a nearly equal role in atmospheric regional circulation. In case of circulation classes the hybrid types are added to the directional ones, resulting in 10 main circulation classes.

Mean sea level pressure fields for 10 JCCs centered at the Danish Straits (55° N, 10° E) are represented in Fig. 4. The center over the Danish Straits was selected in relation to previous studies (Matthäus and Schinke, 1994; Lass and Matthäus, 1996).

JCTs and JCCs were calculated for all days during the period of 1948–2013 using gridded MSLP data of NCEP/NCAR reanalysis (Kalnay et al., 1996) at 12:00 UTC. The software package cost733class (Philipp et al., 2014) was used as it offers a flexible way to calculate several circulation classifications with variable number of types over arbitrarily selected areas.

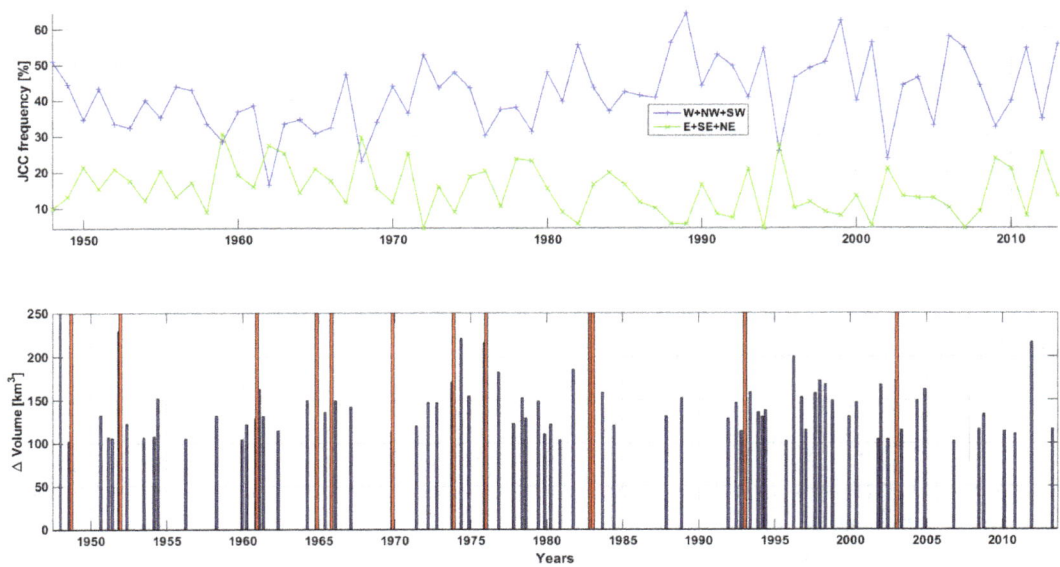

Figure 3. Upper panel: relative frequencies (%) of selected JCCs: W+NW+SW (blue line) and E+SE+NE (green line) for the half-year period October to March (ONDJFM), lower panel: detected LVCs (blue bars) based on smoothed sea surface elevation at Landsort for the period 1948–2013. Red bars mark MBIs (Matthäus et al., 2008).

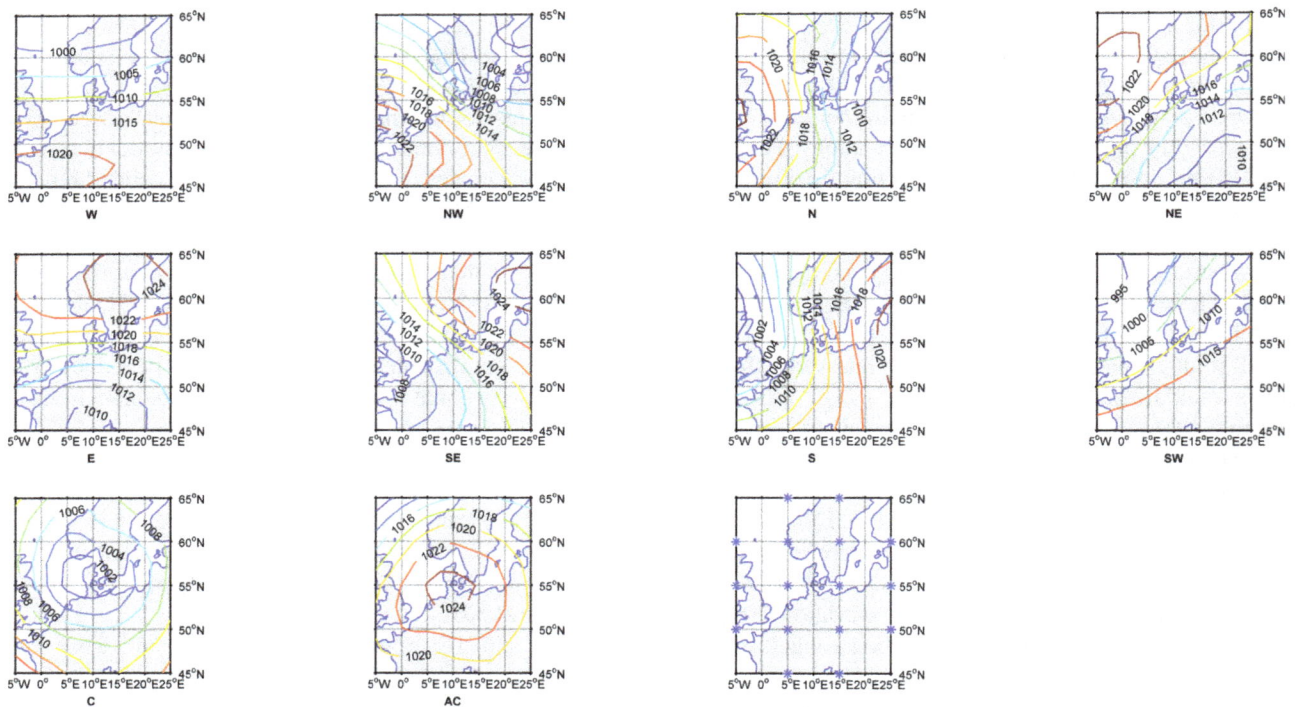

Figure 4. MSLP fields (in hPa) for 10 Jenkinson–Collison circulation classes (JCC) centered over the middle of the Danish Straits at 10° E, 55° N. In the right lower panel the geographical positions of 16 data points which are used for the JCC-calculations is shown.

3 Results

3.1 Atmospheric circulation patterns prevailing during LVC periods

Mean annual and seasonal frequencies of circulation classes were calculated to compare the special circulation during the 74 LVC periods with the mean atmospheric circulation in this domain. For each LVC, periods of 121-days were under closer inspection. Based on Landsort SSE data it is reasonable to center the periods either on the day of the minimum or the maximum sea level, or the day in the middle between them. We call this reference time 0-day. At first we tested the sensitivity of JCC sequences with respect to the 0-day selection. Sequences of circulation classes for 74 large volume changes of the Baltic Sea in the period 1948–2013 with 0-days at (a) minimum and (b) maximum Landsort SSE are presented in Fig. 5. The images appear to be somewhat noisy because of the high variability of the circulation classes during the LVC periods. There is obviously not just one sequence of patterns that results in large volume changes. The first structure, that could be noticed is a greenish yellow stripe before the 0-day (Fig. 5a), that demonstrates dominating easterly and southeasterly atmospheric flow (class) before the minimum of SSE at Landsort. A similar noisy pattern can also be found in Fig. 5b associated with dominating westerly and north-westerly atmospheric flow (class) before the maximum of SSE at Landsort. The patterns in Fig. 5a are shifted to the left with respect to the chosen reference but the structure of the patterns are very similar. The high variability in the length of LVC periods makes the synthesizing of the atmospheric circulation difficult. Thus, we have calculated frequency histograms for 30-day sub-periods, for which the circulation patterns are more homogeneous. These are presented in Fig. 6 for three 30-day sub-periods taking the 0-day at SSE maximum. The first sub-period corresponds to the pre-inflow period (60 to 31 days before the 0-day), the second sub-period corresponds to the inflow period (30-days until the 0-day), and the third sub-period, after the inflow, is chosen as a control period (1–30 days after the 0-day). The relative frequencies of JCCs in dependency of varying LVC thresholds are also displayed in Fig. 6. The threshold has been varied in the range of 24 to 44 cm. The significance of the difference in frequency distributions between the 1948–2013 average JCCs (bright green columns in Fig. 6) and the corresponding sub-period is checked by the Kolmogorov-Smirnov test. The JCCs histogram for the third sub-period never differed from the long-term mean. For the whole 121-day period and for the second sub-period the histogram differed significantly from the long period mean in case of all used thresholds. The first sub-period was significantly different from the mean for 24–38 cm LVC thresholds, but failed the test in case of higher SSE values. In case of larger LVCs, the duration of the inflow is generally longer (more than 30 days) and in case of higher SSEs the separation of the pre-

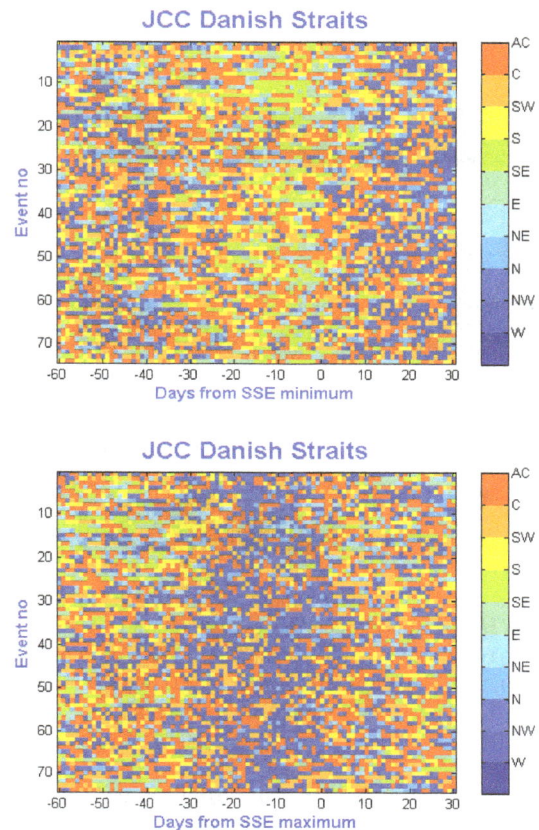

Figure 5. Sequences of Jenkinson–Collison circulation classes for 91-days periods during 74 LVC events. Upper panel: the 0-day is centered at the minimum sea surface elevation at Landsort, lower panel: the 0-day is centered at the maximum sea surface elevation at Landsort.

inflow period and the main inflow period is not always possible (therefore, the percentage of W, NW and SW grows at higher thresholds, Fig. 6a).

For the chosen 100 km^3 (29 cm) threshold, during the precursory period (30-days before the minimum in Landsort SSE) the largest increase compared to long-term mean frequency show SE and E types of JCCs. During the inflow period (30-days before the maximum in Landsort SSE) the dominant JCCs are W and NW. From Fig. 6, it could be deduced that the frequency of these specific circulation classes grows if the threshold is increased.

To check if it is possible to associate the temporal changes in the number of LVCs and MBIs with the changes in atmospheric circulation, annual and seasonal occurrences of separate and combined JCCs were calculated for every year. As season, we defined the main period when most of the MBIs occurred (October to March, ONDJFM). The Mann–Kendal nonparametric test (Kendall, 1975) was applied to check the significance ($p = 0.05$) of the trends in JCCs frequencies. For two classes (W, SW) positive trends and also for two

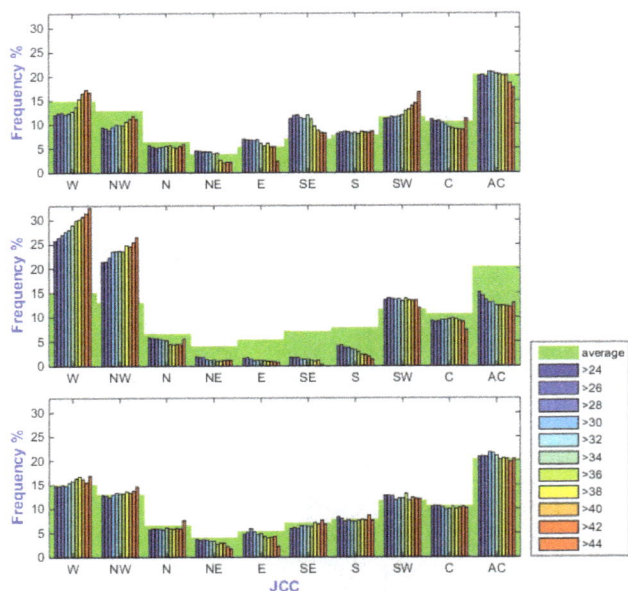

Figure 6. Histograms of occurrences of JCCs for LVC periods: (upper panel) 60–31 days before (upper panel), 30–0 days before (middle panel), and 1–30 days after (lower panel) the SSE maximum at Landsort. The threshold of LVCs varies between 24 to 44 cm.

classes negative trends (NE, E) during the 66 seasons were found. Trends determined for time series of seasonal occurrences of NE+E+SE and SW+W+NW were also significant (see Fig. 3). The occurrence of these easterly types has decreased since 1970s about 20 days, while for the westerly types the frequency has grown about twice. Slopes of trends were calculated by the method of Sen (1968).

4   Conclusions and discussion

A new term "Large Volume Changes (LVCs)" has been introduced to indicate changes in the Baltic Sea volume independent of the salinity content of the inflowing water mass. The idea is based on the assumption that the atmospheric forcing that causes the inflows to the Baltic Sea does not depend on salinity. Large volume changes have been calculated for the period 1948–2013 filtering Landsort sea surface elevation anomalies daily time series. The cases with local minimum and maximum difference of at least 29 cm which correspond to $100\,\text{km}^3$ of Baltic Sea volume change have been chosen for a closer study to reveal atmospheric circulation patterns that force large changes in volume. The total number of such cases is 74.

For the first time, atmospheric circulation classification has been used to characterize the atmospheric forcing before, during and after the LVC events. From earlier studies, it could be deduced that these events are driven by synoptic scale atmospheric forcing, which could easily be represented and interpreted by Lamb automated weather types (Jenkin-

son and Collison, 1977). During different phases of the inflow, the abundance of certain Jenkinson Collison' classes grow or decrease compared to the average frequency of classes. The frequency distributions of Jenkinson Collison' classes 60–30 days before the maximum SSE is significantly different from the average and the same is valid for the period of 30–0 days before the maximum, but not for the 30 days period after the maximum. The first period is characterized by a higher number of E and SE classes, which confirms the idea of Matthäus and Schinke (1994) about the pre-inflow period for which easterly winds prevail which increase the Baltic Sea water outflow, lower the mean sea level and hinder the inflow of North Sea water through the Danish Straits. The associated anticyclonic circulation is related to dry periods with less precipitation and increased saline stratification in the Belt Sea area. An immediate period (next 30 days) of strong to very strong westerly winds trigger the inflow and force effective LVCs/MBIs.

If the threshold for a LVC is raised, the amount of certain aforementioned types increases during both 30 days periods. It is difficult to attribute the lack of MBIs to the increase or decrease of certain JCCs, but the trends in time series of NE+E+SE and SW+W+NW show that the reason can be the negative trend in eastern types after 1980s, as in the western types for which a significant increase occurred. Thus, one possible reason might be the increased atmospheric zonal circulation linked with intensified precipitation in the Baltic region and increased river runoff to the Baltic Sea (Schinke and Matthäus, 1998; Lehmann et al., 2002, 2011). An alternative explanation has been recently proposed by Soomere et al. (2015). They found a significant abrupt change in the meridional component of the airflow direction over the southern Baltic Sea around 1987. This change was established by a substantial increase in northwestern wind events at the expense of other wind directions (e.g. western winds that are critical for MBIs to occur).

We showed that easily interpretable circulation types in synoptic scale offer a complex tool to study the atmospheric forcing of large volume changes to the Baltic Sea. But such a simple circulation classification as JCC with only 10 classes does not reveal very special circulation cases, that could be connected to MBIs. An inverted approach could be used, at first classifying only pressure fields occurring during such events, and then in a second step determining their individual frequency. The main idea is that events like these depend on the sequence or accumulation of forcing which is not easy to follow by an Euler type approach. Therefore the next steps will be the tracking of cyclone pathways and classifying of pressure field sequences which will help to overcome this kind of problems.

Acknowledgements. Circulation weather types have been calculated using software cost733class that has been worked out during COST action 733 at Augsburg University (http://cost733.geo.uni-augsburg.de/cost733class-1.2).

The study was supported by the Estonian Ministry of Education and Research (IUT20-11 and ETF9134) and by the EU Regional Development Foundation, Environmental Conservation and Environmental Technology R & D Program Project no. 3.2.0801.12-0044.

References

Andersson, H. C.: Influence of long-term and large-scale atmospheric circulation on the Baltic Sea level, Tellus, 54A, 76–88, 2002.

Franck, H. and Matthäus, W.: Sea level conditions associated with major Baltic inflows, Beitr. Meereskunde, 63, 65–90, 1992.

Huth, R., Beck, C., Philipp, A., Demuzere, M., Ustrnul, Z., Cahynova, M., Kyseli, J., and Tveito, O. E.: Classification of atmospheric circulation patterns, Ann. N. Y. Acad. Sci., 1146, 105–152, doi:10.1196/annals.1446.019, 2008.

Jacobsen, T. S.: The physical oceanography of the open Danish waters, in: The Belt Project evaluation of the physical, chemical and biological measurements, edited by: Aetebjerg Nielsen, G., National Agency of Environmental Protection, Denmark, 122 pp., 1981.

Jenkinson, A. F. and Collison, F. P.: An initial climatology of gales over the North Sea, Synoptic Climatology Branch Memorandum, No. 62, Meteorological Office, Bracknell, 18 pp., 1977.

Jones, P. D., Harphan, C., and Briffa, K.: Lamb weather types derived from reanalysis products, Int. J. Climatology, 33, 1129–1139, doi:10.1002/joc.3498, 2013.

Kalnay, E., Kanamitsu, M., Kistler, R., Collins, W., Deaven, D., Gandin, L., Iredell, M., Saha, S., White, G., Woollen, J., Zhu, Y., Leetmaa, A., Reynolds, R., Chelliah, M., Ebisuzaki, W., Higgins, W., Janowiak, J., Mo, K. C., Ropelewski, C., Wang, J., Jenne, R., and Joseph, D.: The NCEP/NCAR 40-Year Reanalysis Project, B. Am. Meteorol. Soc., 77, 437–472, 1996.

Kendall, M. G.: Rank Correlation Methods, London, UK, Charles Griffin, 1975.

Lamb, H. H.: British Isles weather types and a register of daily sequence of circulation patterns, 1861–1971, Geophys. Mem., 110, Meteorol. Office, London, 85 pp., 1972.

Lass, H. U. and Matthäus, W.: On temporal wind variations forcing salt water inflows into the Baltic Sea, Tellus, 48A, 663–671, 1996.

Lass, H. U. and Schwabe, R.: An analysis of the salt water inflow into the Baltic in 1975 to 1976, Dt. Hydrogr. Z., 43, 97–125, 1990.

Lehmann, A., Krauss, W., and Hinrichsen, H.-H.: Effects of remote and local atmospheric forcing on the circulation and upwelling in the Baltic Sea, Tellus, 54A, 299–316, 2002.

Lehmann, A., Getzlaff, K., and Harlass, J.: Detailed assessment of climate variability in the Baltic Sea area for the period 1958 to 2009, Clim. Dynam., 46, 185–196, 2011.

Leppäranta, M. and Myrberg, K: The physical oceanography of the Baltic Sea, Springer-Verlag, Berlin-Heidelberg-New York, 378 pp., 2009.

Linderson, M. L.: Objective classification of atmospheric circulation over Southern Scandinavia, Int. J. Climatol., 21, 155–169, 2001.

Matthäus, W., Nehring, D., Feistel, R., Nausch, G., Mohrholz, V., and Lass, H.-U.: The Inflow of Highly Saline Water into the Baltic Sea, In State and Evolution of the Baltic Sea, 1952–2005: A Detailed 50-Year Survey of Meteorology and Climate, Physics, Chemistry, Biology, and Marine Environment, 265–309, 2008.

Matthäus, W. and Schinke, H.: Mean Atmospheric Circulation Patterns Associated with Major Baltic Inflows, Dt. Hydrogr. Z., 46, 321–339, 1994.

Mohrholz, V., Naumann, M., Nausch, G., Krüger, S., and Gräwe, U.: Fresh oxygen for the Baltic Sea – An exceptional saline inflow after a decade of stagnation, J. Marine Syst., 148, 152–166, 2015.

Pasanen, L., Launonen, I., and L. Holmström: A scale space multiresolution method for extracting of time series features, Stat, 2, 273–291, 2013.

Philipp, A., Beck, C., Huth, R., and Jacobeit, J.: Development and Comparison of Circulation Type Classifications Using the COST 733 Dataset and Software, Int. J. Climatol., doi:10.1002/joc.3920, in press, 2014.

Post, P., Truija, V., and Tuulik, J.: Circulation Weather Types and Their Influence on Temperature and Precipitation in Estonia, Boreal Environ. Res., 7, 281–289, 2002.

Schimanke, S., Dieterich, C., and Meier, H. E. M.: An algorithm based on sea-level pressure fluctuations to identify major Baltic inflow events, Tellus A, 66, 23452, doi:10.3402/tellusa.v66.23452, 2014.

Schinke, H. and Matthäus, M.: On the Causes of Major Baltic Inflows – an Analysis of Long Time Series, Cont. Shelf Res., 1, 67–97, 1998.

Sen, P. K.: Estimates of Regression Coefficient Based on Kendall's Tau, J. Am. Stat. Assoc., 63, 1379–1389, 1968.

Soomere, T., Bishop, S. R., Viska, M., and Räämet, A.: An abrupt change in winds that may radically affect the coasts and deep sections of the Baltic Sea, Clim. Res., 62, 163–171, 2015.

Trigo, R. M. and DaCamara, C. C.: Circulation Weather Types and Their Influence on the Precipitation Regime in Portugal, Int. J. Climatol., 20, 1559–1581, 2000.

Väli, G., Meier, H. E. M., and Elken, J.: Simulated halocline variability in the Baltic Sea and its impact on hypoxia during 1961–2007, J. Geophys. Res., 118, 6982–7000, doi:10.1002/2013JC009192, 2013.

20

Quality control of the RMI's AWS wind observations

Cédric Bertrand, Luis González Sotelino, and Michel Journée

Royal Meteorological Institute of Belgium, Brussels, Belgium

Correspondence to: Cédric Bertrand (cedric.bertrand@meteo.be)

Abstract. Wind observations are important for a wide range of domains including among others meteorology, agriculture and extreme wind engineering. To ensure the provision of high quality surface wind data over Belgium, a new semi-automated data quality control (QC) has been developed and applied to wind observations from the automated weather stations operated by the Royal Meteorological Institute of Belgium. This new QC applies to 10 m 10 min averaged wind speed and direction, 10 m gust speed and direction, 2 m 10 min averaged wind speed and 30 m 10 min averaged wind speed records. After an existence test, automated procedures check the data for limits consistency, internal consistency, temporal consistency and spatial consistency. At the end of the automated QC, a decision algorithm attributes a flag to each particular data point. Each day, the QC staff analyzes the preceding day's observations in the light of the assigned quality flags.

1 Introduction

This paper describes the quality control (QC) procedures developed at the Royal Meteorological Institute of Belgium (RMI) to ensure the accuracy and reliability of the wind observations performed within the Automatic Weather Stations (AWS) network operated by RMI. Indeed, while high quality wind measurements are critical for many fields of science and engineering, only little attention has been paid in literature to quality control (QC) of wind-related variables. DeGaetano (1997) was the first to introduce complex automated QC checks for hourly wind records. In a subsequent study (DeGaetano, 1998), the author went a step further by proposing a distinct treatment for calm and non-calm wind speed values in the detection of wind speed bias. Later, Graybeal (2006) proposed to evaluate the reliability of extreme wind values using a relationship between daily wind speed and daily wind gust peaks (Weggel, 1999). In more recent years, Jiménez et al. (2010) extended the automated QC procedures of wind speed and wind direction values to wind data collected at higher temporal resolutions of 10 or 30 min. Lastly, Chávez-Arroyo and Probst (2015) presented a set of eleven QC procedures applied to the wind velocity records of the automated surface observation network of the Mexican National Weather Service. In the present approach, the

automated QC functions are included in a larger QC protocol involving manual inspections.

Wind data are evaluated daily by automated screening and manual inspection. Every morning a comprehensive suite of QC algorithms is applied to the previous day's data and a report summarizing the results for each station is produced for the RMI's QC staff. The purpose of the automated data screening is to objectively identify anomalous data values for subsequent review by an experienced data analyst. Note that false positives (i.e. type I error) increase the burden on the manual QC and false negatives (i.e. type II error) reduce the data quality. The review is necessary to determine whether an anomaly results from a problem with the instrument – and what maintenance action may be necessary – or whether it accurately reflects unusual meteorological conditions.

The paper is organized as follows. In Sect. 2, we briefly describe the wind measurements performed within the AWS operated by RMI. In Sect. 3, the automated quality control procedures are presented. Manual QC is discussed in Sect. 4. Finally, conclusion and perspective are given in Sect. 5.

2 Wind measurements within the RMI's AWS network

Wind speed and direction are recorded in 14 AWSs operated by RMI (see Fig. 1 for the stations' location). Wind

Table 1. List of the 14 RMI's Automatic Weather Stations performing wind observations and the associated measurements. Group of stations with similar measurements are defined by "QC group".

AWS		Wind measurement(s)					QC group
		Wind (gust) speed			Wind (gust) direction		
Synop code	Name	2 m	10 m	30 m	10 m	30 m	
6414	Beitem	X	X (X)		X (X)		2
6418	Zeebrugge		X (X)		X (X)		3
6431	Zelzate		X (X)		X (X)		3
6434	Melle	X	X (X)	X	X (X)		1
6438	Stabroek	X	X (X)		X (X)		2
6439	Sint Katelijn Waver	X	X (X)		X (X)		2
6447	Uccle	X		X (X)		X (X)	4
6455	Dourbes	X	X (X)		X (X)		2
6459	Ernage	X	X (X)		X (X)		2
6464	Retie	X	X (X)		X (X)		2
6472	Humain	X	X (X)	X	X (X)		1
6477	Diepenbeek	X	X (X)		X (X)		2
6484	Buzenol	X	X (X)		X (X)		2
6494	Mont Rigi	X	X (X)		X (X)		2

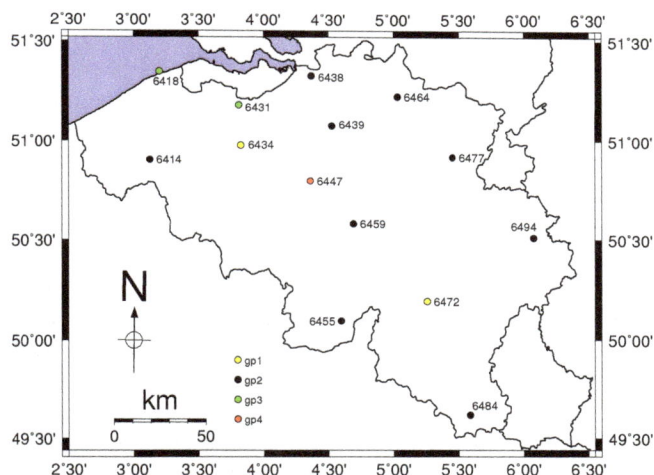

Figure 1. Location of wind measurements performed within the RMI's AWS network and the associated QC groups (see Table 1 for QC group definition).

speed is measured by the Siggelkow's wind velocity sensor LISA excepted for the 2 m wind speed in Uccle (AWS 6447), Ernage (AWS 6459) and Beitem (AWS 6414) where the Wind Transmitter "First Class" Advanced sensor of Thies clima is used. For the wind direction, all AWS are equipped with the Siggelkow's wind direction sensor RITA i WR08 Gray/Analogoutput except in Humain (AWS 6472) and Zeebrugge (AWS 6418) where the Wind Direction Transmitter "First Class" sensor of Thies clima is used. At a given station, wind speed can be measured at up to 3 different levels (i.e. 2, 10 and 30 m high). By contrast, wind direction is

recorded at 10 m high in all stations excepted in Uccle where it is measured at 30 m high. Table 1 provides an overview of the wind measurements performed in each AWS and the associated QC group. Indeed, due to the large heterogeneity within the RMI's AWS, four groups based on the recorded wind parameters have been distinguished for the automated data QC.

RMI's AWS are built around a programmable data logger that acquires the sensors' measurements, then processes, stores and transmits the data to the central RMI database (DB) in Uccle, Brussels. Once converted to digital values a first processing is performed at the raw data level allowing calculation of 10 min wind speed and direction averages from the 1 s measurements together with the computation of the gust speed and direction (the gust speed being defined as the maximum 3 s wind speed running average over the 10 min time period).

3 Automated data quality control

Similarly to what is done for the air and soil temperatures measurements from the RMI's AWS (Bertrand et al., 2013, 2015), a first basic QC is performed on all wind records once acquired centrally to ensure that gross errors are found before being further transmitted in the central DB. Automated procedures monitor the data to make sure they are collected and that the system performance is acceptable. After an existence test, a module checks for physical limits and flags the data violating these limits (erroneous when data lie outside physical limits and suspect when lying outside basic long-term climatological extremes that do not take into account the time of

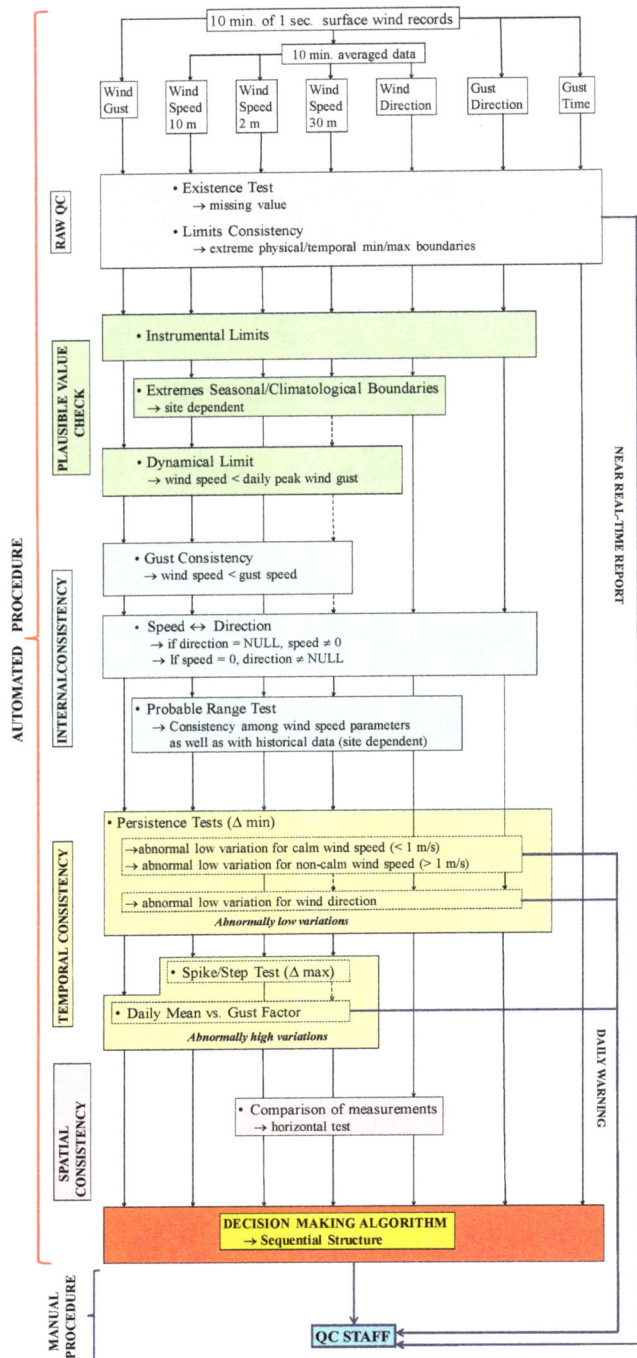

Figure 2. Flowchart of the wind quality assurance process implemented at RMI. (Dashed arrow only applies to 30 m wind measurements in Uccle – AWS6447.)

year and location). A list of missing and flagged data is automatically produced after each control cycle and transmitted to the AWS network maintenance team for further intervention (see Fig. 2). Note that values flagged as erroneous fail immediately and do not need to undergo additional checks.

Second, each night automated procedures check the previous day wind records for more subtle errors. Based on previous works by, e.g., DeGaetano (1997), Graybeal (2006), Liljegren et al. (2009), Jiménez et al. (2010) and Chávez-Arroyo and Probst (2015), the developed data quality tests fall into four categories ranging from simple to complex, from less restrictive to more restrictive. More specifically, the quality assessment process (see Fig. 2) starts with the identification of the records of all wind parameters which show readings beyond plausible values, both fixed and defined dynamically. Afterwards, wind speed, wind direction and gust speed and direction are evaluated for internal consistency. Thereafter, a temporal consistency check is applied. It aims at detecting abnormally low and high variations in wind speed and direction records. Finally, a spatial check is applied to the wind direction records.

To interpret the results of the automated tests, a decision algorithm has been developed that is applicable to all wind parameters and all sites. The algorithm proceeds sequentially through each step until a failure mode is identified. If no failure mode is identified, the measurement is judged to be valid. At the end of the process, a report is automatically generated for each AWS and sent to the QC staff.

3.1 Plausible value check

Implausible values are first defined according to the ranges specified by the manufacturer of the measurement equipment. Here, the fixed range is from 0 to 360° for the average 10 min wind direction and gust direction, and from 0 to 60 or 75 m s^{-1} for the Siggelkow or Thies clima systems, respectively, both for the average 10 min wind speed and wind gust parameters. Second, to verify whether the 10 min wind speed values are within acceptable limits depending on the climatic conditions of the measurement site, individual values are compared with upper seasonal bounds derived at each station from each measuring height from several years of manually controlled wind speed records. To minimize the possibility of a false positive identification, the decision algorithm does not report a range anomaly in case where 10 min wind speed measurements performed at different heights in the same site are larger than their corresponding seasonal limit. Finally, based on DeGaetano (1997), the daily peak gust (computed as the maximum value of the 10 min gust speed records over the given day), u_g, is used as daily limit for the averaged 10 min wind speed measurements recorded at the same height than the gust speed. Similarly to what is done in Chávez-Arroyo and Probst (2015), this dynamical limit test is performed only if more than 85 % of the daily 10 min gust speed records (i.e. more than 122 values a day) are available.

3.2 Internal consistency check

The internal consistency check consists of three different stages; in the first stage gust consistency is verified while in the second stage it is required that zero wind (gust) speed records must have zero and non-changing associated wind (gust) direction records. The third stages involve vertical comparisons of the 10 min wind speed measurement at different heights on the same AWS/site. This test provides a more stringent constraint than simple valid maximum/minimum limit tests by requiring consistency among the measurements as well as consistency with historical data. In order to implement such a QC procedure, 10 min wind speed record measured at a given height is related to 10 min wind speed value at another height using a simple linear regression model. At each station location, parameters of the regression model were estimated using the resistant least trimmed squares (LTS) regression method (e.g., Rousseeuw, 1984) due to the expected existence of outliers in the historical station data considered to fit the model. The biweight mean and standard deviation (Lanzante, 1996) were then used to calculate the confidence intervals around the regression line. Note that prediction intervals were constructed on the basis of a target-flagging rate of 1 per 1000 (e.g. a 99.9 % interval) for erroneous and of 10 per 1000 (e.g. a 99 % interval) for suspicious, respectively (e.g., Eischeid et al., 1995; Graybeal et al., 2004; Graybeal, 2006; Liljegren et al., 2009). Because two comparisons are necessary to unambiguously identify which level is problematic, at least two vertical tests (comparing three levels) must fail for the decision algorithm to report an anomaly. Consequently the decision algorithm will never report a vertical anomaly for AWS where only two wind speed measurement levels are available. However, a single vertical test is still valuable because a single failed vertical test can confirm a range test failure and cause the decision algorithm to report a range anomaly. For this reason, wind speed vertical comparisons are performed not only at stations of QC group 1 (where the test is the more efficient as it involves three measuring levels) but also at stations of QC groups 2 and 4 (see Table 1 for group QC definition).

3.3 Temporal consistency check

The temporal consistency check aims at detecting abnormally low and high variations in wind speed and direction records. Because the frequency distribution of repetitive readings under calm conditions for a given AWS has a far heavier tail than the distribution for non calm condition (i.e. both true calms and total sensor failures produce a sequence of repetitive values, while a similar situation is quite improbable for valid non-calm values) a distinct treatment is applied to calm and non-calm wind speed records, respectively. Ideally the limit between calm and non-calm wind speeds should be given by the anemometers cut-in wind speed (typically in the order of $0.3 \, \mathrm{m \, s^{-1}}$ for both the Siggelkow's LISA sensor

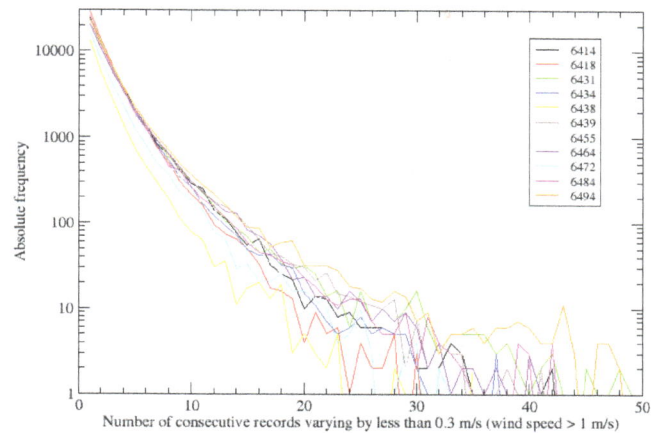

Figure 3. Frequency of occurrence of different consecutive 10 m wind speed repetitions at given AWS (non-calm wind speed) gathered over five years (i.e., 2010–2014) of 10 min wind speed records (see Table 1 for the AWS names).

and the Thies Clima wind transmitter). Here, for the sake of simplicity, we conservatively assume a typical cut-in value of $1 \, \mathrm{m \, s^{-1}}$ similarly to what is done in Jiménez et al. (2010) and Chávez-Arroyo and Probst (2015).

To identify the maximum number of consecutive unchanging records that can be assumed to be valid, an analysis of the frequency counts for different numbers of consecutive repetitions was performed at each station's location for each wind parameter at each recording height using manually quality controlled historical data. For short durations, constant wind periods are reported with a high frequency of occurence. As duration increases, an abrupt decrease in frequency of constant wind periods appears in all stations although each site has a different decay rate. As an example, Fig. 3 displays the number of consecutive records of 10 m wind speed measurements varying by less than $0.3 \, \mathrm{m \, s^{-1}}$ for non-calm wind speed situation at various AWS locations. The threshold value for suspect (erroneous) periods of constant wind speed was selected as being the number of repetitions reached by 99 % (99.9 %) of cumulated frequencies. A methodology similar to the one used for abnormally low variations of wind speed was used for the wind direction (for non-calm wind speed situation).

Besides the persistence test, a spike/step test compares the magnitude of change between 10 min wind speed records with the maximum probable change for a 10 min, 1h, 2h, 3h, and 6h time step periods. As for the persistence test, the maximum probable change is based on the 99th (99.9th) percentile change for several years of quality controlled prior data. The maximum probable change for wind speed depends on the location and sensor mounting height. To minimize the possibility of a false positive identification, a given 10 min wind speed record must fail in at least three of the five tested time steps prior to be flagged as suspect or erroneous. Moreover, because wind speed associated with thunderstorms can produce large changes in successive data values, the decision algorithm does not report a spike/step anomaly if more than one sensor mounted at different height on the same location fails the spike/step test. Note that the wind speed spike/step test is performed in all our AWS. Indeed, even if a failure in only one height is insufficient to report an anomaly, it can support other kind of detected failure mode during the algorithm decision making process.

The step/spike test is however not well suited when missing values are found in the time series as the difference between records cannot be calculated. Using different time steps allow to partly overcome such a limitation but does not solve the handling of abnormally high wind speed records surrounded by missing values (which tend to be systematically detected as invalid; Jiménez et al., 2010). Therefore, based on Graybeal (2006) an additional QC procedure evaluates the reliability of extreme wind values using the empirical relation of Weggel (1999) between daily mean wind speed, U, and the gust factor, G, to fit a linear regression model and establish prediction intervals around the regression line:

$$G = A\,U^n \quad \text{or} \tag{1}$$
$$\log G = \log A + n \log U,$$

where A is a constant and log the natural logarithm. G the gust factor is defined as

$$G = \frac{u_g}{U} - 1 \tag{2}$$

with u_g the daily peak gust (i.e., the maximum daily gust value). For each location, the parameters of the linear regression in Eq. (1) were estimated at the sensor mounting height where both 10 min wind speed and gust speed are recorded. As previously done for the vertical comparisons of the same measurement (see Sect. 3.2), the LTS regression method was used to fit the model and the biweight mean and standard deviation to calculate the variance of the predictions. The procedure is illustrated in Fig. 4 which presents the regression plot for the logarithm of the gust factor, G, versus the logarithm of the daily mean wind speed, U, for the station of Dourbes (AWS 6455). The LTS fit is given in blue and the 99 % (99.9 %) confidence level region in green (red), respectively.

It is worth pointing out that when abnormal low or high (excepted for the spike/test) variations are detected by the automated procedures, a daily warning is sent to the QC staff for requesting a visualization of the entire daily time evolution of the identified problematic wind parameter during the manual QC.

3.4 Spatial consistency check

In the horizontal test, the differences between a measurement and the corresponding measurements on other locations are

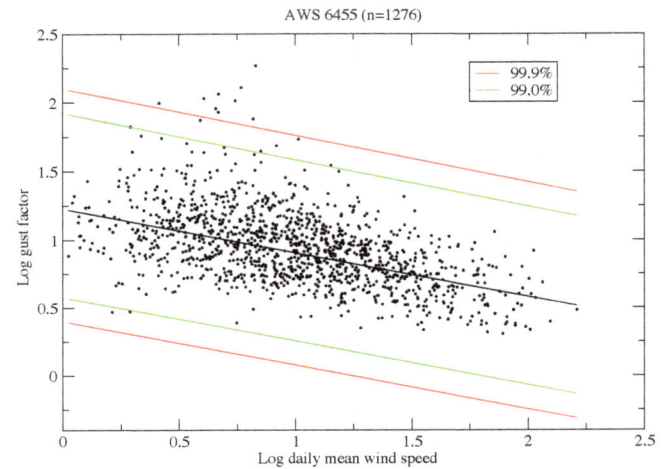

Figure 4. Log daily peak-gust factor plotted against daily mean wind speed, for an arbitrary sample of $n = 1276$ consecutive days (AWS 6455), with regression (blue line) and resistant prediction bounds (99 % in green and 99.9 % in red, respectively).

compared. Here, the spatial check only applies to the 10 min averaged wind direction records (for non-calm wind speed situations). It compares the station's direction to the mean wind direction in a radius of 75 to 100 km around the analyzed station's direction value. Basically, the station fails the neighboring test if the difference between the recorded station's direction and the computed direction in the radius is larger than 100°. It is worth pointing out that this test does not apply to the wind direction records performed at AWS 6484 and 6494 (see Fig. 1 for the stations location with the Belgian territory) as a minimum of five neighboring stations in radius is required to perform the test. An illustration of the spatial consistency check is given in Fig. 5. In this example, the wind direction recorded at the Zelzate's station (AWS 6431) and represented by the red arrow differs by more than 100° from the mean direction of the five nearest stations directions (represented by the black arrows).

4 Manual QC

Each day, the QC staff analyses the preceding day's wind records in the light of the assigned quality flags from the automated system. Results of the automated system can be graphically plotted on the operator terminal screen. In that case, all the analyzed wind speed records (including the gust speed) of the inspected day at a given station are presented in a graphic window with erroneous or suspect values indicated in the corresponding parameter daily time series. Similarly, the daily time series of the wind and gust directions are reported in separate window together with wind direction recorded at neighboring stations (e.g. daily time series of wind direction measurements from stations included in a domain surrounding the analyzed station – domain delimited by

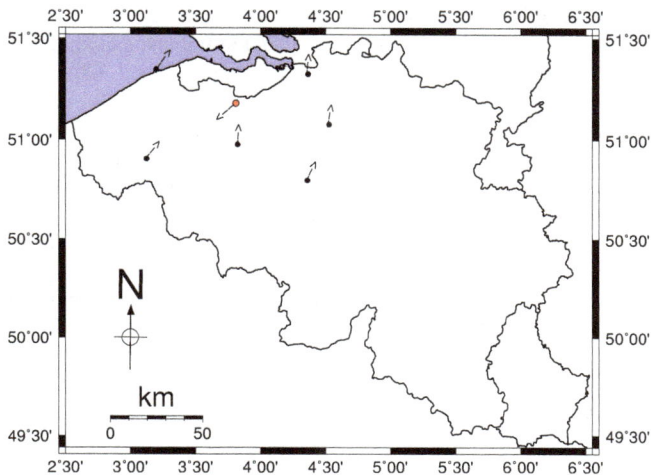

Figure 5. Illustration of the spatial consistency check applied to the 10 min averaged direction records. Date: 2015-08-23 15:00 UTC. Station flagged: Zelzate (AWS 6431). Station's speed: $2.5\,\text{m s}^{-1}$. Station's direction: $52.74°$. Radius: 75 km. Stations in the radius: 6. Speed in the radius: $3.02\,\text{m s}^{-1}$. Direction in radius: $201.70°$.

the operator). Visual inspection of all records flagged by the automated decision making algorithm is done to distinguish instrumental problems from plausible behaviors. It is the human decision whether or not a value is accepted. When errors are verified or visually detected, faulty records are eliminated and "trouble tickets" are issued where needed to the maintenance team so that sensors can be replaced or repaired. More than simply deleting erroneous measurements, human operators supply corrections and estimations (i.e., when values are missing) where possible. They have the opportunity to visualize different automated corrections on the problematic time series in order to determine the most appropriate in their specific case while it is always possible for individuals to apply their own corrections. When the correction/estimation process is completed, all modifications introduced by the operator are automatically recorded in the central RMI's DB. Note that the original parameters values are kept in the database and still accessible by the QC staff if required.

5 Conclusions

Automation of the RMI's AWS data quality control is in progress. After the automated quality control of 10 min air and soil temperatures records (Bertrand et al., 2013, 2015), automated quality assurance procedures devoted to wind records have been operationally implemented to support the QC staff in their work. Validation exercises have revealed that unsurprisingly the automatic QC system performs better for stations of the QC group 1 than for those of the QC group 3 as the increased wind speed recording heights allow to refine the final decision of the algorithm. Nevertheless, it has been found that the automated QC is able to correctly

identify problematic parameters in a particular station on a given day irrespectively of the AWS QC group. However, the spatial consistency check applied to the 10 min wind direction tends to produce type I error (i.e. false positives) at some stations (located in the North-East part of Belgium). This occurs when the station's direction while being close to the direction recorded at nearest neighboring station differs by more than 100° from the mean wind direction in radius. Enlarging the acceptable direction difference threshold does not satisfactory solve the problem as the direct counterpart is an increasing number of type II error (false negatives). Current investigations tend to indicate that incorporating the wind measurements performed in the 7 AWS operated by the Belgian army on military airports (three being located within the problematic area) into the automated QC and limiting the number of neighboring in the radius to the five or four nearest stations in the spatial consistency check will substantially reduce the occurrence of type I error. Another advantage of using the wind records from the military AWS is that the increased stations density will authorize to extend the spatial consistency check to the wind directions recorded in our eastern AWS (Mont Rigi station, AWS 6494).

Finally, while the validation exercise has not revealed a particular weakness in the step/spike test, we are planning to investigate if it could be relevant or not to adapt the procedure using the daily mean wind speed and the gust factor to detect abnormally high variations in presence of missing data records (Eq. 1 in Sect. 3.3). The idea is that similarly to Jiménez et al. (2010) the daily peak gust (u_g) could be replaced by the daily maximum 10 min wind speed to calculate the gust factor (G) in Eq. (2). In that case, only historical daily time series with no missing 10 min values will be considered to determine the linear regression parameters in Eq. (1).

References

Bertrand, C., González Sotelino, L., and Journée, M.: Quality control of 10 min air temperature data at RMI. Adv. Sci. Res., 10, 1–5, 2013.

Bertrand, C., González Sotelino, L., and Journée, M.: Quality control of 10 min soil temperatures data at RMI, Adv. Sci. Res., 12, 23–30, 2015.

Chávez-Arroyo, R. and Probst, O.: Quality assurance of near-surface wind velocity measurements in Mexico, Meteorol. Appl., 22, 165–177, 2015.

DeGaetano, A. T.: A quality-control routine for hourly wind observations, J. Atmos. Ocean. Technol., 14, 308–317, 1997.

DeGaetano, A. T.: Identification and implications of biases in U.S. surface wind observation, archival, and summarization methods, Theor. Appl. Climatol., 60, 151–162, 1998.

Eischeid, J. K., Baker, C. B., Karl, T. R., and Diaz, H. F.: The quality control of long-term climatological data using objective data analysis, J. Appl. Meteor., 34, 2787–2795, 1995.

Graybeal, D. Y.: Relationships among daily mean and maximum wind speeds, with application to data quality assurance, Int. J. Climatol., 26, 29–43, 2006.

Graybeal, D. Y., DeGaetano, A. T., and Eggleston, K. L.: Complex quality assurance of historical hourly surface airways meteorological data, J. Atmos. Oceanic Technol., 21, 1156–1169, 2004.

Jiménez, P. A., González-Rouco, J. F., Montávez, J. P., and Garcia-Bustamante, E: Quality assurance of surface wind observations from automated weather stations. J. Atmos. Ocean. Technol., 27, 1101–1123, 2010.

Lanzante, J. R.: Resistant, robust and nonparametric techniques for the analysis of climate data: theory and examples, including applications to historical radiosonde station data, Int. J. Climatol., 16, 1197–1226, 1996.

Liljegren, J. C., Tschopp, S., Rogers, K., Wasmer, F., Lijegren, L., and Myirski, M: Quality control of meteorological data for the chemical stockpile emergency preparedness program, J. Atmos. Ocean. Technol., 26, 1510–1526, 2009.

Rousseeuw, P. J.: Least median of squares regression. J. Am. Statist. Assoc., 79, 871–880, 1984.

Weggel, J. R.: Maximum daily wind gustrelated to mean daily wind speed, J. Struct. Eng., 125, 465–468, 1999.

21

The impact of clouds, land use and snow cover on climate in the Canadian Prairies

Alan K. Betts[1], Raymond L. Desjardins[2], and Devon E. Worth[2]

[1]Atmospheric Research, Pittsford, Vermont, USA
[2]Science and Technology Branch, Agriculture and Agri-Food Canada, Ottawa, Ontario, Canada

Correspondence to: Alan K. Betts (akbetts@aol.com)

Abstract. This study uses 55 years of hourly observations of air temperature, relative humidity, daily precipitation, snow cover and cloud cover from 15 climate stations across the Canadian Prairies to analyze biosphere-atmosphere interactions. We will provide examples of the coupling between climate, snow cover, clouds, and land use. Snow cover acts as a fast climate switch. With the first snow fall, air temperature falls by $10\,°C$, and a similar increase in temperature occurs with snow melt. Climatologically, days with snow cover are $10\,°C$ cooler than days with no snow cover in Alberta. However the interannual variability has a larger range, so that for every $10\,\%$ decrease in days with snow cover, the mean October to April climate is warmer by 1.4 to $1.5\,°C$. Snow cover also transforms the coupling between clouds and the diurnal cycle of air temperature from a boundary layer regime dominated by shortwave cloud forcing in the warm season to one dominated by longwave cloud forcing with snow cover. Changing agricultural land use in the past thirty years, specifically the reduction of summer fallowing, has cooled and moistened the growing season climate and increased summer precipitation. These hourly climate data provide a solid observational basis for understanding land surface coupling, which can be used to improve the representation of clouds and land-surface processes in atmospheric models.

1 Introduction

This paper reviews the analysis of the diurnal climatology of the Canadian Prairies (Betts et al., 2013a, b, 2014a, b, 2015; Betts and Tawfik, 2016). There are 15 climate stations with more than 55 years of hourly temperature, relative humidity, wind, precipitation and snow depth, as well as a unique set of hourly estimates of opaque, reflective cloud cover, made by trained observers (Environment Canada, 2013). We have used snow cover and the daily mean opaque cloud cover, $OPAQ_m$, to stratify this large dataset (660 station-years), and study the diurnal and seasonal climate. The land surface, the boundary layer (BL) and the overlying atmosphere are a tightly coupled system, with two distinct climates above and below the freezing point of water. Below freezing, precipitation falls as snow and the highly reflective snow cover acts as a fast climate switch, which drops the mean surface temperature by $10\,°C$ within days, and changes the climatology (Sect. 2). One result is the mean cold season temperature is linearly related to the fraction of days with snow cover. An-

other impact of snow cover is to transform the coupling of the diurnal cycle to cloud cover (Sect. 3). There have been large land use changes on the Prairies over the past 30 years, as 5 million hectares of summer fallow has been converted to continuous cropping. Increased transpiration has cooled and moistened the growing season climate and increased precipitation (Sect. 4). These observational studies provide a baseline for model evaluation.

Figure 1 shows the location of the climate stations, Canadian ecozones, regional zones, agricultural regions and boreal forest, with 50 km radius circles around each station. We generated daily means for all variables, such as mean temperature and humidity, T_m and RH_m and opaque cloud, $OPAQ_m$, and merged daily total precipitation and daily snow depth. From the hourly data we computed the diurnal temperature range between maximum temperature, T_x, and minimum temperature, T_n, as

$$DTR = T_x - T_n. \tag{1}$$

Figure 1. Climate station locations, Canadian ecozones, regional zones, agricultural regions and boreal forest.

At the time of maximum temperature, T_x, typically in the afternoon in the warm season, we derived the mixing ratio, Q_{tx}, and the corresponding equivalent potential temperature and pressure-height to the lifting condensation level (LCL), which is close to cloud-base in the warm season (Betts et al., 2013a).

2 Impact of snow cover on climate

Snow/ice albedo feedback is well-known on global scales, but global models show a wide variation in their representation of this key process (Qu and Hall, 2007; Bony et al., 2006; Xu and Dirmeyer, 2013). Early case studies showed that snow cover reduces surface temperatures by about 5 °C on both the short-term timescale and on monthly timescales (Namias, 1960, 1985; Cohen and Rind, 1991; Dewey, 1977; Wagner, 1973). However there has been surprisingly little quantitative observational analysis of the impact of snow on local and regional climate. Betts et al. (2015) found that there are two very distinct Prairie climates, sharply separated by the freezing point of water: one for the cold season with $T_m < 0$ °C and surface snow cover, and one for the warm season with $T_m > 0$ °C with no snow cover.

Figure 2 shows the annual mean climatology (black) and the partition for each month into the days with no snow cover (red) and with snow cover (blue). The mean climatology is the weighted mean of the no-snow and snow values, using the number of days in each class as weights. We show three stations in Alberta, which have an average of 36 % of the days in winter with no snow cover. For October to April, we plot the temperature difference ΔT between these snow and no-snow climatologies. There is a clear separation of the climatologies for these cold season months for all three stations with a mean value of

$$\Delta T = -9.8(\pm 0.8)°C.$$

That is, it is climatologically about 10 °C cooler when there is reflective snow cover with a surface albedo ≈ 0.7 (Betts et al., 2014b). Other stations across the Prairies show a similar cooling of the climatology with snow, although some have too few snow-free days in January and February to make this comparison.

Figure 3 shows the temporal perspective (adapted from Betts et al., 2014b): the climatological transition across the first lasting snow in the fall (left panel) and the final melt of the snowpack in the spring (center panel). The left panel is a composite across the fall transitions from no snow to lasting snow cover for 1955–2004 for six stations in Saskatchewan. The average date of this first snowfall is 15 November, and there is a 10 °C drop of mean temperature within less than a week, as the surface albedo changes from about 0.2 to 0.7 with snow. From a climatological perspective this means that the transition into winter is not a smooth process lasting 4–6 weeks, but it typically occurs in less than a week with the first lasting snow fall. In spring there is a longer reverse transition with a rise of 10 °C in about 10 days as the snowpack melts.

Betts et al. (2014b) described the role of reflective snow cover as a fast climate switch which drops the surface temperature by about 10 °C, as seen in both Figs. 2 and 3. Conceptually, temperatures fall as the solar zenith angle increases with the approach of winter, until precipitation falls as snow, once temperatures fall below freezing in a favorable synoptic situation. With snow cover, two feedbacks cool the surface making it much harder for transient synoptic advection to melt the snow cover. The primary one is the reflection of sunlight by snow, but Betts et al. (2014b) showed that a drop in the incoming long wave radiation is also significant in cooling the surface. Snowmelt in spring is initiated by decreasing solar zenith angle driving warmer temperatures, and with the loss of the feedbacks associated with snow cover, temperatures rise about 10 °C. Snow cover has a further feedback, as it insulates the ground, reducing the ground heat flux, which is typically upward in fall and downward in spring. We will see in the next section that there is also a profound change in the coupling between cloud cover and surface temperature with snow, which may also play a role in the longwave coupling.

In Alberta in the lee of the Rocky Mountains the winter snowpack is often transient, so there is a wide interannual variation in the number of days with snow cover. The right panel plots mean temperature for the period October to April, against the fraction of days with snow cover for five Alberta stations, each with almost fifty years of data. There is a strong linear dependence (with $R^2 = 0.79$) of the interannual variation in the cold season mean temperature on the fraction of days with snow cover. The temperature range of 14.6 ± 0.6 °C from zero to 100 % snow cover is larger than

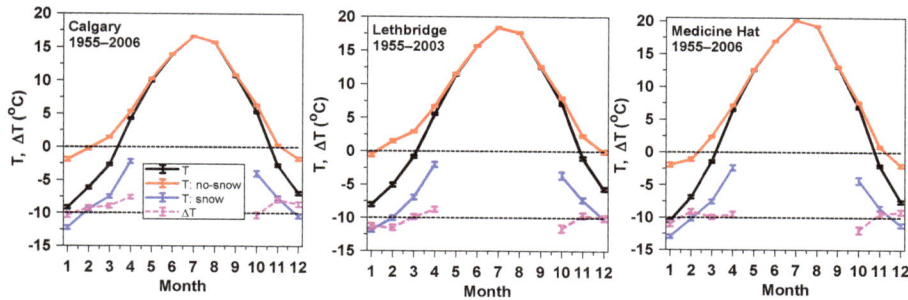

Figure 2. Annual Climatology for Calgary, Lethbridge and Medicine Hat, with the separation into days with and without snow cover.

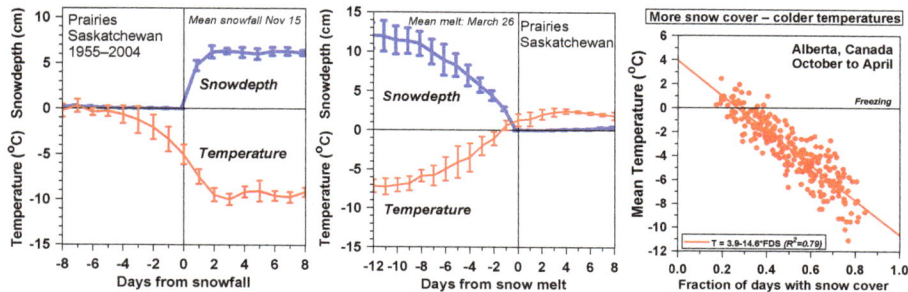

Figure 3. Fall (left panel) and spring (center panel) snow transition for six Saskatchewan stations, and (right panel) relation of cold season temperature to fraction of days with snow cover for five Alberta stations.

the local change with snow cover of order 10 °C. This apparent amplification suggests that there might be a further positive feedback between the local impact of the snowpack and the larger-scale regional climate.

3 Impact of snow cover and cloud cover on the diurnal cycle

The coupling between cloud cover, temperature and the diurnal cycle is a central feature of the Earth's climate (Betts, 2015). However, the representation in global models is a challenge because of uncertainties in their representation of clouds (Groisman et al., 2000). So observational studies play a key role (Dai et al., 1999). The long time-series of opaque reflective cloud data at climate stations on the Canadian Prairies used for our analyses has been transformative, because it can be calibrated against baseline surface radiation measurements to give the longwave and shortwave cloud forcing (Betts et al., 2015).

Figure 4 shows the climatology of the diurnal cycle for January, July and November stratified by opaque cloud fraction and by surface snow cover for all the Canadian Prairie data (adapted from Betts and Tawfik, 2016). In January, which is representative of all the winter months with snow (not shown), mean air temperature increases with opaque cloud cover, and sunrise minimum temperatures plunge under clear skies. The amplitude of the diurnal cycle increases as cloud cover decreases in response to the daytime solar

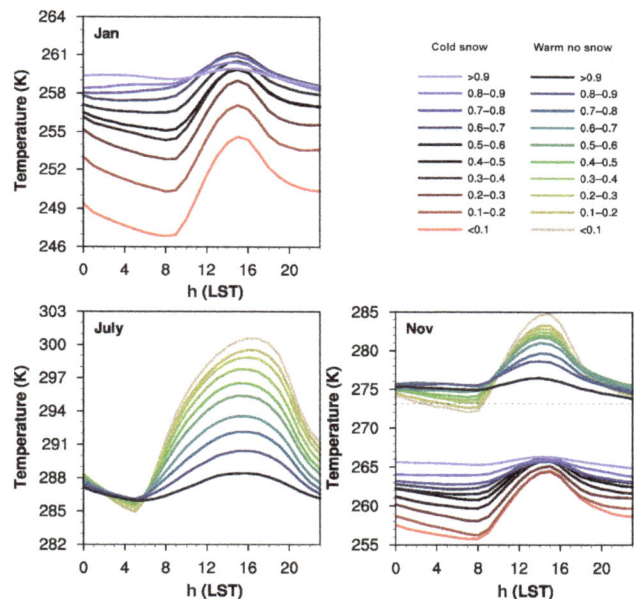

Figure 4. Diurnal cycle of temperature stratified by opaque cloud for January, July and November (adapted from Betts and Tawfik, 2016).

forcing. This winter stable boundary layer regime is dominated by long-wave cloud forcing (Betts et al., 2015), for which cloud cover reduces the cooling of the surface to

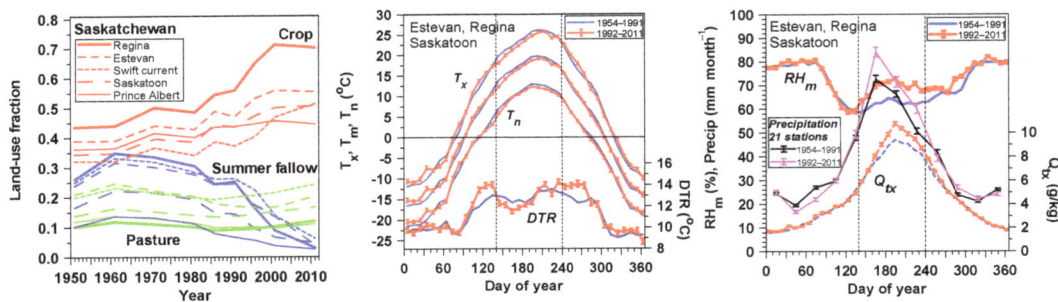

Figure 5. Long term trends in total cropland, pasture, and summer fallow around five climate stations in Saskatchewan (left panel); mean changes in annual cycle of DTR, T_x, T_m and T_n for Saskatoon, Regina and Estevan (center panel) and RH$_m$, Q_{tx} and mean precipitation for 21 stations in southern Saskatchewan (right panel).

space. In contrast, in July and indeed all months with no snow cover (not shown), minimum temperatures vary little with cloud cover, and the diurnal range to the afternoon maximum is largest under clear skies. This is because in the warm season, the shortwave reflection by clouds dominates (Betts et al., 2015), so that mean temperatures are warmest under clear skies. This is characteristic of an unstable daytime convective BL (Betts et al., 2013a). For November (as well as March and April, not shown), there is sufficient data to show both climatologies, with and without snow, and the clear separation in temperature between them that increases as cloud cover falls (Betts and Tawfik, 2016). We see that snow cover completely transforms the coupling between opaque cloud and the diurnal cycle of temperature.

4   Impact of land use on climate

In the past thirty years there has been a major change in land use across the Canadian Prairies, specifically the conversion of more than five million hectares of summer fallow to continuous cropping. The large increase in the area of cropland has increased summer transpiration, which has reduced maximum temperatures in the growing season over the Prairies (Gameda et al., 2007). Other analyses of US Midwest summer temperature maxima also show a cooling from land-use change to cropland (Bonan, 2001) and cropland intensification (Mueller et al., 2016).

Figure 5 summarizes the long-term climate impact of the reduction of summer fallow in Saskatchewan (Betts et al., 2013b). The left panel shows the land-use trends in total cropland, pasture, and summer fallow around five climate stations. The 50 km radius circles around each station in Fig. 1 were used to generate local averages of the ecodistrict crop data. We split the climate station timeseries into two periods: a longer historic period, 1954–1991, when summer fallow cover was large (although slowly decreasing) and a recent 20-year period, 1992–2011, when summer fallow has fallen rapidly to its present low value. We did not attempt any analysis of decadal trends.

Saskatoon, Regina and Estevan in Saskatchewan have complete datasets and show similar changes between the two time periods, so we averaged them as 10-day means (Betts et al., 2013b). The center panel shows the mean changes in the annual cycle of DTR, T_x, T_m and T_n and the right panel the mean changes in RH$_m$ and mixing ratio Q_{tx} at the time of the afternoon maximum temperature. We show the standard errors of the difference between the two mean time series as an indication of significance. Precipitation has much more variability than temperature and humidity, so we used the 21 stations in Saskatchewan south of 53.22° N in the second generation adjusted precipitation dataset (Mekis and Vincent, 2011). The right panel shows the difference in the monthly mean time-series of precipitation for the two time periods, with the standard error of each monthly mean.

It is clear that there are significant changes in the growing season climate between the historic period, 1953–1991, and the more recent period since 1992. Betts et al. (2013b) considered the period $140 \leq DOY < 240$ (20 May–27 August) to be representative of the growing season (where Day of Year is abbreviated DOY). Since 1992, the growing season is significantly cooler, with a drop of (T_x, T_m, T_n) of (-0.93 ± 0.09, -0.82 ± 0.07, -0.68 ± 0.06 °C), and significantly moister with a rise of (RH$_m$, Q_{tx}) of (6.9 ± 0.2 %, 0.70 ± 0.04 g kg^{-1}). There is a corresponding fall of the LCL of 22.3 ± 1.1 hPa, and a small rise of equivalent potential temperature of 1.1 ± 0.2 K, at the time of afternoon maximum temperature. There is also an increase of summer (June, July and August) precipitation of 25.9 ± 4.6 mm. The change of DTR in the growing season has a more complex structure with a decrease in the early part of the growing season ($140 \leq DOY < 200$) of -0.60 ± 0.09°C that is coupled to an increase in cloud cover (Betts et al., 2013b).

We conclude that more intensive agriculture has increased transpiration, which has cooled and moistened the growing season climate with a lowered cloud-base, increased equivalent potential temperature, and an increase of summer precipitation.

5 Conclusions

The long time-series of hourly climate data including opaque cloud measurements provides insight into the fully coupled land–atmosphere–cloud interactions in the Canadian Prairies. Reflective surface snow cover acts as a climate switch, since surface albedo changes from around 0.2 to 0.7. The climatology with and without snow cover shows that near-surface temperature drops by $10\,^\circ$C with snow cover, and the coupling between opaque cloud and the diurnal cycle of temperature is transformed. During the warm season, the shortwave reflection by clouds dominates and temperatures are coolest under cloudy skies. Minimum temperatures vary little with cloud cover and the diurnal range to the afternoon maximum is largest under clear skies. The reverse is true during the cold season with snow cover, when the longwave warming by clouds dominates. The near-surface air is warmest under cloudy skies, while sunrise minimum temperatures plunge under clear skies. This snow-albedo coupling is so strong and rapid that the mean temperatures of the cold season are linearly related to the fraction of days with snow cover. Increased transpiration related to the shift to more intensive agriculture over the past 30 years, from summer fallow to continuous cropping on 5 million hectares of the agricultural land, has cooled and moistened the growing season climate, and increased precipitation. The on-going analyses of these data are providing new understanding of the climate and land–atmosphere–cloud system for northern latitudes. We expect that these observational data sets will also be extremely valuable for testing and evaluating atmospheric models.

Acknowledgements. This research was supported by Agriculture and Agri-Food Canada. We thank the civilian and military technicians of the Meteorological Service of Canada and the Canadian Forces Weather Service, who have made reliable cloud observations hourly for 60 years.

References

Betts, A. K.: Diurnal cycle, in: Encyclodedia of Atmospheric Sciences, 2nd Edn., edited by: North, G. R., Pyle, J., and Zhang, F., Elsevier, 319–323, doi:10.1016/B978-0-12-382225-3.00135-3, 2015.

Betts, A. K. and Tawfik, A. B.: Annual climatology of the diurnal cycle on the Canadian Prairies, Front. Earth Sci., 4, 1–23, doi:10.3389/feart.2016.00001, 2016.

Betts, A. K., Desjardins, R. L., and Worth, D. E.: Cloud radiative forcing of the diurnal cycle climate of the Canadian Prairies, J. Geophys. Res.-Atmos., 118, 8935–8953, doi:10.1002/jgrd.50593, 2013a.

Betts, A. K., Desjardins, R. L., Worth, D. E., and Cerkowniak, D.: Impact of land use change on the diurnal cycle climate of the Canadian Prairies, J. Geophys. Res.-Atmos., 118, 11996-12011, doi:10.1002/2013JD020717, 2013b.

Betts, A. K., Desjardins, R. L., Worth, D. E., and Beckage, B.: Climate coupling between temperature, humidity, precipitation, and cloud cover over the Canadian Prairies, J. Geophys. Res.-Atmos., 119, 13305–13326, doi:10.1002/2014JD022511, 2014a.

Betts, A. K., Desjardins, R. L., Worth, D. E., Wang, S., and Li, J.: Coupling of winter climate transitions to snow and clouds over the Prairies, J. Geophys. Res.-Atmos., 119, 1118–1139, doi:10.1002/2013JD020717, 2014b.

Betts, A. K., Desjardins, R. L., Beljaars, A. C. M., and Tawfik, A.: Observational study of land–surface–cloud–atmosphere coupling on daily timescales, Front. Earth Sci., 3, 13, doi:10.3389/feart.2015.00013, 2015.

Bonan, G. D.: Observational evidence for reduction of daily maximum temperature by croplands in the Midwest United States, J. Climate, 14, 2430–2442, 2001.

Bony, S., Colman, R., Kattsov, V. M., Allan, R. P., Bretherton, C. S., Dufresne, J.-L., Hall, A., Hallegatte, S., Holland, M. M., Ingram, W., Randall, D. A., Soden, B. J., Tselioudis, G., and Webb, M. J.: How well do we understand and evaluate climate change feedback processes?, J. Climate, 19, 3445–3482, doi:10.1175/JCLI3819.1, 2006.

Cohen, J., and Rind, D.: The impact of snow cover on the climate, J. Climate, 4, 689–706, doi:10.1175/1520-0442(1991)004<0689:TEOSCO>2.0.CO;2, 1991.

Dai, A., Trenberth, K. E., and Karl, T. R.: Effects of clouds, soil moisture, precipitation and water vapor on diurnal temperature range, J. Climate, 12, 2451–2473, 1999.

Dewey, K. F.: Daily maximum and minimum temperature forecasts and the influence of snow cover, Mon. Weather Rev., 105, 1594–1597, 1977.

Environment Canada: MANOBS, Chapter 1, Sky: http://www.ec.gc.ca/manobs/default.asp?lang=En&n=A1B2F73E-1, last access: 21 March 2013.

Gameda, S., Qian, B., Campbell, C., and Desjardins, R. L.: Climatic trends associated with summerfallow in the Canadian Prairies, Agr. Forest Meteorol., 142, 170–185, 2007.

Groisman, P. Y., Bradley, R. S., and Sun, B.: The relatioship of cloud cover to near-surface temperature and humidity: Comparison of GCM simulations with empirical data, J. Climate, 13, 1858–1878, 2000.

Mekis, É. and Vincent, L. A.: An overview of the second generation adjusted daily precipitation dataset for trend analysis in Canada, Atmos.-Ocean, 49, 163–177, 2011.

Mueller, N. D., Butler, E. E., McKinnon, K. A., Rhines, A., Tingley, M., Holbrook, N. M., and Huybers, P.: Cooling of US Midwest summer temperature extremes from cropland intensification, Nat. Clim. Change, 6, 317–322, doi:10.1038/nclimate2825, 2016.

Namias, J.: Snowfall over Eastern United States: Factors leading to its monthly and seasonal variations, Weatherwise, 13, 238–247, doi:10.1080/00431672.1960.9940990, 1960.

Namias, J.: Some empirical evidence for the influence of snow cover on temperature and precipitation, Mon. Weather Rev., 113, 1542–1553, doi:10.1175/1520-0493(1985)113<1542:SEEFTI>2.0.CO;2, 1985.

Qu, X. and Hall, A.: What controls the strength of the snow-albedo feedback?, J. Climate, 20, 3971–3981, 2007.

Wagner, J. A.: The influence of average snow depth on monthly mean temperature anomaly, Mon. Weather Rev., 101, 624–626, doi:10.1175/1520-0493(1973)101<0624:TIOASD>2.3.CO;2, 1973.

Xu, L. and Dirmeyer, P. A.: Snow-atmosphere coupling strength. Part I: Effect of model biases, J. Hydrometeorol., 14, 389–403, doi:10.1175/JHM-D-11-0102.1, 2013.

22

Comparison of regional and global reanalysis near-surface winds with station observations over Germany

A. K. Kaiser-Weiss[1], **F. Kaspar**[1], **V. Heene**[1], **M. Borsche**[1], **D. G. H. Tan**[2], **P. Poli**[2], **A. Obregon**[1], and **H. Gregow**[3]

[1]Deutscher Wetterdienst, Frankfurter Straße 135, 63067 Offenbach, Germany
[2]European Centre for Medium-Range Weather Forecasts, Shinfield Park, Reading, RG2 9AX, UK
[3]Finnish Meteorological Institute, P.O. Box 503, 00101 Helsinki, Finland

Correspondence to: A. K. Kaiser-Weiss (andrea.kaiser-weiss@dwd.de)

Abstract. Reanalysis near-surface wind fields from multiple reanalyses are potentially an important information source for wind energy applications. Inter-comparing reanalyses via employing independent observations can help to guide users to useful spatio-temporal scales. Here we compare the statistical properties of wind speeds observed at 210 traditional meteorological stations over Germany with the reanalyses' near-surface fields, confining the analysis to the recent years (2007 to 2010). In this period, the station time series in Germany can be expected to be mostly homogeneous. We compare with a regional reanalysis (COSMO-REA6) and two global reanalyses, ERA-Interim and ERA-20C. We show that for the majority of the stations, the Weibull parameters of the daily mean wind speed frequency distribution match remarkably well with the ones derived from the reanalysis fields. High correlations (larger than 0.9) can be found between stations and reanalysis monthly mean wind speeds all over Germany. Generally, the correlation between the higher resolved COSMO-REA6 wind fields and station observations is highest, for both assimilated and non-assimilated (i.e., independent) observations. As expected from the lower spatial resolution and reduced amount of data assimilated into ERA-20C, the correlation of monthly means decreases somewhat relative to the other reanalyses (in our investigated period of 2007 to 2010). Still, the inter-annual variability connected to the North Atlantic Oscillation (NAO) found in the reanalysis surface wind anomalies is in accordance with the anomalies recorded by the stations.

We discuss some typical examples where differences are found, e.g., where the mean wind distributions differ (probably related to either height or model topography differences) and where the correlations break down (because of unresolved local topography) which applies to a minority of stations. We also identified stations with homogeneity problems in the reported station values, demonstrating how reanalyses can be applied to support quality control for the observed station data.

Finally, as a demonstration of concept, we discuss how comparing feedback files of the different reanalyses can guide users to useful scales of variability.

1 Introduction

Atmospheric reanalysis is "a consistent reprocessing of archived weather observations using a modern forecasting system" (Dee et al., 2014). The past two decades have brought forth remarkable advances in the quality and quantity of the resulting reanalysis datasets (Dee et al., 2014; Bosilovich et al., 2013; Hartmann et al., 2013). These advances have been accompanied (and in many respects catalysed) by rapid growth in uptake of reanalysis datasets for a diverse and growing range of applications. But it remains challenging for individual users to know which (if any) of the available datasets (whether from reanalysis or other procedures) are appropriate for their applications (see, e.g., Gregow et al., 2015). Assessing "fitness-for-purpose" requires synthesising knowledge about (a) the relative strengths/limitations of the available datasets, (b) the extent to which such characteristics affect the fidelity of derived results, and (c) the user's own application-specific tolerance of uncertainty; it is important to note that fitness for one purpose does not guarantee fitness for all purposes. Here, we describe the inter-comparisons that were undertaken to elucidate the relative properties of the near-surface winds over Germany in two global and one regional reanalysis datasets on the time scale of days to several years. The inter-comparisons employ independent wind observations from German weather stations and also illustrate how reanalyses can support quality control of the observations. The COSMO-REA6 regional reanalysis is driven by the global ERA-Interim reanalysis, and covers Europe with increased spatial and temporal resolution. We demonstrate that inter-comparison of reanalyses can help to guide users to useful scales of variability.

For the purpose of wind energy production, for example, there are dedicated wind mast measurements, which measure at a certain point over a short period of time and at heights typical for wind energy production. This is highly valuable for wind energy applications, but it remains desirable to add to this short-term information an estimate of the inter-annual temporal variability, as is typically available in the time series of station measurements made for traditional weather/climate applications (approximately 10 m above ground). In Germany, these are relatively abundant and evenly distributed and cover many decades. On the other hand, observation practices, instruments, and the height of the sensor above ground might have changed over time. This information should ideally be found in the metadata accompanying the historical station records. It is also valuable to track changes in the surroundings (like growing trees or changes in land cover) and measurement or processing errors, but this is not uniformly available worldwide given historical variations in the rigour of observing practice around the globe.

Reanalyses also provide wind information, here derived from a physically consistent state of the atmosphere using a state-of-the-art NWP model, constrained by the whole observing system. While larger wind energy companies might be able to build their own wind measurement masts and develop sophisticated analyses, other users like smaller enterprises, district managers, government agencies, or interested individual citizens rely on products based on station measurements at 10 m above ground, and statistical models build upon those, like e.g., wind climatologies at levels of interest for wind energy production (Gerth and Christoffer, 1994). All these users might draw benefit from our study.

Reanalysis wind fields have been found to differ from station records on the multi-decadal scale. The so-called "stilling effect", i.e., a decrease of wind speeds deduced from station measurements since the 1970s, is reported in many papers on mid-latitude wind observations (see McVicar et al., 2012, for a comprehensive review). This is in general contrast to various reanalyses, which do not show a stilling effect (Smits et al., 2005; McVicar et al., 2008, 2012, and references therein). Several reasons have been suggested. Long-term changes in surface roughness could explain, at least partly, the difference (Vautard et al., 2010; Wever, 2012). Changes in aerosols, sea surface temperature and greenhouse gas concentrations were found to be unlikely causes (Bichet et al., 2012). There still remain the possibilities that measurement artefacts, or processes not modelled in the reanalysis, may contribute to the discrepancies.

In this work we do not investigate the multi-decadal stilling effect directly, but report on a preparatory multi-annual study. We focus on the area of Germany and the period 2007 to 2010. Although the dataset is short, it is best fitted for the inter-comparison since all stations are known to operate automatically, instrumental problems are unlikely, and changes in surface roughness are negligible. This allows us to characterize the recent variability captured in regional and global reanalysis near surface wind fields, and to compare their statistical characteristics with the traditional station measurements. Several approaches have been identified as useful for reanalysis comparisons (see Fig. 1) and applied below for the near-surface wind fields, namely direct comparisons (Sect. 3.1 and 3.2), thematic comparisons (Sect. 3.3) and internal metrics comparisons (Sect. 3.4).

2 Data

We used the 10 m wind fields and feedback files containing the radiosonde data from the global reanalysis ERA-Interim (Dee et al., 2011). The data of this reanalysis are available from the ECMWF website http://apps.ecmwf.int/datasets/. ERA-Interim is based on the ECMWF Integrated Forecast System, IFS (Cy31r2), 4-dimensional variational analysis (4D-Var) with a 12-h analysis window, and assimilation of a wide range of surface and upper-air observational data from both in-situ and satellite instruments. The spatial resolution of the data set is approximately 80 km (T255 spectral) on 60

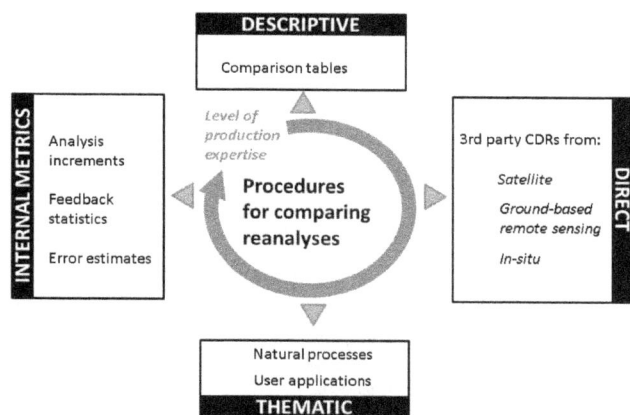

Figure 1. Useful approaches for intercomparing reanalyses include: (1) descriptive, e.g., a comparison table of characteristics, (2) direct comparisons with in-situ, ground-based remote sensing or satellite observations, (3) thematic comparison, and (4) internal metrics comparison.

vertical levels from the surface up to 0.1 hPa. For comparison with station data, the interpolated fields (interpolated to 0.125°) as obtained from the ECMWF website are taken.

We also used the ERA-20C reanalysis (Poli et al., 2013; Dee et al., 2014), also accessed through http://apps.ecmwf.int/datasets/. This reanalysis provides a global realization at spectral resolution T159 which corresponds to the horizontal resolution of the reduced Gaussian grid of ∼ 125 km. The data assimilation method applied is an updated version of the 4D-Var scheme (IFS Cy38r1) and used a 24-h assimilation window but it assimilated observations of surface pressure and surface marine winds only. From ERA-20C, the 10 m winds as pre-processed by ECMWF and interpolated to 0.125° horizontal resolution were taken for this comparison study.

The third reanalysis product used is the COSMO-based regional reanalysis of DWD's Hans-Ertel Centre for Weather Research (HErZ), University of Bonn (Bollmeyer et al., 2014, http://www.herz-tb4.uni-bonn.de/) called COSMO-REA6. The regional reanalysis COSMO-REA6 is driven via the global reanalysis boundary conditions (ERA-Interim) every 3 h. It is based on the COSMO-model and applies nudging as the data assimilation technique. Here we use data from a configuration that was run for the European CORDEX domain at a nominal resolution of 0.055°. From COSMO-REA6, the wind speed at the ground layer (10 m) as well as feedback files containing radiosonde data were used in the work below.

The station data and associated metadata were taken from the freely accessible climate data archive of Deutscher Wetterdienst (DWD) which is available through ftp://ftp-cdc.dwd.de/pub/CDC/observations_germany/climate/ (Kaspar et al., 2013). Available are the station data and associated metadata describing the particularities of the station instrumentation and measurement time series. From this archive (August 2014, version v002), the example station observations below were chosen. The station observations of wind speeds are not assimilated within ERA-Interim and ERA-20C, i.e., are independent measurements. For COSMO-REA6, station data below 100 m are assimilated, above 100 m they are not, i.e., the former may be dependent, the latter are independent measurements.

The North Atlantic Oscillation (NAO) index was taken from http://www.cru.uea.ac.uk/~timo/datapages/naoi.htm which is an extension of Jones et al. (1997).

For our comparison, we took the grid cells matching the location of the station from the regional reanalysis, and from the global reanalyses interpolated by 0.125° (provided by ECMWF). Because it cannot be expected that processes on the sub-daily scale can be resolved in the archived global reanalysis datasets, we compare frequency distributions of daily means. The correlation coefficients were calculated with monthly means, which were calculated from daily means, to avoid the issue of calculating correlations with incomplete time series. Here we restrict ourselves to the time period 2007 to 2010, which is the overlap period of COSMO-REA6 at our disposal. In this time span, we are relatively confident no large changes in surroundings of the stations occurred, the method of measurements remained constant, and no gross errors are expected. As users would especially be interested in information about the past spanning many decades, we include ERA-20C into our preparatory comparisons because of its length. ERA-Interim is used as a benchmark. Based on the comparison, we discuss the potential of reanalyses to enhance the traditional approaches for assessing wind variability based on the 10 m station winds.

3 Results

For an illustration of spatial variability, the 10 m surface winds from the regional reanalysis COSMO-REA6 and the interpolated global reanalysis ERA-20C for February 2007 are shown in Fig. 2. The higher wind speeds over the North and Baltic Seas are prominent in both reanalyses as well as the sharp decline at the coast line. As expected, due to the much finer horizontal resolution of the COSMO-REA6 regional reanalysis, higher spatial variability is present here. Note the coincidences of wind features with topographical features, e.g., mountain ridges in the South of East Germany (compare to Fig. 3). The features related to topography are more pronounced in the regional reanalysis, e.g., the land-sea transition and the wind speed variability over the heterogeneous low mountain ranges in Mid-Germany. Note, however, the close match of the average absolute values of both reanalysis datasets.

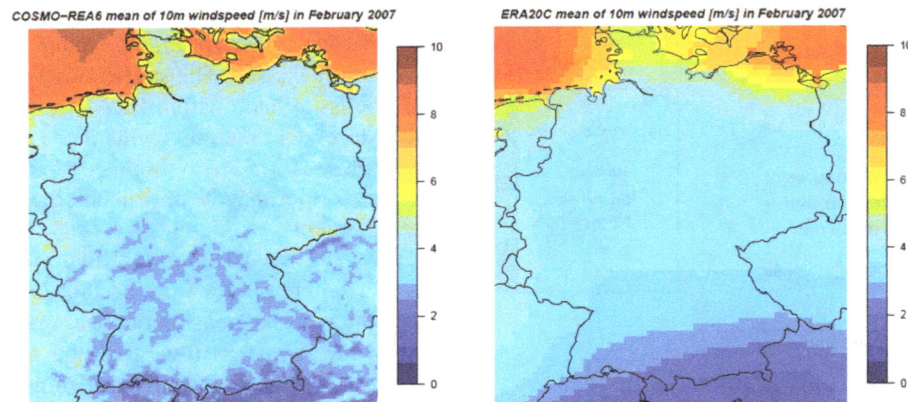

Figure 2. 10 m surface winds from the COSMO-REA6 regional reanalysis (left) and the ERA-20C interpolated wind fields (right) illustrate the different spatial resolution for the month of February 2007.

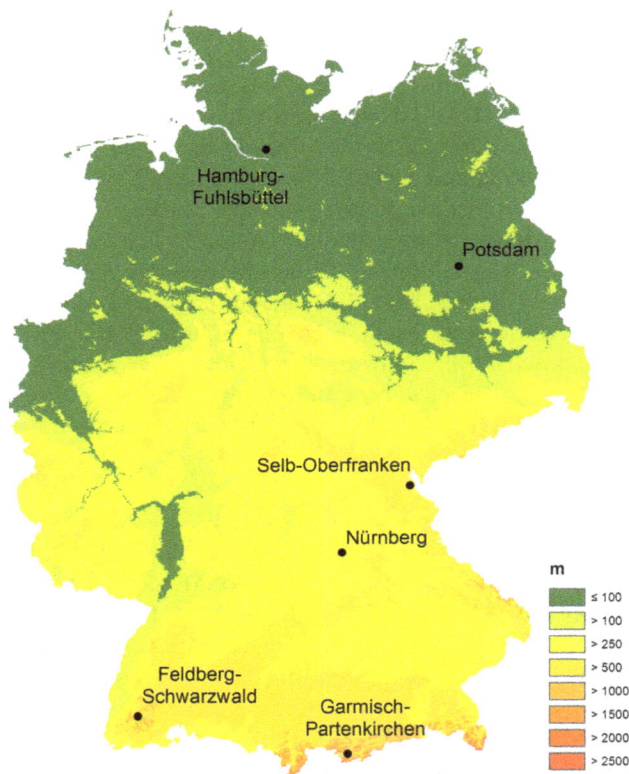

Figure 3. Topography of Germany with colour-coded height above sea level, and location of selected stations.

3.1 Frequency distributions

We compare the frequency distributions of daily-mean wind speed from the reanalyses with the distribution from the station observations, using the respective interpolated reanalysis grid cells containing the geographical location of the station. Daily-mean diagnostics were chosen to accommodate differ-

ences in temporal sampling of the reanalyses (3 vs. 6 h for ERA-20C and ERA-Interim respectively).

For most station locations, the frequency distributions derived from the reanalyses match well with the one from the observations (e.g., Nuremberg, see Fig. 4). This might come as a surprise, given the rather coarse grid resolutions, and that the 10 m winds are expected to be strongly influenced by the local scale topography. Some stations show an offset in wind speed, due to the model height differing from the real topographic height, e.g., mountain stations like Feldberg-Schwarzwald (Fig. 5). For some applications, fitness-for-purpose would therefore be improved by determining a more representative model height in cases of complex topography. Local influences of topography can result in pronounced differences in the frequency distributions. As an extreme example, the station Garmisch-Partenkirchen (Fig. 6) is shown, which is located in a valley for which the restricted representativity of the station cannot match the scales resolved in the reanalyses. Generally, only few histogram mismatches are found (and less for the higher resolved COSMO-REA6 reanalysis). An unexpected outcome was that some dubious station data could be identified: Fig. 7 shows an unusually large proportion of zero wind-speed values; it seems that these were in fact reported on occasions where no observations were performed. This illustrates the use of reanalysis data to support quality control for observed wind data.

Wind energy applications often require Weibull parameters, thus Weibull distributions with two parameters had been fitted to the frequency distributions:

$$P(v) = \frac{k}{c}\left(\frac{v}{c}\right)^{k-1} e^{-\left(\frac{v}{c}\right)k}, \tag{1}$$

where v is the wind speed ($v \geq 0$), k is the shape parameter ($k > 0$), and c is the scale parameter of the distribution ($c > 0$). The Weibull parameters and error estimates are calculated with the maximum likelihood method implemented in the Cran R-project package *fitdistrplus* (Venables and Rip-

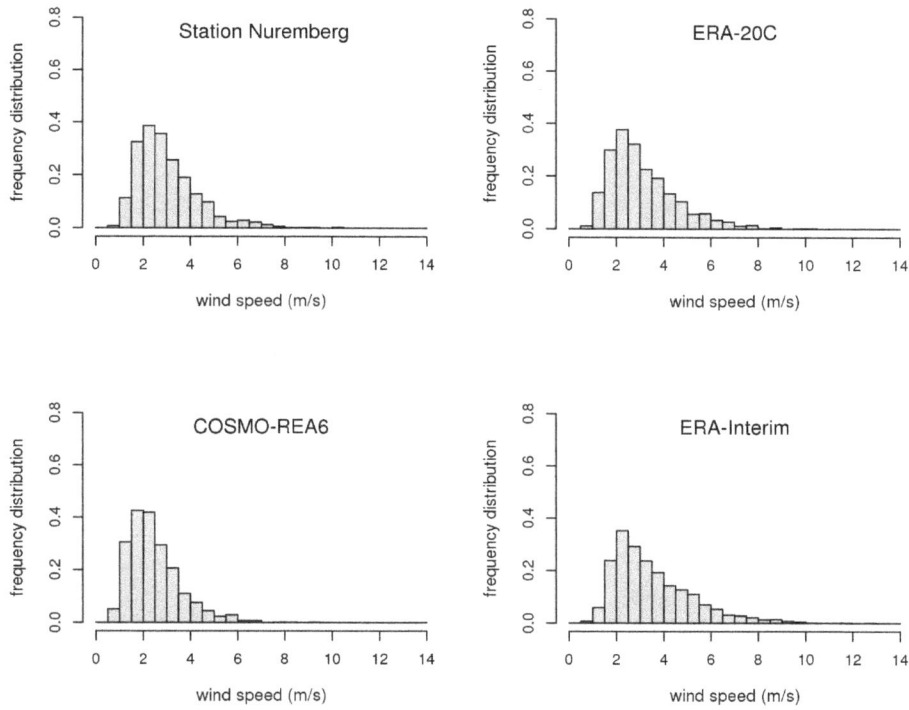

Figure 4. Frequency distribution of daily mean wind speeds in the period 2007 to 2010 for the station Nuremberg (top left) and matching grid cells of ERA-20C (top right), COSMO-REA6 (bottom left), and ERA-Interim wind fields (bottom right).

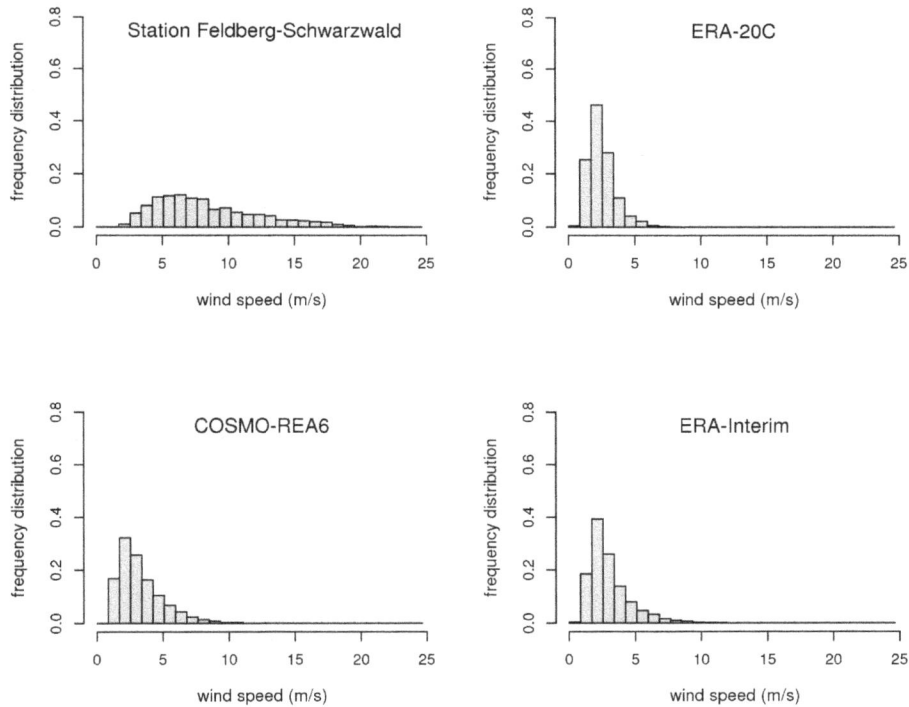

Figure 5. Frequency distribution of daily mean wind speeds in the period 2007 to 2010 for the station Feldberg-Schwarzwald (top left) and matching grid cells of ERA-20C (top right), COSMO-REA6 (bottom left), and ERA-Interim wind fields (bottom right).

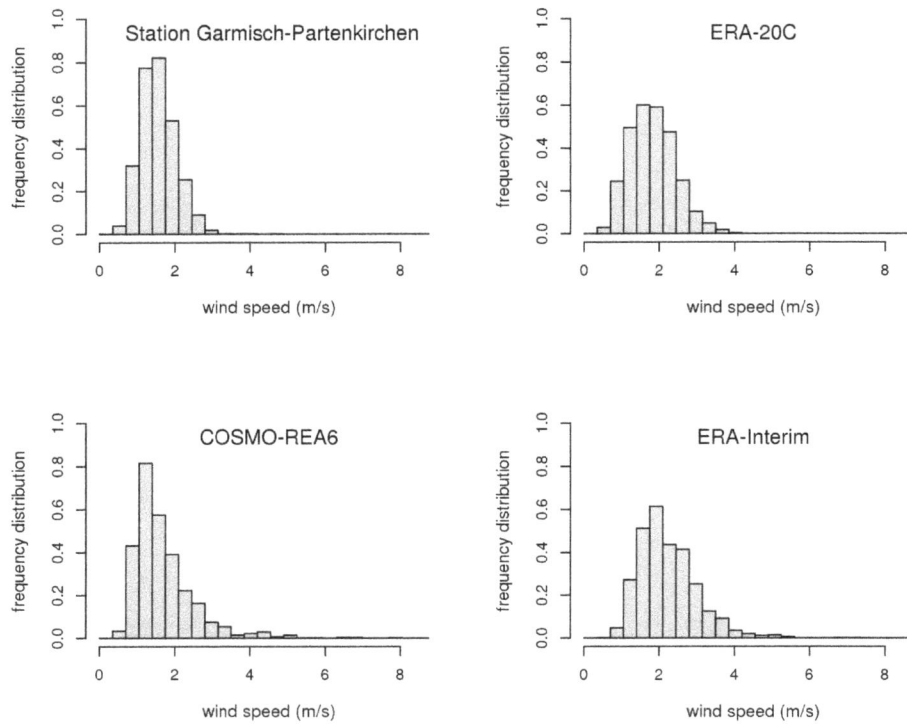

Figure 6. Frequency distribution of daily mean wind speeds in the period 2007 to 2010 for the example station Garmisch-Partenkirchen (top left) and matching grid cells of ERA-20C (top right), COSMO-REA6 (bottom left), and ERA-Interim wind fields (bottom right).

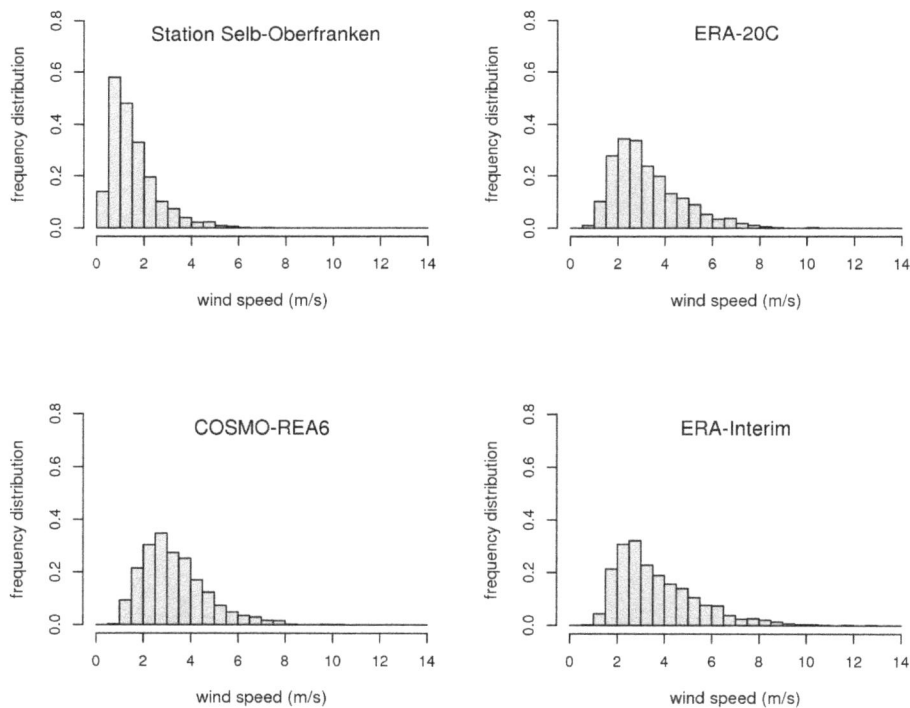

Figure 7. Frequency distribution of daily mean wind speeds in the period 2007 to 2010 for the station Selb-Oberfranken (top left) and matching grid cells of ERA-20C (top right), COSMO-REA6 (bottom left), and ERA-Interim wind fields (bottom right).

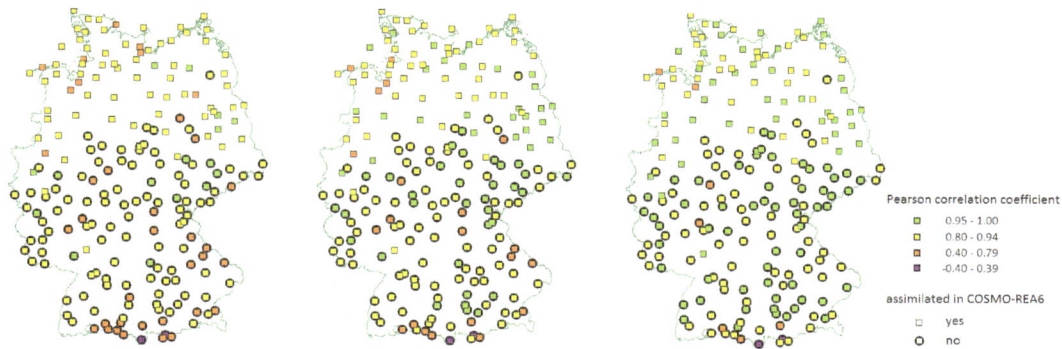

Figure 8. Pearson correlation coefficient of monthly mean wind speeds (calculated from daily wind speeds) between station observations and reanalyses: ERA-20C (left), ERA-Interim (middle), and COSMO-REA6 (right). Stations not assimilated in COSMO-REA6 are highlighted with grey circles.

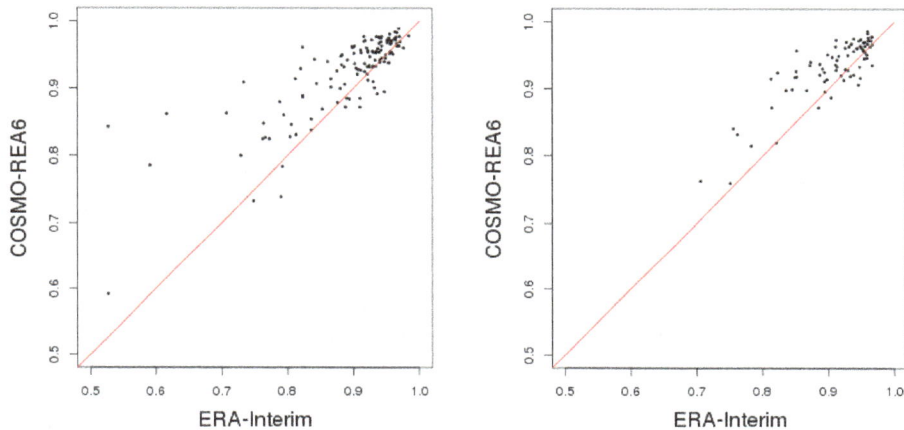

Figure 9. Scatterplots of Pearson correlation coefficients for stations that are not assimilated in COSMO-REA6 (left) and stations that are assimilated (right). Correlations are calculated between monthly mean wind speeds of station observations and COSMO-REA6 (*y* axis) and compared to correlations between station observations and ERA-Interim (*x* axis), respectively. Three of the total 210 stations are outliers (correlation below 0.5) and are not shown.

ley, 2010; Delignette-Muller et al., 2014). The Weibull parameters for selected stations are compared in Tables 1 and 2, with the dependence on station altitude (listed in Table 3) clearly prominent. For Hamburg-Fuhlsbüttel, the Weibull parameters k_s (shape) and c_s (scale) of the station observations match well with the ones from reanalyses, also for Potsdam and Nuremberg. Mountain stations like Feldberg-Schwarzwald have significantly different c (scale) parameters. Here a more representative height would be needed to be determined for each reanalysis. The situations in Garmisch-Partenkirchen and Selb-Oberfranken are different: The suspect histogram of the station data from Selb-Oberfranken (Fig. 7) causes a deviation of k_s (shape) and c_s (scale) due to issues with the missing values of the station recordings. For stations in complex topography, we expect significant differences in Weibull parameters due to the difference of real topography and model topography and the coarse reanalysis resolution. A close match of Weibull parameters (derived

from stations and from reanalyses) could still be due to a favourable match of height above ground and above model ground. Thus we need to examine the correlation coefficients to judge whether reanalysis winds and station observations are in accordance. For instance, the Weibull parameters of the station Garmisch-Partenkirchen are not close, but still comparable to the ones from reanalysis (see Tables 1 and 2), but there is no correlation found between the time series of monthly means (see Table 3) which can be explained by its location in a valley.

3.2 Correlation of reanalysis data with monthly station data

The correlation of monthly wind speed from the ERA-20C, ERA-Interim, and COSMO-REA6 reanalyses with the 210 stations over Germany were calculated with the Pearson correlation for the period 2007 to 2010. In case of ERA-20C,

Table 1. Weibull parameters k (shape) derived from selected station daily mean wind observations (k_s), and from ERA-20C, ERA-Interim, and COSMO-REA6 interpolated grid cells matching the station locations, together with their 1 σ standard deviations (SD).

	k_s (SD)	$k\_$ERA-20C (SD)	$k\_$ERA-Interim (SD)	$k\_$COSMO-REA6 (SD)
Hamburg-Fuhlsbüttel	2.65 (0.05)	2.61 (0.05)	2.56 (0.05)	2.62 (0.05)
Potsdam	2.89 (0.05)	2.56 (0.05)	2.48 (0.05)	2.73 (0.05)
Nuremberg	2.43 (0.04)	2.32 (0.04)	2.25 (0.04)	2.35 (0.04)
Selb-Oberfranken	1.68 (0.03)	2.36 (0.04)	2.26 (0.04)	2.56 (0.05)
Garmisch-Partenkirchen	3.40 (0.06)	3.26 (0.06)	2.94 (0.05)	2.17 (0.04)
Feldberg-Schwarzwald	2.26 (0.05)	2.55 (0.05)	2.07 (0.04)	2.07 (0.04)

Table 2. Weibull parameters c (scale) derived from selected station daily mean wind observations (c_s), and from ERA-20C, ERA-Interim, and COSMO-REA6 interpolated grid cells matching the station locations, together with their 1 σ standard deviations (SD).

	c_s (SD)	$c\_$ERA-20C (SD)	$c\_$ERA-Interim (SD)	$c\_$COSMO-REA6 (SD)
Hamburg-Fuhlsbüttel	4.46 (0.05)	4.49 (0.05)	4.77 (0.05)	3.83 (0.04)
Potsdam	4.78 (0.05)	3.90 (0.04)	4.44 (0.05)	3.40 (0.03)
Nuremberg	3.39 (0.04)	3.50 (0.04)	3.99 (0.05)	2.79 (0.03)
Selb-Oberfranken	1.72 (0.03)	3.66 (0.04)	4.17 (0.05)	3.73 (0.04)
Garmisch-Partenkirchen	1.73 (0.01)	2.03 (0.02)	2.49 (0.02)	1.92 (0.02)
Feldberg-Schwarzwald	9.50 (0.12)	2.80 (0.03)	3.40 (0.05)	3.69 (0.05)

82 % of stations show a correlation coefficient greater than or equal to 0.8 and 47 % greater than or equal to 0.9. For ERA-Interim, these correlations are achieved at more stations (89 and 66 %, respectively), and yet more in COSMO-REA6 (96 and 80 %, respectively). The COSMO-REA6 reanalysis assimilates station observations below 100 m height; whether locations contribute assimilated observations or not can be distinguished in Figs. 8 and 9. In our case, 130 stations are not assimilated, i.e., are independent from COSMO-REA6. These independent stations show generally an improvement of correlation with COSMO-REA6 (Fig. 9). In Fig. 8 it can be seen that the regional reanalysis improves monthly correlations where it can be expected, namely in the areas with more complex topography (South of Germany, towards the Alps and Bayrischer Wald in the South-West). Three stations in complex terrain (Garmisch-Partenkirchen is one of them) have no correlation with the reanalyses and remain outside the scales of Fig. 9. Still, the correlations over time do hold (see Table 3) for several exposed mountain stations, like Feldberg-Schwarzwald.

Single stations only sometimes show statistically significant differences between the correlations with the three reanalyses, but the combined effect of the 210 stations becomes clear in Fig. 8, and especially in Fig. 9.

3.3 Interannual variability related to NAO

For many users, notably renewable energy users, the spatial characteristics and inter-annual variability of weather and cli-

mate and its spatial extent are of special interest. The North Atlantic Oscillation (NAO) (Walker and Bliss, 1932; Hurell, 1995), is the main synoptic mode of atmospheric circulation and climate variability for Germany. Thus it is of interest, what the magnitude of the NAO-related effect is for the station data and to what extend the NAO-related effect is captured in the surface fields of the reanalysis over Germany.

For the recent years, the correlations of the stations and reanalysis fields are high. The question is, whether the correlation caused by NAO-related effects holds also for the past, keeping in mind the reanalysis systems might have long-term variability connected to changes of the observing system (Dee, 2005; Dee and Uppala, 2009). To describe the magnitude of the effect, we select seasonal averages for selected seasons corresponding to high, neutral, and low values of NAO index, and calculate the anomalies relative to the average of the period 2007 to 2010. This is done with the interpolated ERA-20C near-surface fields. The station anomalies are calculated analogously, i.e., their reference is also derived from the station averages. In Fig. 10, we show examples for seasons from 2000 and 2002. The NAO-related anomalies show a west-east gradient (left), or a north-south gradient (middle and right) over Germany. The first panel (left) corresponds to a neutral NAO index autumn with higher than usual wind speeds over the western part and lower wind speeds in the East. With a negative NAO index, the average wind pattern is shifted southwards, leaving the North Sea area with lower wind speeds on average (middle panel). For the case of a positive NAO index, the wind pattern is shifted towards

Table 3. Pearson correlation coefficient together with its 1σ-confidence interval, between mean monthly wind speeds from selected station observations and matching grid cells from interpolated ERA-20C, interpolated ERA-Interim, and COSMO-REA6. Note that for ERA-20C and ERA-Interim, all station observations are independent. For COSMO-REA6, station observations are assimilated (nudged) as indicated in the last column.

Station	Altitude	r_ERA-20C (conf. int.)	r_ERA-Interim (conf. int.)	r_COSMO-REA6 (conf. int.)	nudged
Hamburg-Fuhlsbüttel	11	0.93 (0.90, 0.94)	0.96 (0.95, 0.97)	0.98 (0.97, 0.98)	yes
Potsdam	81	0.94 (0.92, 0.96)	0.96 (0.95, 0.97)	0.98 (0.98, 0.99)	yes
Nuremberg	314	0.93 (0.90, 0.94)	0.92 (0.89, 0.94)	0.97 (0.96, 0.98)	no
Selb-Oberfranken	535	0.86 (0.82, 0.90)	0.89 (0.86, 0.92)	0.89 (0.85, 0.91)	no
Garmisch-Partenkirchen	719	−0.23 (−0.36, −0.09)	−0.11 (−0.38, 0.18)	−0.04 (−0.18, 0.12)	no
Feldberg-Schwarzwald	1490	0.90 (0.87, 0.92)	0.89 (0.85, 0.91)	0.93 (0.91, 0.95)	no

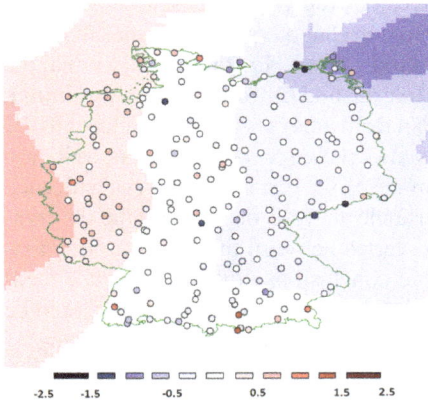 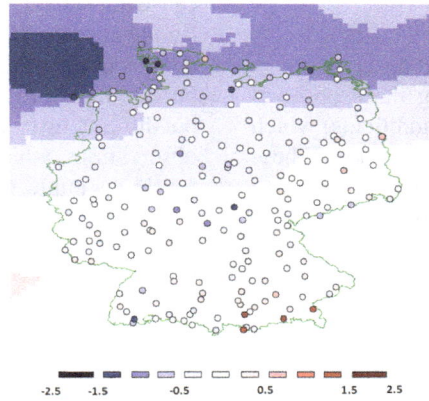 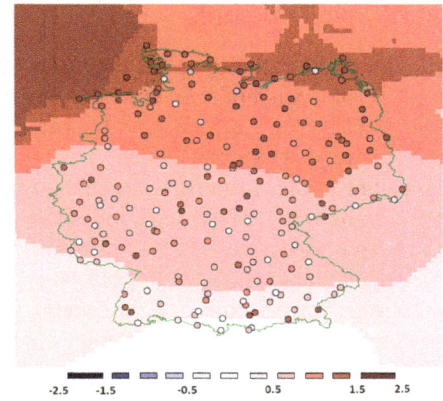

Figure 10. Seasonal anomalies of the ERA-20C 10 m wind speeds compared to the reference period 2007 to 2010 for September, October, November (SON) of 2000 at neutral NAO-index (left panel), SON2002 at negative NAO-index (middle), and December, January, February (DJF) 1999/2000 at positive NAO-index (right).

the North (left panel). Note that, allowing for some scatter among the stations, they generally match the spatial extent and the magnitude of the anomaly present in the ERA-20C fields. The examples picked here have months with a similar NAO-index, and a rather pronounced NAO anomaly. Of course, the inter-annual wind variability is dependent not only on the NAO strength, but also on details of the jet stream position, and the individual storm tracks, when moving to shorter time scales and selected station positions. Our conclusion here is that the inter-annual variability in the near surface winds captured in the reanalyses is consistent with the station time series, considering a time scale of several years. We can detect this common NAO-pattern (in both reanalysis and station data) in recent years. To extend our studies to longer time scales, we would have to separate the effect

of instrumental changes at the stations from possible effects of NAO-related variability. For instance, Welker and Martius (2015) found a positive correlation between the NAO-index and occurrence of high wind speeds over Switzerland, deduced from the global 20CR reanalysis (Compo et al., 2011) on the inter-annual time scale. Welker and Martius (2015) found an increase of the correlation at the decadal time scale; and point to a change of strength of this correlation over time due to an eastward shift of the NAO pattern.

3.4 Comparing feedback statistics

The opportunities to compare reanalyses against stable time series of fully-independent observations covering a long time-period are limited. Fortunately, the reanalysis process

RMSE

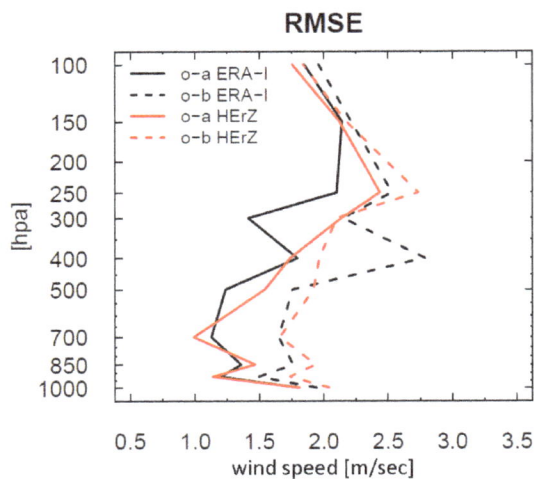

Figure 11. Root mean square errors of analysis departures (o-a, solid line) and background departures (o-b, dashed line) for Lindenberg radiosonde observations of wind speed at standard pressure levels, in January 2011: feedback statistics for the ERA-Interim (black) and COSMO-REA6 (red).

provides opportunities to use the assimilated observations in a quasi-independent manner, by comparing these observations against the free forecasts (or background fields) which are started from the preceding (re-)analysis window. These so-called feedback statistics are in fact routinely produced by the data assimilation system, and relate assimilated observations, free forecasts (i.e., background fields), analysis increments, and analysis fields to each other. They yield valuable additional information, e.g., estimates of the analysis error or diagnostics on systematic changes in increments which, if traced to biases in observations or model (or both), indicate deficiencies in the system. Favourable statistics may show that the frequency distribution and time series of observed and reanalysed parameters are matching. Thus, it is potentially of high practical value for the user to take into account the results of feedback statistics.

As a demonstration of concept, here we analyse the differences between observations (o), background (b), and (re-)analysis (a) of wind speed for a global (ERA-Interim) and the COSMO-REA6 regional reanalysis, for the Lindenberg radiosondes launched at 06Z (see Fig. 11). Care must be taken to make the differences arising from different reanalysis systems as comparable as possible. In particular, the uncertainties of the background fields from the reanalysis systems should be made as similar as possible. These uncertainties depend on the initial condition uncertainty and the growth of this uncertainty with forecast lead time. The forecast lead times of both systems should be comparable. In our example, COSMO background fields valid at 06Z are taken from forecasts initialized at 0Z and propagated for a lead time of 6 h. The most comparable ERA-Interim background fields are from forecasts initialized at 3Z and prop-

agated for a lead time of 3 h. In addition, the sounding data that have influenced the initial conditions for the background forecasts in both systems are comparable. At leading order, the regional and global reanalyses both match the sounding observations to the same extent, for the statistical sample investigated here. The match is not uniform, for example ERA-Interim o-b exhibits large values around 400 hPa while COSMO o-b exceeds ERA-Interim o-b between 1000 and 700 hPa. While the limited sample shown here does not permit such discrepancies to be definitively traced to specific system deficiencies, this example highlights a more general issue: namely, that relatively few users have the tools and skills to identify such discrepancies, or to make informed assessments of any consequent impacts in their particular application, and hence lack clarity on whether a dataset is fit for their specific purpose. We believe that these are important issues that warrant attention in the context of current efforts to develop climate services.

To draw conclusion on the performance of the different systems, a comparison of observations to background (o-b) is easier to interpret than observations to analysis (o-a), because the background can be regarded, in first approximation (in the absence of time-correlated errors), as rather independent from the observations. From Fig. 11 we can see that the background of ERA-Interim and COSMO-REA6 are similar, but differ to up to 25 %. Both analyses are drawn closer to the observations, which shows the effect of the data assimilation for this example. Comparing the resulting analyses with each other is harder, because the analysis depends on the observations to a degree that varies between the different assimilation systems. For more conclusions, longer periods of feedback statistics with comparable forecast lead times should be analysed, which were not available at the time of writing.

4 Summary and conclusions

Here we inter-compared the frequency distributions, Weibull parameters and monthly to inter-annual variability of regional and global reanalysis near surface wind fields of the recent years (2007–2010). We illustrated three approaches, namely direct comparison (with independent station measurements), thematic comparison (the magnitude and spatial extend of NAO-related mean wind speed anomalies) and internal metrics (feedback statistics).

The 2σ confidence intervals of the Weibull parameters derived from the reanalysis wind fields overlap the ones derived from the stations in most cases for the Weibull shape parameter (k), whereas the Weibull scale parameters (c) differ more. With respect to correlations, generally the ERA-20C fits the station data well, ERA-Interim fits better, and the regional COSMO-REA6 reanalysis is closest to the station data. This is valid for the daily, monthly, and seasonal scale, for the period investigated here (2007 to 2010). It has to be kept in mind that the station observations situated below 100 m a.s.l. are assimilated within COSMO-REA6. The increased correlation with COSMO-REA6 holds true regardless whether or not a station contributes observations that are assimilated in COSMO-REA6. Noticeable improvement for COSMO-REA6 correlations are found across Germany, for instance at stations in Southern Germany, closer to the Alps (not assimilated in COSMO-REA6), and at the coastline (assimilated in COSMO-REA6).

As expected from the fact that the regional COSMO-REA6 reanalysis is driven by the global ERA-Interim reanalysis via boundary condition, the feedback statistics exhibit a similar fit to the Lindenberg radio soundings for both reanalysis systems for our example month of January 2011.

We demonstrated that although local and regional effects can be expected to determine the variability in wind fields measured at the stations (10 m over ground), the frequency distributions of mean wind speeds match quite well with the ones of the reanalysis fields for which such effects are at the sub-grid scale. The correlation of ERA-20C, ERA-Interim and COSMO-REA6 monthly means with station observations is high (> 0.8) for the majority of the German stations. Thus we conclude that the monthly and seasonal anomalies recorded at these stations can be understood as representative for a wider spatial area, comparable to the resolution of the reanalyses, at least for the recent years. Due to the shortness of period, we cannot make such a statement concerning inter-annual variability, though the NAO-related anomalies indicate there is also coherence at the inter-annual scale.

The correlation holds for ERA-20C, even though fewer observations are assimilated than in ERA-Interim and the spatial resolution is significantly reduced. The correlation with COSMO-REA6 reanalysis is highest, because of several, and possibly combined, reasons: the higher resolution of the model, the regional data assimilation (nudging), and possibly also the higher temporal resolution of the output. It should be kept in mind that COSMO-REA6 is forced every 3 h by ERA-Interim, and we compare aggregated daily means from hourly COSMO-REA6 output with daily means from 6-hourly ERA-Interim output. While skill is inherited from ERA-Interim via boundary conditions, we get the additional benefit from COSMO-REA6, possibly also from the temporal resolution. Particular the stations with low correlations show a strong improvement, pointing to the expected difference in the reanalysis performance, namely that the smaller scales which are resolved with COSMO-REA6 are of importance for the monthly mean near surface wind speeds, i.e., hinting to that a simple scaling of global reanalysis would not give the same information (on the daily to monthly time scale).

Further analysis would be needed to identify under which circumstances the differences are most pronounced, and to what extent the low-frequency information from global reanalyses is represented or improved with regional reanalyses for regional-scale parameters.

Author contributions. The ideas and methodologies were developed by Andrea Kaiser-Weiss, Frank Kaspar, David Tan, Paul Poli and Hilppa Gregow during the CORE-CLIMAX project. Analyses in the manuscript were conducted by Andrea Kaiser-Weiss, Frank Kaspar, Vera Heene, Michael Borsche and Andre Obregon. Andrea Kaiser-Weiss prepared the manuscript with contribution from all co-authors.

Acknowledgements. Andre Obregon and Vera Heene were supported through the CORE-CLIMAX (grant no. 313085 within the EU Seventh Framework Programme). Michael Borsche is supported by the UERRA project (grant no. 607193 within the EU Seventh Framework Programme). We thank Christoph Bollmeyer and Liselotte Bach from University of Bonn for the provision of the COSMO-REA6 feedback files. We thank Karsten Friedrich (DWD) for his help with the figures. We would like to acknowledge helpful discussions with Hermann Mächel (DWD). We thank our two reviewers for their constructive criticism which improved the manuscript.

References

Bichet, A., Wild, M., Folini, D., and Schär, C.: Causes for decadal variations of wind speed over land: Sensitivity studies with a global climate model. Geophys. Res. Lett., 39, L11701, doi:10.1029/2012GL051685, 2012.

Bollmeyer, C., Keller, J. D., Ohlwein, C., Wahl, S., Crewell, S., Friederichs, P., Hense, A., Keune, J., Kneifel, S., Pscheidt, I., Redl, S., and Steinke, S.: Towards a high-resolution regional reanalysis for the European CORDEX domain, Q. J. Roy. Meteorol. Soc., 141, 1–15, doi:10.1002/qj.2486, 2014.

Bosilovich, M. G., Kennedy, J., Dee, D., Allan, R., and O'Neill, A.: On the Reprocessing and Reanalysis of Observations for Climate. Climate Science for Serving Society: Research, Modelling and Prediction Priorities, edited by: Asrar, G. and Hurrell, J. W., Springer Netherlands, 51–71, 2013.

Compo, G. P., Whitaker, J. S., Sardeshmukh, P. D., Matsui, N., Allan, R. J., Yin, X., Gleason, B. E., Vose, R. S., Rutledge, G., Bessemoulin, P., Brönnimann, S., Brunet, M., Crouthamel, R. I., Grant, A. N., Groisman, P. Y., Jones, P. D., Kruk, M. C., Kruger, A. C., Marshall, G. J., Maugeri, M., Mok, H. Y., Nordli, Ø., Ross, T. F., Trigo, R. M., Wang, X. L., Woodruff, S. D., and Worley, S. J.: The Twentieth Century Reanalysis Project. Q. J. Roy. Meteorol. Soc., 137, 1–28, doi:10.1002/qj.776, 2011.

Dee, D. P.: Bias and data assimilation, Q. J. Roy. Meteorol. Soc., 131, 3323–3343, doi:10.1256/qj.05.137, 2005.

Dee, D. P. and Uppala, S.: Variational bias correction of satellite radiance data in the ERA-Interim reanalysis, Q. J. Roy. Meteorol. Soc., 135, 1830–1841, doi:10.1002/qj.493, 2009.

Dee, D. P., Uppala, S. M., Simmons, A. J., Berrisford, P., Poli, P., Kobayashi, S., Andrae, U., Balmaseda, M. A., Balsamo, G., Bauer, P., Bechtold, P., Beljaars, A. C. M., van de Berg, L., Bidlot, J., Bormann, N., Delsol, C., Dragani, R., Fuentes, M., Geer, A. J., Haimberger, L., Healy, S. B., Hersbach, H., Hólm, E. V.,

Isaksen, L., Kållberg, P., Köhler, M., Matricardi, M., McNally, A. P., Monge-Sanz, B. M., Morcrette, J.-J., Park, B.-K., Peubey, C., de Rosnay, P., Tavolato, C., Thépaut, J.-N., and Vitart, F.: The ERA-Interim reanalysis: configuration and performance of the data assimilation system, Q. J. Roy. Meteorol. Soc., 137, 553–597, doi:10.1002/qj.828, 2011.

Dee, D. P., Balmaseda, M., Balsamo, G., Engelen, R., Simmons, A. J., and Thepaut, J.-N.: Towards a consistent reanalysis of the climate system, B. Am. Meteorol. Soc., 95, 1235–1248, doi:10.1175/BAMS-D-13-00043.1, 2014.

Delignette-Muller, M., Pouillot, R., Denis, J., and Dutang, C.: _tdistrplus: Help to Fit of a Parametric Distribution to Non-Censored or Censored Data, R package version 1.0-2, available at: http://cran.r-project.org (last access: 1 April 2015), 2014.

Gerth, W. P. and Christoffer, J.: Windkarten von Deutschland, Meteorolog. Zeitschrift, N.F., 3, 67–77, 1994.

Gregow, H., Poli, P., Mäkelä, H. M., Jylhä, K., Kaiser-Weiss, A. K., Obregon, A., Tan, D. G. H., Kekki, S., and Kaspar, F.: User awareness concerning feedback data and input observations used in reanalysis systems, Adv. Sci. Res., 12, 63–67, doi:10.5194/asr-12-63-2015, 2015.

Hartmann, D. L., Klein Tank, A. M. G., Rusticucci, M., Alexander, L. V., Brönnimann, S., Charabi, Y., Dentener, F. J., Dlugokencky, E. J., Easterling, D. R., Kaplan, A., Soden, B. J., Thorne, P. W., Wild, M., and Zhai, P. M.: Observations: Atmosphere and Surface, in: Climate Change 2013: The Physical Science Basis. Contribution of Working Group I to the Fifth Assessment Report of the Intergovernmental Panel on Climate Change. Cambridge University Press, Cambridge, United Kingdom and New York, NY, USA, 2013.

Hurrell, J. W.: Decadal trends in the North Atlantic oscillation: regional temperature and precipitation, Science, 269, 676–697, 1995.

Jones, P. D., Jónsson, T., and Wheeler, D.: Extension to the North Atlantic Oscillation using early instrumental pressure observations from Gibraltar and South-West Iceland, Int. J. Climatol., 17, 1433–1450, 1997.

Kaspar, F., Müller-Westermeier, G., Penda, E., Mächel, H., Zimmermann, K., Kaiser-Weiss, A., and Deutschländer, T.: Monitoring of climate change in Germany – data, products and services of Germany's National Climate Data Centre, Adv. Sci. Res., 10, 99–106, doi:10.5194/asr-10-99-2013, 2013.

McVicar, T. R., Van Niel, T. G., Li, L. T., Roderick, M. L., Rayner, D. P., Ricciardulli, L., and Donohue R. J., Wind speed climatology and trends for Australia, 1975–2006: Capturing the stilling phenomenon and comparison with near-surface reanalysis output, Geophys. Res. Lett., 35, L20403, doi:10.1029/2008GL035627, 2008.

McVicar, T. R., Roderick, M. L., Donohue, R. J., Li, L. T., Van Niel, T. G., Thomas, A., Grieser, J., Jhajharia, D., Himri, Y., Mahowald, N. M., Mescherskaya, A. V., Kruger, A. C., Rehman, S., and Dinpashoh, Y.: Global review and synthesis of trends in observed terrestrial near-surface wind speeds: Implications for evaporation, J. Hydrol., 416–417, 182–205, 2012.

Poli, P., Hersbach, H., Tan, D., Dee, D., Thépaut, J.-N., Simmons, A., Peubey, C., Laloyaux, P., Komori, T., Berrisford, P., Dragani, R., Trémolet, Y., Holm, E., Bonavita, M., Isaksen, L., and Fisher, M.: The data assimilation system and initial performance evaluation of the ECMWF pilot reanalysis of the 20th-century assimilating surface observations only (ERA-20C), ERA report series No. 14, European Centre for Medium-Range Weather Forecasts, Reading, England, 2013.

Smits, A., Klein-Tank, A. M. G., and Können, G. P.: Trends in storminess over the Netherlands, 1962–2002, Int. J. Climatol., 25, 1331–1344, doi:10.1002/joc.1195, 2005.

Vautard, R., Cattiaux, J., Yiou, P., Thépaut, J.-N., and Ciais, P.: Northern Hemisphere atmospheric stilling partly attributed to an increase in surface roughness, Nature GeoScience, 3, 756–761, 2010.

Venables, W. N. and Ripley, B. D.: Modern Applied Statistics with S, 4th Edn., Springer-Verlag, 2010.

Walker, G. T. and Bliss, E. W.: World Weather V, Mem. R. Meteorol. Soc., 4, 53–84, 1932.

Welker, C. and Martius, O.: Large-scale atmospheric flow conditions and sea surface temperatures associated with hazardous winds in Switzerland, Clim. Dynam., 44, 1857–1869, doi:10.1007/s00382-014-2404-1, 2015.

Wever, N.: Quantifying trends in surface roughness and the effect on surface wind speed observations, J. Geophys. Res., 117, D11104, doi:10.1029/2011JD017118, 2012.

23

Representation of the grey zone of turbulence in the atmospheric boundary layer

Rachel Honnert

CNRM-Météo-France, CNRM/GMAP, Toulouse, France

Correspondence to: Rachel Honnert (rachel.honnert@meteo.fr)

Abstract. Numerical weather prediction model forecasts at horizontal grid lengths in the range of 100 to 1 km are now possible. This range of scales is the "grey zone of turbulence". Previous studies, based on large-eddy simulation (LES) analysis from the MésoNH model, showed that some assumptions of some turbulence schemes on boundary-layer structures are not valid. Indeed, boundary-layer thermals are now partly resolved, and the subgrid remaining part of the thermals is possibly largely or completely absent from the model columns. First, some modifications of the equations of the shallow convection scheme have been tested in the MésoNH model and in an idealized version of the operational AROME model at resolutions coarser than 500 m. Secondly, although the turbulence is mainly vertical at mesoscale (> 2 km resolution), it is isotropic in LES (< 100 m resolution). It has been proved by LES analysis that, in convective boundary layers, the horizontal production of turbulence cannot be neglected at resolutions finer than half of the boundary-layer height. Thus, in the grey zone, fully unidirectional turbulence scheme should become tridirectional around 500 m resolution. At Météo-France, the dynamical turbulence is modelled by a K-gradient in LES as well as at mesoscale in both MésoNH and AROME, which needs mixing lengths in the formulation. Vertical and horizontal mixing lengths have been calculated from LES of neutral and convective cases at resolutions in the grey zone.

1 Introduction

The grey zone of turbulence is defined by Wyngaard (2004) as the scales on the order of the energy-containing turbulence scale. At these resolutions, the turbulence structures are neither entirely subgrid scale (as in global and mesoscale models) nor largely resolved (as in large-eddy simulations – LESs). Honnert et al. (2011) used LES coarse-graining to produce similarity functions linking the subgrid or resolved part of the turbulent fluxes and the horizontal resolution of the model out of the height of the thermals. They indicated that the grey zone exists from resolutions smaller than 2 times the boundary-layer height in convective boundary layers (CBLs). Regional models are now approaching the sub-kilometre scales, and Honnert et al. (2011) showed that neither unidirectional (1-D) non-local mesoscale boundary-layer (BL) turbulence scheme nor isotropic (3-D) LES schemes are appropriate at these scales. That is why the turbulence schemes have to be adapted to the grey zone of turbulence.

Boutle et al. (2014) blended a 3-D-Smagorinsky with a 1-D non-local BL scheme with the help of the similarity functions proposed by Honnert et al. (2011). Ito et al. (2015) extended the Mellor and Yamada scheme by modifying the length scales using statistics obtained from LES coarse-graining. Shin and Hong (2015) quantified the local and non-local turbulence at scales in the grey zone to adjust the vertical profiles resulting from their non-local K-gradient scheme.

These adaptations strongly depend on the schemes which are currently used at mesoscale or LES. At Météo-France, the turbulence in the atmospheric BL is represented by an eddy-diffusivity/mass-flux parameterization (EDMF; Hourdin et al., 2002; Soares et al., 2004). The updraughts are represented by the mass-flux scheme which starts at the ground (hereafter PM$_{09}$; Pergaud et al., 2009) and represents the shallow convection, while the rest of the turbulence is represented by a K-gradient scheme (hereafter CBR; Cuxart et al., 2000). Both parts of this scheme are being modified to adapt Météo-France models to the grey zone of turbulence. In this article, modifications of PM_{09} are presented in the second section as well as preliminary results in the third section. As perspective, the "true" CBR mixing lengths in the grey zone are presented.

2 A new mass-flux scheme

As many mass-flux schemes, PM_{09} is based on several assumptions which are valid at large scales. It assumes in particular that the thermal surface is small, the resolved vertical velocity is zero, and the thermal field is quasi-stationary.

Honnert et al. (2016) determined the characteristics of the non-local turbulence (BL thermals) in the grey zone by means of a conditional sampling. Figure 1 shows a 16 km long horizontal cross section of an LES. The thermals (in white) and the part of the thermals which impact the subgrid mass-flux scheme at 1 km resolution (in black) have been determined by the conditional sampling of Honnert et al. (2016). The environment of the structures is in red. Figure 1 shows that at 16 km resolution, PM_{09}'s assumptions are valid: the thermal surface is small, the resolved vertical velocity is zero, as the grid cell contains both the updraughts and the compensatory subsidence, and the thermal field is quasi-stationary. However, in the grey zone, they are not verified. Indeed, as seen on the 1 km zoom of Fig. 1, the thermal surface (in black) may be large, the resolved vertical velocity is not zero, as one thermal can fill the grid cell, and the thermal field is probably not quasi-stationary.

However, mass-flux schemes can be developed without the three assumptions presented before. The initial schemes (PM_{09}; Rio and Hourdin, 2008) describe the behaviour of parameters of one unique thermal in the mesh (the vertical velocity w_u, the mass-flux M_u, the total potential temperature θ_{l_u}, the thermal surface area α, the buoyancy inside the thermal B_u and the pressure and the entrainment (ϵ)/detrainment (δ) lateral closure). a_1 and b_1 are constant. Equations (1)–(4) show the modifications (in red) of PM_{09}. The non-negligible resolved vertical velocity ($\overline{w}^{\Delta x}$) is added in Eqs. (1), (3), and (4). The thermal surface is not negligible and appears at the denominator in Eqs. (3)–(4). The surface triggering of the mass flux ($M_{u_{z=0}}$, Eq. 5) depends on the resolution.

$$M_u = \rho\alpha\left(w_u - \overline{w}^{\Delta x}\right) \tag{1}$$

$$\frac{1}{M_u}\frac{\partial M_u}{\partial z} = \epsilon - \delta \tag{2}$$

$$\frac{\partial \theta_{l_u}}{\partial z} = -\frac{\epsilon}{1-\alpha}\left(\theta_{l_u} - \overline{\theta_l}^{\Delta x}\right) \tag{3}$$

$$\frac{1}{2}\frac{\partial\left(w_u - \overline{w}^{\Delta x}\right)^2}{\partial z} = a_1 B_u - b_1\frac{\epsilon}{1-\alpha}\left(w_u - \overline{w}^{\Delta x}\right)^2 \tag{4}$$

$$\frac{M_{u_{z=0}}}{w^*} = 0.075 \times \left(1 + \tanh\left(\ln\left(\frac{\Delta x}{h+h_c}\right)+0.8\right)\right) \tag{5}$$

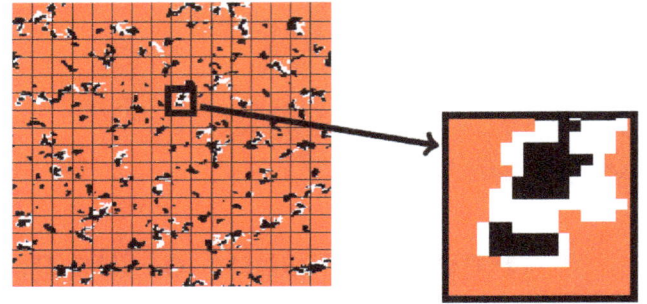

Figure 1. 16 km long horizontal cross section of an LES (IHOP case, 14:00 LT, 500 m altitude) and 1 km long zoom. The thermal fraction is in white, the core of the thermals (strong vertical velocity) is in black, and the environment is in red (see Honnert et al., 2016).

Then, the finer the resolution or the smaller the BL height, the smaller the subgrid turbulent flux. The scheme produces less subgrid turbulence. Consequently, resolved BL thermals are created.

3 Results

The results presented in this section compare model simulations and coarse-grained LES of a dry CBL (hereafter IHOP). This case has been performed using radio soundings collected during the International H_2O Project ($IHOP_{2002}$) campaign (Weckwerth et al., 2004). This field experiment took place in the US Southern Great Plains from 13 May to 25 June 2002. Here, we used the 14 June 2002 case corresponding to a growing CBL near Homestead, Oklahoma (cf. Couvreux et al., 2015). This day was characterized by high pressure (1016 hPa or more) and light wind (less than $5\,\text{m s}^{-1}$). The vertical shear was weak. The well-mixed boundary layer reached 1.5 km in the beginning of the afternoon. The radio soundings were made in the morning from 14:00 to 18:00 UTC (09:00 to 13:00 LT – local time). This case was chosen as it presented a relatively uniform site topography and a typical development of continental convective boundary layer. The simulations lasted for 7 h.

Méso-NH (Lafore et al., 1998) is the research model at Météo France. It can be used in various configurations of the turbulence scheme (from LES to synoptic), in idealized cases, as well as in real cases. In Fig. 2a and b, the resolved turbulent kinetic energy (TKE) of the new parameterization (in green) is compared at 500 m and 1 km resolution with the results of the LES coarse-graining (in black), those of simulations with PM_{09} (in blue), and without mass flux at all (in red). The new parameterization is scale-adaptive and produces the resolved TKE calculated from the LES, even if it produces a bit too much TKE at 500 m and not enough at 1 km resolution.

Resolved TKE IHOP, 12h, PMMC09-No-convection-HRIO-LES

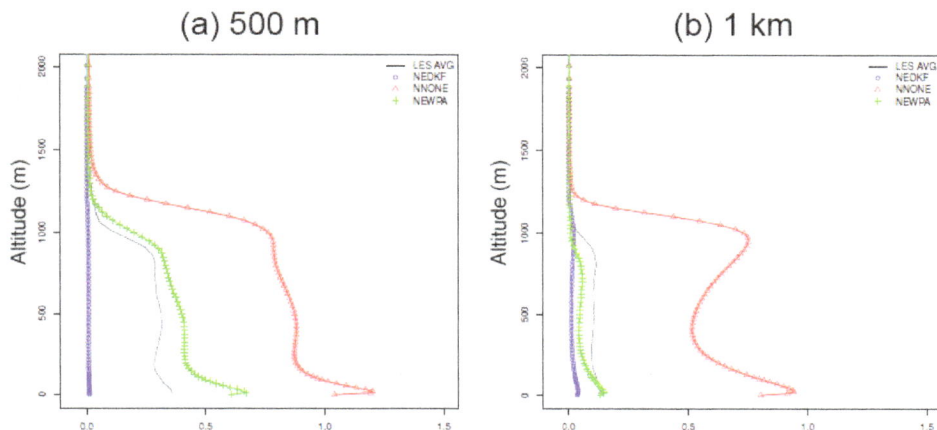

(a) 500 m   (b) 1 km

Figure 2. Resolved TKE in Méso-NH at (**a**) 500 m and (**b**) 1 km resolution in IHOP. The reference (coarse-grained LES) is in black, and the parameterization in green. The blue lines are results of simulations with PM_{09}, and the red ones result from simulations without shallow convection.

AROME (Seity et al., 2010) is the operational regional model at Météo France. Its turbulence scheme is the same as Méso-NH, but the configuration is fixed (for mesoscale simulations) and it simulates real cases only. It is challenging to test a new turbulence parameterization in the operational AROME as there is no reference. That is why, in these tests, we used idealized-AROME. This model has no land surface or lateral coupling and no surface scheme. The imposed boundary conditions allow us to reproduce in AROME the idealized cases previously studied in LES.

Figure 3 shows subgrid TKE produced by the new parameterization at resolutions from 500 m to 2 km (dotted lines) and the subgrid TKE at resolutions from 62.5 m to 8 km calculated from LES. In the middle of the BL, the parameterization follows the LES reference. However, in the surface layer, the turbulence is underestimated. This default is not due to the mass-flux parameterization, but it results from limits of the K-gradient scheme. Indeed, it is purely 1-D in AROME, while in the surface layer, a 3-D dynamical turbulence is required.

Comparison of Ma_Modif/LES AVG for SBG_TKE

Figure 3. Subgrid TKE in AROME at 500 m and 1 km resolution in the IHOP. "MNH" means the reference LES (in full lines) and AROME means the new parameterization (in dotted lines).

4 Limits of the mass-flux modifications: from 1-D to 3-D turbulence scheme

A second problem appears in the grey zone of turbulence: the dimensionality of the scheme. At mesoscale, the horizontal homogeneity assumption allows the computation of the vertical (1-D) turbulent flux only. On the contrary, in LES, the turbulence is assumed isotropic, thus 3-D. The limit resolution at which the horizontal turbulent movements are not negligible is in the grey zone. Honnert and Masson (2014) quantified the production terms of the TKE at grey-zone resolutions from LES coarse-graining. At mesoscale, the turbulence is

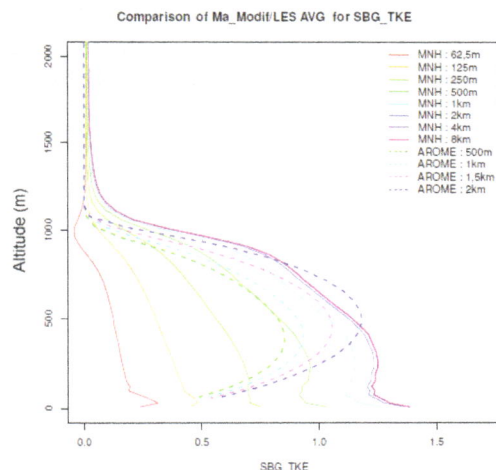

mainly produced by thermals; thus the (vertical) mass-flux scheme has the most impact. However thermal production of TKE is reduced in the grey zone, as the thermals are partly resolved, and the vertical and horizontal components of the dynamical production of TKE become larger; thus the 3-D K-gradient scheme becomes critical.

Honnert and Masson (2014) have proved that the limit resolution at which the horizontal turbulent movements are not negligible is about 0.5 times the size of the energy-containing structures in free CBL (about 500 m resolution). So, the modifications made in the mass-flux part impact the scheme until about 500 m resolution. The K-gradient scheme has to

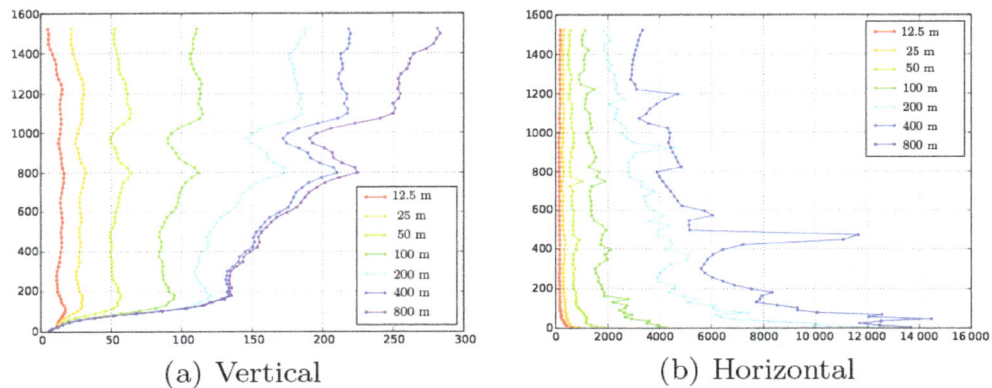

(a) Vertical   (b) Horizontal

Figure 4. (**a**) Vertical and (**b**) horizontal mixing lengths computed at resolutions from 12.5 to 800 m. CASES-99 (neutral BL).

be modified for finer resolutions. Firstly, a 3-D K gradient is necessary. AROME, for instance, has no 3-D turbulence scheme. This limits the modelling of the smallest scales of the grey zone of the turbulence with AROME. Secondly, the subgrid turbulence is not isotropic in the grey zone (as proved in Honnert and Masson, 2014). Méso-NH, for instance, possesses a 3-D turbulence scheme. However, the turbulence is always assumed isotropic. In particular, the LES scheme uses a unique mixing length on the horizontal and on the vertical. Finally, the mixing length is on the order of the boundary-layer height at mesoscale (Bougeault and Lacarrère, 1989) and on the order of the mesh size in LES.

The size of the vertical and horizontal mixing lengths is studied in the grey zone. The eddy diffusivity (K) is calculated from the fluxes ($\overline{u'v'}, \overline{v'w'}, \overline{u'w'}$ in Eqs. 6–8) and gradients (e.g. $\frac{\partial \overline{u}}{\partial y}$ in Eqs. 6–8) computed by LES coarse-graining of IHOP and an additional neutral case at several resolution in the grey zone (Eqs. 6–8). The mixing lengths (L) are computed from the eddy diffusivity and the TKE (e) (Eqs. 9–11) in the CBR equations (C is a constant).

$$\overline{u'v'} = -K_{u,v}\left(\frac{\partial \overline{u}}{\partial y} + \frac{\partial \overline{v}}{\partial x}\right) \tag{6}$$

$$\overline{u'w'} = -K_{u,w}\left(\frac{\partial \overline{u}}{\partial z} + \frac{\partial \overline{w}}{\partial x}\right) \tag{7}$$

$$\overline{v'w'} = -K_{v,w}\left(\frac{\partial \overline{v}}{\partial z} + \frac{\partial \overline{w}}{\partial y}\right) \tag{8}$$

$$K_{u,v} = -CL_{u,v}\sqrt{e} \tag{9}$$

$$K_{u,w} = -CL_{u,w}\sqrt{e} \tag{10}$$

$$K_{v,w} = -CL_{v,w}\sqrt{e} \tag{11}$$

Figure 4 shows the mixing lengths in the neutral BL (CASES-99) based on the CASES-99 experiment which took place from 1 to 31 October 1999 near Leon, Kansas. It was first designed to study stable BL, morning and evening transitions periods. Measurements were taken during neutral conditions. Drobinski et al. (2007) first described this LES. The rugosity length of the site is 0.1 m, and the friction ve-

locity is 0.42 m s^{-1}. A constant (293.15 K) potential temperature is imposed until 750 m altitude (Drobinski et al., 2007) and then a constant adiabatic gradient until 1500 m altitude. The stable boundary layer kills the turbulence above 750 m altitude and limits the size of the eddies. Thus, the eddies remain small enough to be contained in the LES domain. The heat flux at the surface is zero along the simulations, as well as the humidity flux at the surface. The simulations are dry. Thus, the buoyancy flux is zero during the simulations and the production of turbulence is purely dynamical. The geostrophic wind is zonal and prescribed at 10 m s^{-1}.

Figure 4 shows vertical and horizontal mixing lengths computed by Eqs. (6)–(11). Both vertical and horizontal mixing lengths are larger at mesoscale. Under 1000 m altitude, the vertical mixing lengths are consistent with the literature: at mesoscale, they behave as in Bougeault and Lacarrère (1989) with a maximum of a few hundred metres in the BL, while at small scale, they behave as the Deardorff mixing length (the size of the grid cell) with a relatively constant value in the BL of a few tens of metres. Above the BL, the method reaches its limits as the fluxes and gradients are very small. The horizontal mixing lengths are larger at the surface, where there is a maximum of horizontal movements. At mesoscale, the very large horizontal mixing lengths may result from small horizontal gradients at these scales. The fine resolutions present horizontal mixing lengths on the order of the vertical lengths as turbulence is isotropic at those scales.

5 Conclusions

Numerical weather prediction model forecasts at horizontal grid lengths in the range of 100 m to 1 km are now possible. This range of scales is in the "grey zone of turbulence". Previous studies, based on LES analysis from the MésoNH model, showed that some assumptions of turbulence schemes on BL structures are not valid. Indeed, BL thermals are now partly resolved and the subgrid remaining part of the thermals is possibly largely or completely absent from the model columns. Moreover, although the turbulence is mainly vertical at mesoscale, it is isotropic in LES. It has been proved by LES analysis that, in CBL, the turbulence is neither 1-D nor isotropic in the grey zone.

At Météo-France, the turbulence scheme is an EDMF. In this study, in order to model the turbulence at all scales, both the mass flux and the K-gradient part of the scheme are examined. Firstly, the equations of the shallow convection (mass flux) scheme have been modified in order to remove the mesoscale assumptions, which are not valid in the grey zone. These modifications have been tested in the MésoNH model and in an idealized version of the operational AROME model. Secondly, horizontal and vertical mixing lengths have been calculated from LES of neutral and CBL cases at resolutions in the grey zone. These mixing lengths will be introduced in 3-D CBR, which will amend the K-gradient scheme in the grey zone.

References

Bougeault, P. and Lacarrère, P.: Parametrisation of Orography-Induced Turbulence in a Mesobeta-Scale Model, Mon. Weather Rev., 117, 1872–1890, 1989.

Boutle, I. A., Eyre, J. E. J., and Lock, A. P.: Seamless Stratocumulus SImulation across the Turbulent Grey Zone, Mon. Weather Rev., 142, 1655–1668, 2014.

Couvreux, F., Guichard, F., Redelsperger, J.-L., Kiemle, C., Masson, V., Lafore, J.-P., and Flamant, C.: Water Vapour variability within a convective boundary-layer assessed by large-eddy simulations and IHOP2002 observations, Q. J. Roy. Meteorol. Soc., 131, 2665–2693, 2015.

Cuxart, J., Bougeault, P., and Redelsperger, J.-L.: A turbulence scheme allowing for mesoscale and large-eddy simulations, Q. J. Roy. Meteorol. Soc., 126, 1–30, 2000.

Drobinski, P., Carlotti, P., Redelsperger, J.-L., Banta, R., Masson, V., and Newsom, R. K.: Numerical and Experimental Investgation of the neutral Atmospheric surface layer, J. Atmos. Sci., 64, 137–156, 2007.

Honnert, R. and Masson, V.: What is the smallest physically acceptable scale for 1D turbulence schemes?, Front. Earth Sci., 27, 2, 2014.

Honnert, R., Masson, V., and Couvreux, F.: A diagnostic for Evaluating the Representation of Turbulence in Atmospheric Models at the Kilometric Scale, J. Atmos. Sci., 68, 3112–3131, 2011.

Honnert, R., Couvreux, F., Masson, V., and Lancz, D.: Sampling the structure of convective turbulence and implications for grey-zone parametrizations, Bound.-Lay. Meteorol., doi:10.1007/s10546-016-0130-4, in press, 2016.

Hourdin, F., Couvreux, F., and Menut, L.: Parameterization of the dry convective boundary layer based on a mass flux representation of thermals, J. Atmos. Sci., 59, 1105–1122, 2002.

Ito, J., Niino, H., Nakanishi, M., and Moeung, C.-H.: An extension of Mellor-Yamada model to the terra incognita zone fr dry convective mixed layers int he free convection regime, Bound.-Lay. Meteorol., 157, 23–43, 2015.

Lafore, J. P., Stein, J., Asencio, N., Bougeault, P., Ducrocq, V., Duron, J., Fischer, C., Héreil, P., Mascart, P., Masson, V., Pinty, J. P., Redelsperger, J. L., Richard, E., and Vila Guerau de Arellano, J.: The Méso-NH atmospheric simulation system. Part I: Adiabatic formulation and control simulation, Ann. Geophys., 16, 90–109, 1998.

Pergaud, J., Masson, V., Malardel, S., and Couvreux, F.: A parametrization ofSampling of the structure of turbulence dry thermals and shallow cumuli for mesoscale numerical weather prediction, Bound.-Lay. Meteorol., 132, 83–106, 2009.

Rio, C. and Hourdin, F.: A thermal Plume Model for the Convective Boundary Layer: Representation of Cumulus Clouds, J. Atmos. Sci., 65, 407–425, 2008.

Seity, Y., Brousseau, P., Malardel, S., Hello, G., Benard, P., Bouttier, F., Lac, C., and Masson, V.: The AROME-France convective scale operational model, Mon. Weather Rev., 139, 976–991, 2010.

Shin, H. and Hong, S.: Representation of the Subgrid-Scale Turbulent Transport in Convective Boundary Layers at Gray-Zone Resolutions, Mon. Weather Rev, 143, 250–271, 2015.

Soares, P. M. M., Miranda, P. M. A., Siebesma, A. P., and Teixeira, J.: An eddy-diffusivity/mass-flux parametrization for dry and shallow cumulus convection, Q. J. Roy. Meteorol. Soc., 130, 3365–3383, 2004.

Weckwerth, T. M., Parsons, D. B., Koch, S. E., Moore, J. A., Lemone, M. A., Demoz, B. R., Flamant, C., Geerts, B., Wang, J., and Feltz, W.: An overview of the international H_2O project (IHOP 2002) and some preliminary highlights, B. Am. Meteorol. Soc., 85, 253–277, 2004.

Wyngaard, J. C.: Toward numerical modelling in the "Terra Incognita", J. Atmos. Sci., 61, 1816–1826, 2004.

24

Validation of the McClear clear-sky model in desert conditions with three stations in Israel

Mireille Lefèvre and Lucien Wald

MINES ParisTech – PSL Research University, Sophia Antipolis, Paris, France

Correspondence to: Lucien Wald (lucien.wald@mines-paristech.fr)

Abstract. The new McClear clear-sky model, a fast model based on a radiative transfer solver, exploits the atmospheric properties provided by the EU-funded Copernicus Atmosphere Monitoring Service (CAMS) to estimate the solar direct and global irradiances received at ground level in cloud-free conditions at any place any time. The work presented here focuses on desert conditions and compares the McClear irradiances to coincident 1 min measurements made in clear-sky conditions at three stations in Israel which are distant from less than 100 km. The bias for global irradiance is comprised between 2 and 32 W m^{-2}, i.e. between 0 and 4 % of the mean observed irradiance (approximately 830 W m^{-2}). The RMSE ranges from 30 to 41 W m^{-2} (4 %) and the squared correlation coefficient is greater than 0.976. The bias for the direct irradiance at normal incidence (DNI) is comprised between -68 and $+13$ W m^{-2}, i.e. between -8 and 2 % of the mean observed DNI (approximately 840 W m^{-2}). The RMSE ranges from 53 (7 %) to 83 W m^{-2} (10 %). The squared correlation coefficient is close to 0.6. The performances are similar for the three sites for the global irradiance and for the DNI to a lesser extent, demonstrating the robustness of the McClear model combined with CAMS products. These results are discussed in the light of those obtained by McClear for other desert areas in Egypt and United Arab Emirates.

1 Introduction

The downwelling solar irradiance observed at ground level on horizontal surfaces and integrated over the whole spectrum (total irradiance) is called surface solar irradiance (SSI). It is the sum of the direct irradiance, from the direction of the sun, and the diffuse, from the rest of the sky vault, and is also called the global irradiance. The SSI is an essential climate variable as established by the Global Climate Observing System in August 2010 (GCOS, 2016). Knowledge of the SSI and its geographical distribution is of prime importance for numerous domains where SSI plays a major role as e.g. weather, climate, biomass, and energy.

A model estimating the SSI under clear sky or cloud-free conditions is called a clear-sky model. Oumbe et al. (2014) have demonstrated that computations of the SSI from satellite images can be approximated by the product of the clear-sky SSI and a modification factor due to cloud properties and ground albedo only. Changes in clear-atmosphere properties have negligible effect on this modification factor so that both terms can be calculated independently. These results are im-

portant in the view of an operational system as it permits separating the whole processing into two distinct and independent models, whose input variable types and resolutions may be different. This enforces the importance of the availability of an accurate and easy-to-operate model for the assessment of the clear-sky SSI.

The McClear model (Lefèvre et al., 2013) is such a model. It has been designed to benefit from the recent advances on atmosphere composition made in MACC projects (Monitoring Atmosphere Composition and Climate). The latter were preparing the operational provision of global aerosol properties analyses and forecasts together with physically consistent total column content in water vapour and ozone available every 3 h (Benedetti et al., 2009; Kaiser et al., 2012; Peuch et al., 2009). Such information had not been available so far from any operational numerical weather prediction centre. Since 1 January 2016, the McClear model and its inputs are part of the operational services delivered by the Copernicus Atmosphere Monitoring Service (CAMS) operated by ECMWF on behalf of the European Commission.

Table 1. Geographical coordinates of the three stations. Period is 2006–2011. All data are coincident. Number of samples is 19 849 in G, B and D.

Station	Latitude (positive North, ISO 19115)	Longitude (positive East, ISO 19115)	Elevation a.s.l. (m)
Beer Sheva (BEE)	31.25	34.8	195
Sede Boqer (SBO)	30.905	34.782	500
Yotvata (YOT)	29.879	35.065	66

The CAMS McClear service is available as an interoperable Web processing service (WPS), i.e. an application that can be invoked via the Web and that obeys the OGC (Open Geospatial Consortium) standard for interoperability (Percivall et al., 2011). This service delivers estimates of the global SSI and its direct and diffuse components on horizontal surface as well as the direct SSI at normal incidence, for various durations ranging from 1 min to 1 month.

Since its inception as a pre-operational service, McClear has been increasingly used by academics and practitioners. A lot of attention is paid to the validation of the estimates provided by McClear. The goal is to better establish the domain of validity of McClear, its qualities and drawbacks, and to bring transparency and confidence in the use of this operational service.

McClear has been previously validated with respect to 1 min measurements of global and direct SSI on horizontal surface from the Baseline Surface Radiation Network (BSRN) collected from 11 sites located throughout six continents (Lefèvre et al., 2013). The relative root mean square error (RMSE) for global SSI and direct SSI depends on the station and ranges respectively between 3 and 5 % of the mean of the measurements for the station, and between 5 and 10 %.

This article aims at contributing further to the validation of the McClear model. It focuses on desert conditions encountered in Israel where three close stations measure the global, diffuse and direct SSI. This density of stations permits to study the variability of the performances of McClear in this climate homogeneous area.

2 Measurements and McClear estimates

Measurements of the global G and diffuse D SSI and of the beam irradiation received at normal incidence B_N were collected from three stations (Fig. 1 and Table 1) from the Israel Meteorological Service (IMS), the BSRN network and an undisclosed company for the period 2006–2011. The direct SSI B on horizontal surface is computed from the difference $G−D$. Measurements are integrated over 10 min at Beer Sheva and Yotvata and 1 min at Sede Boqer which belongs to the BSRN network. 1 min measurements at Sede Boqer were averaged over 10 min to match the sampling rate of the two other stations. The solar zenith angle θ_S corresponding to each measurement is computed with the SG2 al-

Figure 1. Map of the three stations. The red line is 100 km in length.

gorithm (Blanc and Wald, 2012). After applying the quality check procedures of Roesch et al. (2011), only the measurements were kept which pass the filters proposed by Lefèvre et al. (2013) to retain reliable clear-sky instants. Finally, only are kept clear-sky instants for which measurements are valid for the three stations simultaneously. The number of samples is 19 849 for each station. This last constraint was imposed in order to be able to compare correlation coefficients computed for data sets, whether measurements or estimates, for two stations.

The three stations are fairly close to each other (Fig. 1). Beer Sheva is 40 km north of Sede Boqer and Yotvata is 120 km south of Sede Boqer.

McClear estimates of G, D, B and B_N for 10 min duration were obtained from the SoDa web site (www.soda-pro.com) for these same instants and for each location. It may be of interest here to underline that the McClear model computes G and B_N, then B, and that D is deduced from $G−B$. Also provided were the corresponding time-series of the irradiance at the top of atmosphere on both horizontal and normal surfaces: E_0 and E_{0N}. The clearness index KT and the direct clearness index KT_{B_N} were computed for both measurements and McClear estimates using the following formula:

$$KT = G/E_0 \tag{1}$$

$$KT_{B_N} = B_N/E_{0N}. \tag{2}$$

Table 2. Comparison between clear-sky global G and diffuse D SSI measured by ground stations and estimated by McClear. Units in W m^{-2}.

	G			D		
	BEE	SBO	YOT	BEE	SBO	YOT
Mean observed SSI	810	838	825	124	114	137
Bias	19	2	32	60	69	48
Relative bias (%)	2	0	4	48	60	35
RMSE	32	30	41	66	74	55
Relative RMSE (%)	4	4	5	53	65	40
Squared correlation coefficient	0.977	0.976	0.980	0.594	0.627	0.633

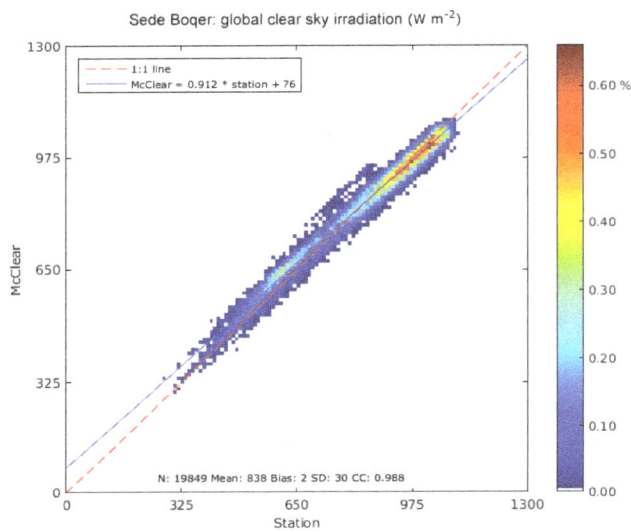

Figure 2. 2-D histogram of measurements (horizontal axis) and McClear estimates (vertical axis) of G for Sede Boqer. The colour represents the frequency of each pair.

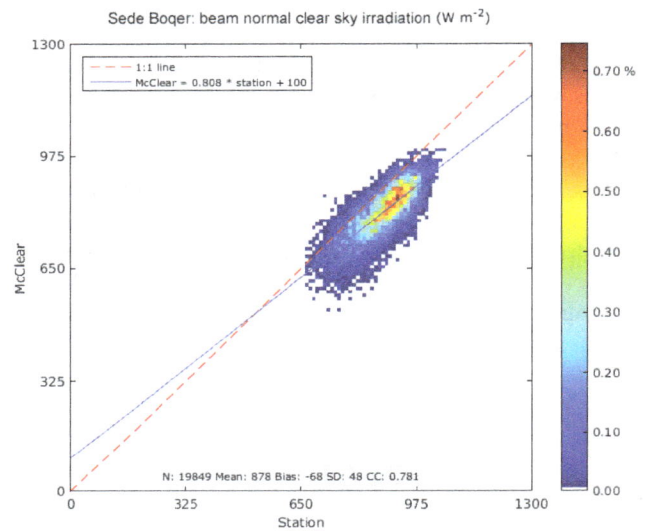

Figure 3. 2-D histogram of measurements (horizontal axis) and McClear estimates (vertical axis) of B_N for Sede Boqer. The colour represents the frequency of each pair.

3 Results

Following the ISO (International Organization for Standardization) standard (1995), the deviations were computed by subtracting measurements for each instant from the McClear estimates and they were summarized by the bias, the root mean square error (RMSE), and the squared correlation coefficient, also known as the coefficient of determination (R^2). Relative values are expressed with respect to the mean observed value. The validations of KT and KT_{B_N} are also included, as they are stricter measures of the performance of a model with respect to the optical state of the atmosphere.

The 2-D histograms of measured and estimated values are presented for Sede Boqer (Figs. 2 and 3). Red, respectively dark blue, dots correspond to regions with great, respectively very low, densities of samples. The plots also present the number of samples, the mean reference value, the bias, the RMSE, the correlation coefficient (CC) and the 1 : 1 line ($y = x$). One may see in Fig. 2 that the points are mostly aligned with the 1 : 1 line with a very limited scattering. The bias and RMSE are respectively 2 and 30 W m^{-2}. The squared correlation coefficient is very large: 0.976, meaning that the temporal changes in G are well reproduced by McClear. The points in Fig. 3 for B_N are less aligned with the 1 : 1 line. The absolute value of the bias and RMSE are much larger: −68 and 83 W m^{-2}. The squared correlation coefficient is 0.610 and a large amount of changes in B_N is unexplained by McClear.

Tables 2–4 present the results of the comparison for respectively G, D, B, B_N, KT and KT_{B_N}. The means of G (Table 2, approximately 830 W m^{-2}), B_N (Table 3, approximately 840 W m^{-2}) and clearness indices (Table 4, 0.75 and 0.64) are large which means that the atmosphere is very often clear and not turbid. Yotvata experiences less B_N – and a lower KT_{B_N} – though it is the southernmost site. It is located 40 km north of the Red Sea in the Negev desert and may be under maritime influence and dust episodes.

Table 3. Comparison between clear-sky beam SSI B and beam at normal incidence B_N measured by ground stations and estimated by McClear. Units in $W\,m^{-2}$.

	B			B_N		
	BEE	SBO	YOT	BEE	SBO	YOT
Mean observed SSI	686	724	688	841	878	809
Bias	−41	−66	−16	−46	−68	13
Relative bias (%)	−6	−9	−2	−6	−8	2
RMSE	59	80	46	69	83	53
Relative RMSE (%)	9	11	7	8	10	7
Squared correlation coefficient	0.931	0.937	0.929	0.576	0.610	0.603

Table 4. Comparison between clear-sky clearness indices KT and KT_{B_N} from ground stations and estimated by McClear.

	KT			KT_{B_N}		
	BEE	SBO	YOT	BEE	SBO	YOT
Mean observed index	0.74	0.76	0.74	0.62	0.66	0.62
Bias	0.02	0.01	0.03	−0.04	−0.06	−0.01
Relative bias (%)	3	1	5	−6	−9	−2
RMSE	0.03	0.03	0.05	0.05	0.07	0.04
Relative RMSE (%)	5	4	6	8	11	7
Squared correlation coefficient	0.524	0.471	0.474	0.625	0.641	0.578

The bias for G is low for Sede Boqer: $2\,W\,m^{-2}$, and is larger for the other sites: 19 and $32\,W\,m^{-2}$, i.e. 2 and 4 % of the mean observed G. The RMSE ranges from 30 to $41\,W\,m^{-2}$ (4 %) and the squared correlation coefficient is greater than 0.976 (Table 2). The influence of θ_S on the SSI creates de facto a correlation between measurements and estimates in clear-sky conditions as θ_S and E_0 can be accurately estimated. The influence of θ_S on KT is much less pronounced and the squared correlation coefficient denotes the ability of McClear to reproduce the optical state of the atmosphere. It ranges between 0.471 and 0.524 (Table 4) and is low. A majority of changes in KT is not reproduced by McClear and improvements should be brought on the McClear model and on the quality of its inputs. The bias and RMSE for KT are similar in relative value to those for G (Table 4). Expectedly, the results for Sede Boqer are fully in line with the bias, RMSE and squared correlation coefficient for both G and KT reported by Lefèvre et al. (2013) for this station though for 1 min SSI: 7 and $30\,W\,m^{-2}$, 0.982, and 0.01, 0.03 and 0.581.

The estimates of D by McClear are inaccurate (Table 2). There is an overestimation ranging between 48 and $69\,W\,m^{-2}$ (35 to 60 % of the mean of D). The RMSE ranges between 55 and $74\,W\,m^{-2}$ (40 to 65 %). The squared correlation coefficient is comprised between 0.594 and 0.633; a large amount of changes in D is unexplained by McClear.

An underestimation is observed for B and B_N (Table 3), except Yotvata for B_N. The bias for B, respectively B_N, is comprised between −66 and $-16\,W\,m^{-2}$, i.e. between −9 and −2 % of the mean B, and between −68 and $+13\,W\,m^{-2}$ (−8 and 2 % of the mean B_N). The RMSE ranges from 46 (7 %) to $80\,W\,m^{-2}$ (11 %) for B, and from 53 (7 %) to $83\,W\,m^{-2}$ (10 %) for B_N. The bias and RMSE for KT_{B_N} are similar in relative value to those for B and B_N (Table 4). The squared correlation coefficient for B_N and KT_{B_N} is close to 0.6; a large amount of changes in B_N or KT_{B_N} is unexplained by McClear. The squared correlation coefficient for B is much larger and close to 0.93 because of the influence of θ_S on the correlation and the accuracy of its estimate.

An additional comparison was performed that dealt with the ability of McClear to reproduce spatial variability. The correlation coefficient between time-series of measurements, respectively McClear estimates, was computed for each pair of stations for G and B_N (Table 5). It is observed (upper right part of the correlation matrix) that the measurements are very much correlated for G (greater than 0.99), which can be explained by the fact that only clear-sky measurements are dealt with. The correlation coefficient is less for B_N, especially between Yotvata and the two others for which it is respectively 0.630 and 0.728. This is in agreement with the remoteness of Yotvata compared to the two others and the above remark on its climate.

The closer the correlation coefficients of the lower part of the matrix to those of the upper part, the more accurately McClear depicts the variability in space. The correlation coefficients for G are almost identical for the measurements and McClear meaning that the actual SSI field is well reproduced by McClear. This is not the case for B_N for which discrep-

Table 5. Correlation matrix between stations for measurements (upper right part of the matrix, in bold) and for McClear (lower left part, in italic) for G and B_N.

G	BEE	SBO	YOT	B_N	BEE	SBO	YOT
BEE	1	**0.995**	**0.991**	BEE	1	**0.884**	**0.630**
SBO	*0.999*	1	**0.994**	SBO	*0.990*	1	**0.728**
YOT	*0.994*	*0.996*	1	YOT	*0.846*	*0.908*	1

ancies may be observed. There is an overestimation of the correlation by McClear which can be attributed to the correlation of its inputs due to the coarse spatial and temporal resolutions. CAMS products on aerosols and total content in water vapour and ozone are available every 3 h. The spatial resolution is $1.125°$, i.e. approx. 120 km along a longitude, for the aerosol properties. This is the same resolution for the total column content of ozone and water vapour before 2014 after which it became $0.8°$. The B_N field estimated by McClear will be smoother than the actual field. Note that the ranking of the correlation coefficients is the same for both the measurements and McClear; the local extrema are respected though the intensity of the variation is decreased.

4 Discussion and conclusion

Like reported in other similar studies, the statistical quantities reported here vary with the period of analysis. A given quantity may change noticeably from one year to another. For example, the bias in B_N at Sede Boqer varies from -63 to $-75\,W\,m^{-2}$ if years are considered separately. This indicates that care must be taken in the analysis of these quantities.

The quantities vary with the month. Trends are more or less marked. There is a tendency for lowest bias – in absolute value – and lowest RMSE in the period May–August. There is a tendency for the bias and the RMSE to increase with θ_S, yielding an increase – in absolute value – of the relative bias and RMSE as the mean G and B_N decrease as θ_S increases. Nevertheless, the changes are limited.

Eissa et al. (2015a, b) have performed similar studies but for respectively Egypt and the United Arab Emirates. Similarly to this study, Aswan and the UAE sites exhibit underestimation of B_N. This underestimation is more pronounced for Beer Sheva and Sede Boqer. On the contrary, Yotvata exhibits an overestimation of $13\,W\,m^{-2}$. The comparison of these different studies shows that the overall picture of the possible causes of the discrepancies between measurements and McClear estimates is still unclear. The underestimation in B_N may be partly caused by overestimation of the aerosol optical depth (AOD). Through comparisons between the AODs measured by AERONET and estimated in CAMS for desert areas in Egypt and UAE, Eissa et al. (2015b) and Oumbe et al. (2012) concluded that one main source of the errors in McClear originates from the CAMS AOD. Therefore, more accurate inputs to McClear would improve its estimates. For

example, Oumbe et al. (2015) have shown that a local empirical correction of the CAMS AOD drastically decreases the bias in the United Arab Emirates.

As for G, if one looks at the results of Eissa et al. (2015a) for Aswan in Egypt – which is located in a desert far from the Cairo megapole, – one would observe that the large overestimation of G by McClear over Aswan: $33\,W\,m^{-2}$, is similar to that observed at Yotvata in this study. Yotvata exhibits the greatest bias of the three sites. The bias at Sede Boqer is $2\,W\,m^{-2}$, expectedly similar to that of $7\,W\,m^{-2}$ reported by Lefèvre et al. (2013) for the same site though for 1 min summarization. The bias for the more turbid sites in the UAE ranges from -5 to $10\,W\,m^{-2}$.

Estimates in G and D – and hence the statistical performances – are sensitive to the type of aerosols that is estimated by the means of the empirical algorithm presented in Lefèvre et al. (2013) applied to the partial aerosol optical depths delivered by CAMS. It is found that in Beer Sheva and Sede Boqer – which are close compared to the size of the CAMS cell, – the most frequent aerosol type is "continental polluted", then "maritime polluted" and finally "desert". The same types are found for Yotvata but "desert" is most frequent than "maritime polluted". An error may arise if the wrong type is selected. Figure 1 in Lefèvre et al. (2013) displays a specific case of daily profile of G in Carpentras (France) with a dramatic change by $30\,W\,m^{-2}$ (approx. 3 %) due to an error in the empirical algorithm. In other cases reported in Eissa et al. (2015a) an overestimation of the fine, strongly scattering pollution particles associated with an underestimation of the coarse, less scattering, mineral dust particles would affect G and D. It should be added that the coarse spatial and temporal resolutions of the CAMS data on aerosols make it difficult to capture the exact atmospheric effects on the incident solar radiation over a specific site. Other causes of uncertainty are the uncertainties in the OPAC model used in McClear (Lefèvre et al., 2013). Zieger et al. (2010) showed noticeable changes in single scattering albedo with relative humidity for the OPAC "continental polluted" and "maritime polluted" types. If relative humidity is assumed too large, then the single scattering albedo is overestimated, yielding an overestimation in D. This may explain the difference between the two sites Beer Sheva and Sede Boqer and the southern one Yotvata where "desert" is more frequent. Simulations performed with the radiative transfer model libRadtran have shown that in case of intense dust storms, i.e. heavy load in dust particles, the single scattering albedo in OPAC "desert" type underestimates that observed in AERONET measurements, which yields an underestimation in D. This is not observed in cases of low or medium loads in dust. This adds to the complexity as intense dust storms may also be observed in the northern sites.

Performances are still far from WMO standards: bias less than $3\,W\,m^{-2}$ and 95 % of the deviations less than $20\,W\,m^{-2}$. Uncertainties in aerosol properties from CAMS are still too large, and more efforts are necessary for a better modelling of the aerosols.

Despite the identified drawbacks and paths for improvements, this validation of the McClear service for the desert conditions in Israel reveals satisfactory results. The comparisons between the McClear estimates and measurements of global horizontal and direct normal irradiances for 3 stations show that a large correlation is attained showing the ability of McClear to capture the temporal and spatial variability of the irradiance field. The performances are similar for the three sites for the global irradiance and for the DNI to a lesser extent, demonstrating the robustness of the CAMS McClear service.

Acknowledgements. The authors thank the Israel Meteorological Service, the operators of the BSRN Sede Boqer station for their valuable measurements and the Alfred-Wegener Institute for hosting the BSRN website. They also thank the anonymous referees whose comments helped in improving this paper. The research leading to these results has received funding from the European Union's Horizon 2020 Programme (H2020/2014–2020) under grant agreement no. 633081 (MACC-III project) and from the European Union's Copernicus Atmosphere Monitoring Service (CAMS).

References

Benedetti, A., Morcrette, J.-J., Boucher, O., Dethof, A., Engelen, R. J., Fisher, M., Flentje, H., Huneeus, N., Jones, L., Kaiser, J. W., Kinne, S., Manglold, A., Razinger, M., Simmons, A. J., and Suttie, M.: Aerosol analysis and forecast in the European Centre for Medium-Range Weather Forecasts Integrated Forecast System: 2. Data assimilation, J. Geophys. Res., 114, D13205, doi:10.1029/2008JD011115, 2009.

Blanc, P. and Wald, L.: The SG2 algorithm for a fast and accurate computation of the position of the Sun, Sol. Energy, 86, 3072–3083, doi:10.1016/j.solener.2012.07.018, 2012.

Eissa, Y., Korany, M., Aoun, Y., Boraiy, M., Abdel Wahab, M., Al-faro, S., Blanc, P., El-Metwally, M., Ghedira, H., and Wald, L.: Validation of the surface downwelling solar irradiance estimates of the HelioClim-3 database in Egypt, Remote Sensing, 7, 9269–9291, doi:10.3390/rs70709269, 2015a.

Eissa, Y., Munawwar, S., Oumbe, A., Blanc, P., Ghedira, H., Wald, L., Bru, H., and Goffe, D.: Validating surface downwelling solar irradiances estimated by the McClear model under cloud-free skies in the United Arab Emirates, Sol. Energy, 114, 17–31, doi:10.1016/j.solener.2015.01.017, 2015b.

GCOS – Global Climate Observing System Essential Climate Variables: available at: www.wmo.int/pages/prog/gcos/index.php?name=EssentialClimateVariables, last access: 20 February 2016.

ISO Guide to the Expression of Uncertainty in Measurement: first edition, International Organization for Standardization, Geneva, Switzerland, 1995.

Kaiser, J. W., Peuch, V.-H., Benedetti, A., Boucher, O., Engelen, R. J., Holzer-Popp, T., Morcrette, J.-J., Wooster, M. J., and the MACC-II Management Board: The pre-operational GMES Atmospheric Service in MACC-II and its potential usage of Sentinel-3 observations, ESA Special Publication SP-708, Proceedings of the 3rd MERIS/(A)ATSR and OCLI-SLSTR (Sentinel-3) Preparatory Workshop, 15–19 October 2012, held in ESA-ESRIN, Frascati, Italy, 2012.

Lefèvre, M., Oumbe, A., Blanc, P., Espinar, B., Gschwind, B., Qu, Z., Wald, L., Schroedter-Homscheidt, M., Hoyer-Klick, C., Arola, A., Benedetti, A., Kaiser, J. W., and Morcrette, J.-J.: Mc-Clear: a new model estimating downwelling solar radiation at ground level in clear-sky conditions, Atmos. Meas. Tech., 6, 2403–2418, doi:10.5194/amt-6-2403-2013, 2013.

Oumbe, A., Bru, H., Hassar, Z., Blanc, P., Wald, L., Fournier, A., Goffe, D., Chiesa, M., and Ghedira, H.: Selection and implementation of aerosol data for the prediction of solar resource in United Arab Emirates, In Proceedings of SolarPACES Conference, 11–14 September 2012, Marrakech, Morocco, PSE AG, Freiburg, Germany, USBKey, Paper#22240, 2012.

Oumbe, A., Qu, Z., Blanc, P., Lefèvre, M., Wald, L., and Cros, S.: Corrigendum to "Decoupling the effects of clear atmosphere and clouds to simplify calculations of the broadband solar irradiance at ground level" published in Geosci. Model Dev., 7, 1661–1669, 2014, Geosci. Model Dev., 7, 2409–2409, doi:10.5194/gmd-7-2409-2014, 2014.

Oumbe, A., Wald, L., Blanc, P., Ghedira, H., and Goffe, D.: Improving the solar resource estimation in the United Arab Emirates using aerosol and irradiance measurements, ISES Solar World Congress 2015, 8–12 November 2015, Daegu, Korea, 2015.

Percivall, G., Ménard, L., Chung, L.-K., Nativi, S., and Pearlman, J.: Geo-processing in cyberinfrastructure: making the web an easy to use geospatial computational platform, in: Proceedings, 34th International Symposium on Remote Sensing of Environment, Sydney, Australia, 10–15 April 2011, available at: www.isprs.org/proceedings/2011/ISRSE-34/211104015Final00671.pdf (last access: 1 March 2016), 2011.

Peuch, V.-H., Rouil, L., Tarrason, L., and Elbern, H.: Towards European-scale Air Quality operational services for GMES Atmosphere, 9th EMS Annual Meeting, EMS2009-511, 9th European Conference on Applications of Meteorology (ECAM) Abstracts, held 28 September–2 October 2009, Toulouse, France, 2009.

Roesch, A., Wild, M., Ohmura, A., Dutton, E. G., Long, C. N., and Zhang, T.: Corrigendum to "Assessment of BSRN radiation records for the computation of monthly means" published in Atmos. Meas. Tech., 4, 339–354, 2011, Atmos. Meas. Tech., 4, 973–973, doi:10.5194/amt-4-973-2011, 2011.

Zieger, P., Fierz-Schmidhauser, R., Gysel, M., Ström, J., Henne, S., Yttri, K. E., Baltensperger, U., and Weingartner, E.: Effects of relative humidity on aerosol light scattering in the Arctic, Atmos. Chem. Phys., 10, 3875–3890, doi:10.5194/acp-10-3875-2010, 2010.

25

Improvement of Solar and Wind forecasting in southern Italy through a multi-model approach: preliminary results

Elenio Avolio[1], **Rosa Claudia Torcasio**[1], **Teresa Lo Feudo**[1], **Claudia Roberta Calidonna**[1],
Daniele Contini[2], **and Stefano Federico**[3]

[1]Institute of Atmospheric Sciences and Climate – Italian National Research Council (ISAC-CNR),
Lamezia Terme, 88046, Italy
[2]Institute of Atmospheric Sciences and Climate – Italian National Research Council (ISAC-CNR),
Lecce, 73100, Italy
[3]Institute of Atmospheric Sciences and Climate – Italian National Research Council (ISAC-CNR),
Rome, 00133, Italy

Correspondence to: Elenio Avolio (e.avolio@isac.cnr.it)

Abstract. The improvement of the Solar and Wind short-term forecasting represents a critical goal for the weather prediction community and is of great importance for a better estimation of power production from solar and wind farms.

In this work we analyze the performance of two deterministic models operational at ISAC-CNR for the prediction of short-wave irradiance and wind speed, at two experimental sites in southern Italy.

A post-processing technique, i.e the multi-model, is adopted to improve the performance of the two mesoscale models.

The results show that the multi-model approach produces a significant error reduction with respect to the forecast of each model. The error is reduced up to 20 % of the model errors, depending on the parameter and forecasting time.

1 Introduction

Solar and wind farm power prediction is of great importance for renewable energy applications (Giebel et al., 2011; Monteiro et al., 2009; Pinson et al., 2009). The models for power prediction often need the output of numerical weather prediction models. Hence, the quality of the power forecast is strictly related to the quality of the radiation and wind prediction over the considered area (Alessandrini et al., 2013; Pinson et al., 2007; Von Bremen, 2007).

A good estimation of power production from solar and wind farms is a significant issue in Southern Italy, due to a large availability of solar and wind energy in spite of a poor integration into the grid (GSE, 2015).

The purpose of this work is to analyse the performance of two deterministic atmospheric models: the WRF (Weather Research and Forecasting Model) (Skamarock et al., 2008)

and the RAMS (Regional Atmospheric Modeling System) (Cotton et al., 2003). Both models were run for six months of 2013 (summer and fall) at 4 km horizontal resolution over Italy. Verification is conducted against two surface stations located in Lamezia Terme and Lecce, and are based on hourly forecasts output and observations.

A multi-model approach is adopted to reduce the forecast errors. The multi-model is trained using a dataset of past forecasts and observations to compute the weights of the models that minimize the Root Mean Square Error; in the forecast phase these weights are used to determine the best estimate of the forecast.

2 Methodology

The models considered in this work are the WRF (Weather Research and Forecasting Model) and the RAMS (Regional

Figure 1. WRF/RAMS domain (same grid at 4 km resolution), and location of the experimental sites.

Table 1. RMSE, MAE and BIAS for Lecce and Lamezia Terme and for each model. Statistics are computed starting from hourly data for the six-month period summer-fall 2013.

	Shortwave RAD [W m^{-2}]		Wind Speed [m s^{-1}]	
LECCE	RAMS	WRF	RAMS	WRF
BIA	18	41	1.3	2.9
MAE	47	53	1.4	2.9
RMSE	96	104	1.6	3.4
LAMEZIA T				
BIA	17	45	0.1	0.4
MAE	49	54	1.0	1.1
RMSE	99	105	1.3	1.4

Atmospheric Modeling System). They were run for 6 months (from 1 June to 30 November 2013) at 4 km horizontal resolution over Italy, using the same domain (Fig. 1). Initial and dynamic boundary conditions are given by the 12:00 UTC deterministic analysis/forecast cycle of the ECMWF-IFS model (0.25° horizontal latitude-longitude grid spacing).

For the WRF model, the PBL scheme adopted is the Mellor–Yamada–Janjic (MYJ) (Janjić, 1994), a 1.5-order prognostic TKE scheme with local vertical mixing. For the shortwave radiation, the Goddard scheme (Chou and Suarez, 1999) is used.

For the RAMS model, the radiation scheme detailed in Chen and Cotton (1983) is used for short-wave radiation; this scheme accounts for the total condensate present in the atmosphere but not for the specific phase of water (i. e. vapour, liquid or ice). The scheme takes into account for the ozone and carbon dioxide on radiative transfer. Unresolved vertical transport is parameterized by the K theory, in which the covariance is evaluated as the product of an eddy mixing coefficient and the gradient of the transported quantity. The turbulent mixing in the horizontal directions is parameterized following Smagorinsky (1963).

Each forecast lasts 36 h and the first 12 h are not considered for the comparison (spin-up time); thus the forecast-time verification is 24 h. Short-wave radiation and 10 m wind speed are simulated and verified.

Verification is carried out against two surface stations located in Southern Italy, Lamezia Terme and Lecce, and are based on hourly output of models forecast; the forecasts are interpolated bilinearly to the positions of the surface stations. Experimental sites are shown in Fig. 1.

The two independent forecasts are regressed towards the measured data during the training period, defining the coefficient of the linear regression to be used in the multi-model forecast phase. More specifically, the two-model forecast at each station location is given by:

$$S = \bar{O} + \sum_{i=1}^{N} a_i(F_i - \bar{F}_i) \tag{1}$$

where N is the number of the models (2); a_i, the weight of the ith model; F_i, the forecast of the ith model; $\bar{F}_i$, its mean value over the training period; and $\bar{O}$, the mean observation over the training period. The calculation of the weights a_i is given by the minimization of the mean square distance D^2:

$$D^2 = \sum_{k=1}^{L} (S_k - O_k)^2 \tag{2}$$

where L is the training period length (80 % of the data).

Therefore, the available dataset is divided in two parts: a training period containing 80 % of the data and a forecasting period with the remaining 20 %. The weights a_i are evaluated over the training period and are then used to compute the forecast ("cross-evaluation"). The methodology is applied 20 times randomly selecting the training dataset to assess the statistical robustness of the results.

3 Results and discussions

First of all, we have computed error statistics for each model, for the two parameters and for both experimental sites. As a second step, the multi-model technique is adopted to improve the forecast performance. Statistics used to quantify the performance of the models are: RMSE (Root Mean Square Error), MAE (Mean Absolute Error) and BIA (Bias).

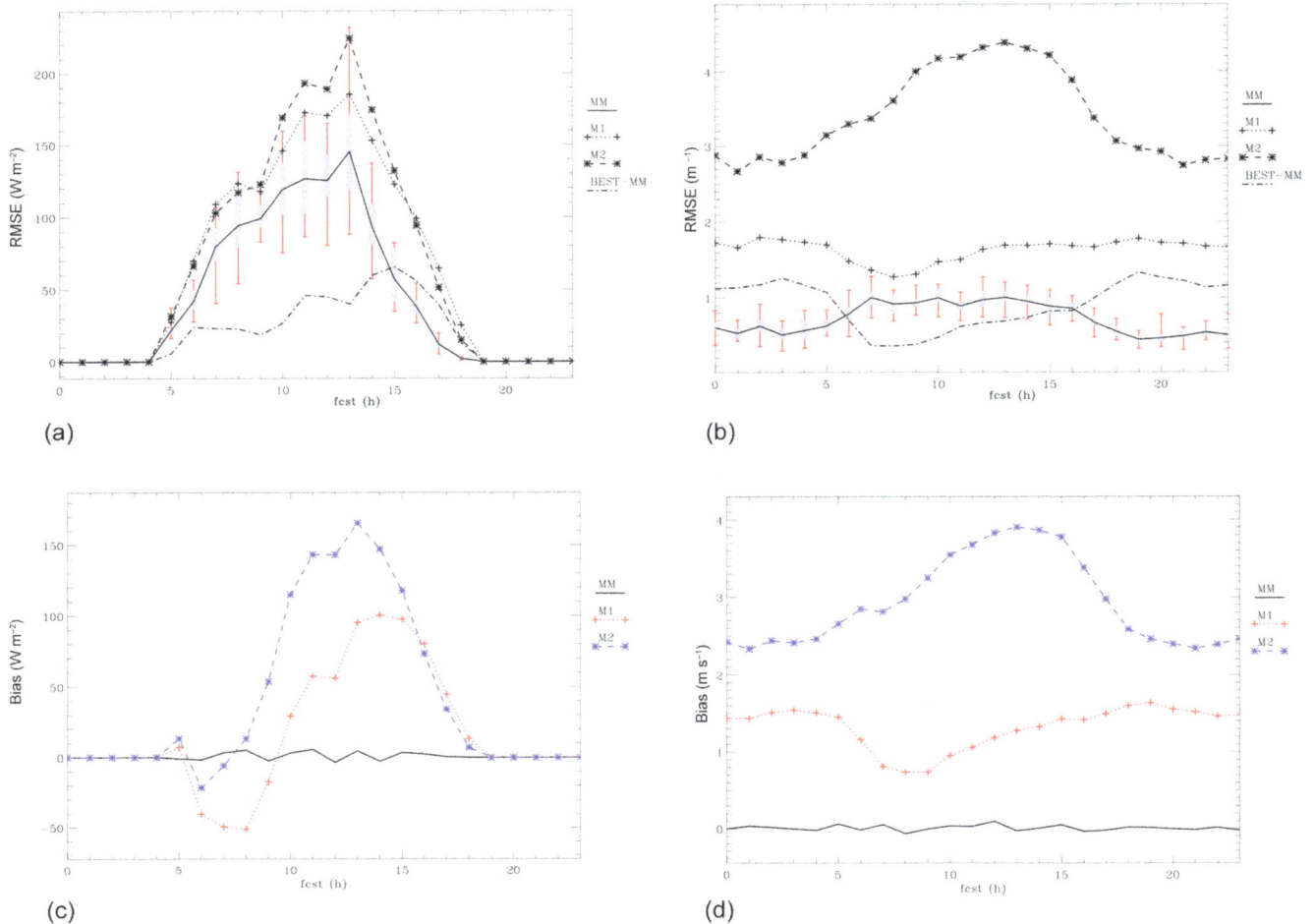

Figure 2. RMSE and BIAS for Lecce site versus forecast time (24 h). M1 and M2 are the models forming the multi-model (MM). **(a)** RMSE for Shortwave Radiation; **(b)** RMSE for Wind speed; **(c)** BIA for Shortwave Radiation; **(d)** BIA for Wind speed. The difference between the RMSE of the best model and that of the MM (BEST-MM) is also shown in **(a)** and **(b)**. The RMSE statistic is computed considering the 20 attempts as a whole. The boxes on the MM RMSE curve show the 25th and 75th percentile of the RMSE distribution for the 20 attempts, while the error bars extent between the maximum and minimum value of the RMSE for the 20 attempts.

3.1   First results: RAMS and WRF performances

Table 1 summarizes the results of the comparison between models and observations.

Initial comparisons show that the minimum hourly radiation errors (RMSE) are 96 W m^{-2} for Lecce and 99 W m^{-2} for Lamezia Terme (RAMS model). WRF performance is quite similar for both experimental sites.

Wind speed is better simulated at Lamezia Terme, where the RMSE is 1.3 m s^{-1} for RAMS (1.4 m s^{-1} for WRF). At Lecce, WRF errors are higher than RAMS and also higher than errors for Lamezia Terme.

For these specific runs, RAMS has slightly better performance compared to WRF and can be considered as the "best model". These results are encouraging because the errors of the models are consistent with the state-of-the-art results obtained with mesoscale models in similar studies (Tiriolo

et al., 2015; Kotroni and Lagouvardos, 2004; Gomez et al., 2014).

3.2   The MULTI-MODEL results

Here we present the results of the multi-model technique adopted to reduce the forecast errors.

In the following these abbreviations are adopted: M1 = RAMS; M2 = WRF; MM = Multi Model; BEST-MM = difference between best model and Multi Model at the specific forecasting time.

Figure 2a and b show that the RMSE of the MM is lower than the RMSE of the best model (M1) for Lecce. The reduction of the error varies depending on the parameter and forecast time. For the solar radiation the error reduction is maximum in correspondence of the middle of the day, when the solar irradiance is at its maximum. Radiation er-

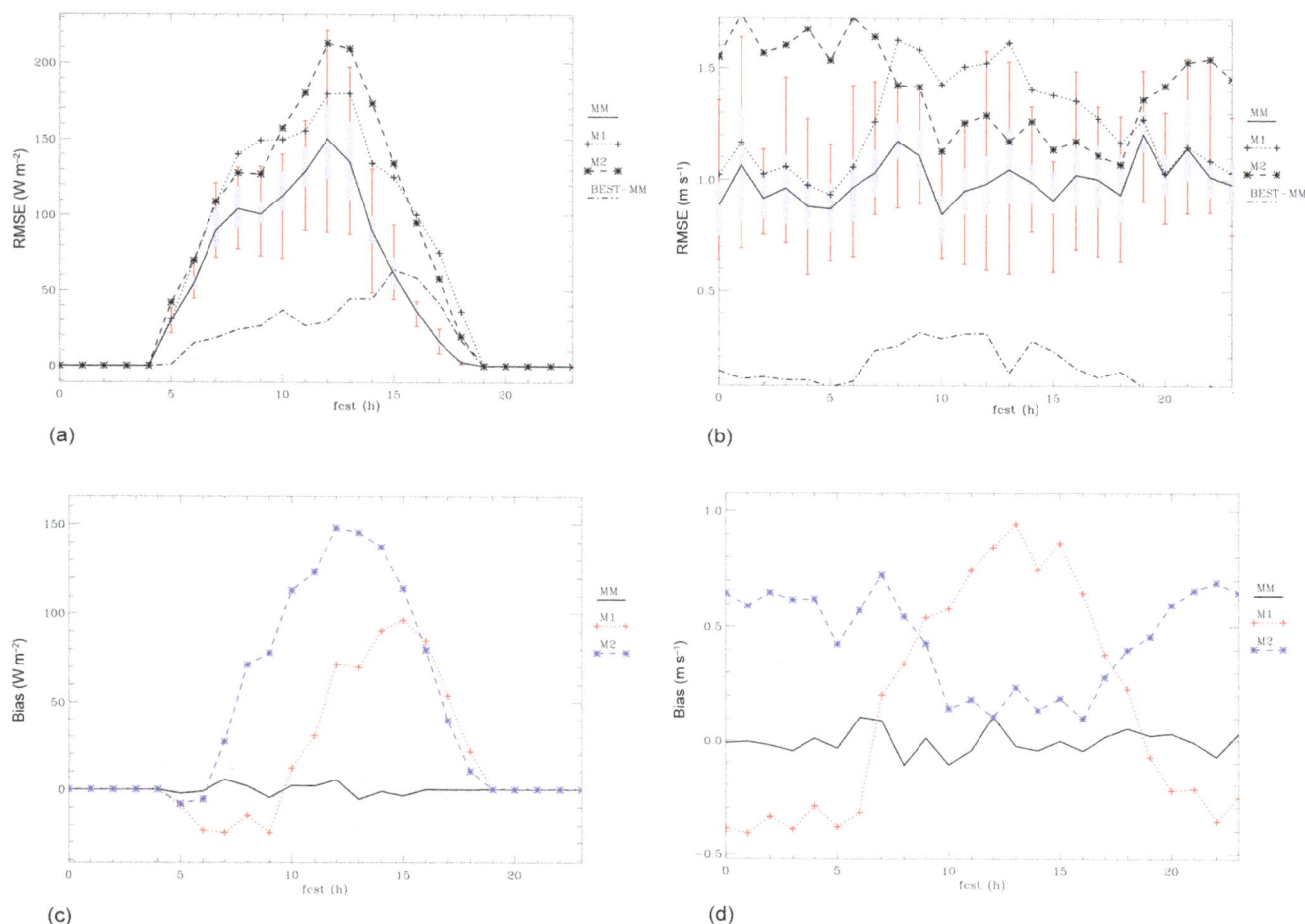

Figure 3. Same as Fig. 2 for the Lamezia Terme site.

rors are reduced up to 30 % compared to the best model and, in some cases, up to 40 % compared to the less accurate model. The maximum RMSE reduction occurs at 15:00 UTC (60 W m^{-2}).

The Wind Speed RMSE shows a clear diurnal cycle for M2, not shown by M1. The MM lowers the RMSE of the best model up to 1.2 m s^{-1} (03:00 and 19:00 UTC), showing the significant impact (more than 50 % reduction of the best model RMSE at 03:00 and 19:00 UTC) of the post-processing technique on the model performance. The behaviour of the MM RMSE over the 20 attempts shows a stable behaviour and the RMSE varies by less than 0.5 m s^{-1} for most forecast times.

Figure 3a and b show the results for Lamezia Terme (RMSE); also for this site it is evident that the RMSE of the MM is less than the RMSE of both models. More in detail the RMSE of the short-wave radiation is reduced up to 60 W m^{-2} (15:00 UTC) and the MM improves the RMSE of the best model by more than 10 % for most diurnal hours.

For the wind speed, the MM improvement is less evident compared to Lecce even if it is still sizable (i.e. larger than 10 % of the best model RMSE) especially for the diurnal hours. Also, the variation of the MM RMSE over the 20 attempts shows a larger variability (0.5–1.0 m s^{-1}) compared to Lecce.

Figure 2c and d (Lecce) and Fig. 3c and d (Lamezia Terme) show that the BIAS has been successfully removed for both parameters and sites because the BIAS of the MM is close to 0 for all forecast times. From the results of Figs. 2 and 3, it follows that the multi-model technique has been able to improve the performance of the best model, regardless to the parameter and forecast hour, showing the significant impact that this method has on the forecast for the two sites. However, the variability of the MM performance for the wind speed RMSE at Lamezia Terme may indicate the need to improve the MM performance.

4   Conclusive remarks

RAMS and WRF models have good performances in predicting the shortwave radiation and wind speed, with RMSE of the order of 95 W m^{-2} and 1.3 m s^{-1}, respectively.

The performance is further improved by the application of the multi-model technique for both wind speed and shortwave radiation both at Lamezia Terme and Lecce, and the RMSE is reduced by a sizeable fraction (almost always larger than 10 % of the model RMSE) depending on the forecasting time and parameter.

These results are important considering the short period used for training the MM; however, the results for the Wind Speed at Lamezia Terme show a large variability, that could be likely reduced by enriching the training period (e.g. considering longer training period, different model's initializations, more members participating to the multi-model).

Acknowledgements. This work was partially supported by the project PONa3_00363 "High Technology Infrastructure for Climate and Environment Monitoring (I-AMICA)", and developed in the framework of the "EERA JP Wind" programme.

References

Alessandrini, A., Sperati, S., and Pinson, P.: A comparison between the ECMWF and COSMO Ensemble Prediction Systems applied to short-term wind power forecasting on real data, Appl. Energ., 107, 271–280, 2013.

Chen, C. and Cotton, W. R.: A One-Dimensional Simulation of the Stratocumulus-Capped Mixed Layer, Bound.-Lay. Meteorol., 25, 289–321, 1983.

Chou, M. D. and Suarez, M. J.: A solar radiation parameterization for atmospheric studies, NASA Tech. Memo, 104606, 15, 40 pp., 1999.

Cotton, W. R., Pielke Sr., R. A., Walko, R. L., Lista, D. E., Tremback, C. J., Jiang, H., McAnelly, R. L., Harrington, J. Y., Nicholls, M. E., Carrio, G. G., and McFadden, J. P.: RAMS 2001: Current status and future directions, Meteorol. Atmos. Phys., 82, 5–29, 2003.

Giebel, G., Brownsword, R., Kariniotakis, G., Denhard, M., and Draxl, C.: The State of the Art in Short-Term Prediction of Wind Power, Anemos.plus, deliverable report, 2011.

Gomez, I., Caselles, V., and Estrela, M. J.: Real-time weather forecasting in the Western Mediterranean Basin: An application of the RAMS model, Atmos. Res., 139, 71–84, 2014.

GSE, Rapporto Statistico "Energia da Fonti Rinnovabili in Italia – 2014", available at: http://www.gse.it/it/salastampa/GSE_Documenti/Rapporto%20statistico%20GSE%20-%202014.pdf (last access: 19 April 2016), 2015.

Janjić, Z. I.: The Step–Mountain Eta Coordinate Model: Further developments of the convection, viscous sublayer, and turbulence closure schemes, Mon. Weather Rev., 122, 927–945, 1994.

Kotroni, V. and Lagouvardos, K.: Evaluation of MM5 High-Resolution Real-Time Forecast over the Urban Area of Athens, Greece, J. Appl. Meteorol., 43, 1666–1678, 2004.

Monteiro, C., Bessa, R., Miranda, V., Botterud, A., Wang, J., and Conzelmann, G.: Wind Power Forecasting: State-of-the-Art 2009, Argonne National Laboratory ANL/DIS-10-1, November 2009.

Pinson, P., Nielsen, H. A., Madsen, H., and Kariniotakis, G.: Skill forecasting from different wind power ensemble prediction methods, J. Phys. Conf. Ser., 75, 012046, doi:10.1088/1742-6596/75/1/012046, 2007.

Pinson, P., Nielsen, H. A., Madsen, H., and Kariniotakis, G.: Skill forecasting from ensemble predictions of wind power, Appl. Energ., 86, 1326–1334, 2009.

Skamarock, W. C., Klemp, J. B., Dudhia, J., Gill, D. O., Barker, D., Duda, M. G., Huang, X.-Y., Powers, J. G., and Wang, W.: A Description of the Advanced Research WRF Version 3, NCAR Technical Note NCAR/TN-475+STR, doi:10.5065/D68S4MVH, 2008.

Smagorinsky, J.: General circulation experiments with the primitive equations. Part I, The basic experiment, Mon. Weather Rev., 91, 99–164, 1963.

Tiriolo, L., Torcasio, R. C., Montesanti, S., and Federico, S.: Verification of a Real Time Weather Forecasting System in Southern Italy, Adv. Meteorol., 2015, 758250, doi:10.1155/2015/758250, 2015.

Von Bremen, L.: Combination of deterministic and probabilistic meteorological models to enhance wind farm forecast, J. Phys. Conf. Ser., 75, 012050, doi:10.1088/1742-6596/75/1/012050, 2007.

26

On the effective solar zenith and azimuth angles to use with measurements of hourly irradiation

P. Blanc and L. Wald

MINES ParisTech – PSL Research University, Sophia Antipolis, France

Correspondence to: L. Wald (lucien.wald@mines-paristech.fr)

Abstract. Several common practices are tested for assessing the effective solar zenith angle that can be associated to each measurement in time-series of in situ or satellite-derived measurements of hourly irradiation on horizontal surface. High quality 1 min measurements of direct irradiation collected by the BSRN stations in Carpentras in France and Payerne in Switzerland, are aggregated to yield time series of hourly direct irradiation on both horizontal and normal planes. Time series of hourly direct horizontal irradiation are reconstructed from those of hourly direct normal irradiation and estimates of the effective solar zenith angle by one of the six practices. Differences between estimated and actual time series of the direct horizontal irradiation indicate the performances of six practices. Several of them yield satisfactory estimates of the effective solar angles. The most accurate results are obtained if the effective angle is computed by two time series of the direct horizontal and normal irradiations that should be observed if the sky were cloud-free. If not possible, then the most accurate results are obtained from using irradiation at the top of atmosphere. Performances show a tendency to decrease during sunrise and sunset hours. The effective solar azimuth angle is computed from the effective solar zenith angle.

1 Introduction

Time-series of measurements of hourly irradiation on horizontal surface are increasingly available from in-situ measurements, satellite retrievals, meteorological numerical models, or combinations of these. They are useful in many aspects in solar energy and other domains, e.g. architecture, building management, agriculture, or biomass. Notably, hourly irradiations are included in Typical Meteorological Year (TMY) data sets that are widely used for simulation of solar conversion and building systems (Kalogirou, 2003; Hall et al., 1978).

In many cases, such data – whether in situ or satellite-derived – are inputs to numerical procedures with different aims, ranging from quality control and gap filling to assessment of the radiation impinging on a tilted plane. Such computations can be performed only if a solar zenith angle θ_S, and azimuth angle Ψ_S in some cases, can be associated to each measurement. However, such angles are seldom given for each hourly irradiation. Currently, only time stamp is given for each measurement.

If measurements are made with integration duration, also called summarization, of 1 min, one may consider that the sun angles are approximately constant, or more correctly that they vary approximately linearly, and that they can be computed for the middle of the corresponding minute. This is not the case for summarization of 15 min or 1 h. Angles are greatly varying within such duration, especially at the beginning and end of the day. In this context, there is a practical request from companies, academics, or researchers: what is the best practice for computing these angles? The article deals with this question. It does not intend to bring definite answers which may be diverse if one considers the final goal of the process requiring solar zenith angle as input. It presents a simple study bringing practical answers to questions brought up by practitioners to the attention of the International Energy Agency (IEA).

Several practices already exist. To the best of the knowledge of the authors, there is no scientific publication supporting these practices and comparing them. The work presented here compares the performances of a few common practices and makes recommendations keeping in mind the practical aspects faced by practitioners, companies, academics, and researchers. Hourly values are dealt with for the sake of the simplicity but the work is applicable to other summarizations.

2 Current practices and new ones

Let B_N denote the direct irradiation received on a plane always normal to the sun rays. Let note G, D and B respectively the global, diffuse and direct irradiation received on a horizontal plane. The direct radiation is also called the beam radiation. Practically, B_N may be measured or B may be deduced from the difference between G and D. In many cases, only G is known. The following relationship holds:

$$B_N = \frac{B}{\cos \theta_S} \qquad (1)$$

and θ_S is the angle to be used in further calculations. The computation of B_N is very sensitive to the solar zenith angle θ_S which for this reason is the quantity dealt with in this work.

In case of summarization greater than 1 min, θ_S varies noticeably and the application of Eq. (1) becomes a problem. Which value is the right one to use? An effective angle θ_S^{eff} must be used to handle measured and modelled hourly irradiation whether global G_h, diffuse D_h, or direct B_h or B_{Nh} irradiations where the subscript "h" means hourly. Practically, Eq. (1) is rewritten with

$$B_{Nh} = \frac{B_h}{\cos \theta_S^{\text{eff}}}. \qquad (2)$$

Six practices to compute θ_S^{eff} have been identified, named from A0 to A5. Let t, expressed in h, define the time of the end of the summarization Δt, equal to 1 h in this case, and assume that the summarization is 1 h.

– *A0*: θ_S^{eff} is taken as θ_S at half-hour, i.e. $(t - 0.5)$

$$\theta_S^{\text{eff,A0}} = \theta_S(t - 0.5). \qquad (3)$$

– *A1*: θ_S^{eff} is taken as the average of θ_S over the hour

$$\theta_S^{\text{eff,A1}} = \frac{1}{\Delta t} \int_{t-1}^{t} \theta_S(u)\mathrm{d}u. \qquad (4)$$

– *A2*: θ_S^{eff} is taken as the average of θ_S over the hour provided $\theta_S < \pi/2$

$$\theta_S^{\text{eff,A2}} = \frac{1}{\Delta t} \int_{\substack{t-1 \\ \theta_S < \pi/2}}^{t} \theta_S(u)\mathrm{d}u. \qquad (5)$$

– *A3*: θ_S^{eff} is taken as the average of θ_S over the hour but limited to the daylight period in the astronomical sense

$$\theta_S^{\text{eff,A3}} = \frac{\int_{\substack{t-1 \\ \theta_S < \pi/2}}^{t} \theta_S(u)\mathrm{d}u}{\int_{\substack{t-1 \\ \theta_S < \pi/2}}^{t} \mathrm{d}u}. \qquad (6)$$

– *A4*: θ_S^{eff} is computed from hourly irradiations B_h^{TOA} and B_{Nh}^{TOA} received at the top of atmosphere. Note that at top of atmosphere, there is no downwelling diffuse component and that the direct irradiation is equal to the global irradiation.

$$\theta_S^{\text{eff,A4}} = \cos^{-1}\left(\frac{B_h^{\text{TOA}}}{B_{Nh}^{\text{TOA}}}\right) \qquad (7)$$

– *A5*: θ_S^{eff} is computed from hourly irradiations given by a clear-sky model

$$\theta_S^{\text{eff,A5}} = \cos^{-1}\left(\frac{B_h^{\text{clear}}}{B_{Nh}^{\text{clear}}}\right) \qquad (8)$$

where a clear-sky model is a model providing estimates of B_h^{clear} and B_{Nh}^{clear} that would be observed if the sky were clear at this instant and location. The McClear model (Lefèvre et al., 2013) is such a model and is used here.

Practices A0 to A4 were discussed during a meeting of the Task #46 of the Solar Heating and Cooling Implementing Agreement of the IEA held in Almeria, Spain, in January 2015. The practice A5 was used by Korany et al. (2015).

3 Methodology for assessing the performances of each practice

Time-series of measurements of the BSRN stations at Carpentras, France, and Payerne, Switzerland, were collected that span 2008 to 2010. Carpentras is located in Provence, in the Southeast of France (Table 1) and experiences Mediterranean climate, i.e. warm temperate climate with dry and hot summer with many days of cloud-free skies throughout the year. Payerne experiences oceanic climate, i.e. warm temperate, fully humid, and warm summer; many small cumulus clouds can be observed during summer days. Measurements are acquired every 1 min for B_N as well as for G and D. Uncertainty requirements for BSRN data are 5 W m^{-2} for global irradiance and 2 W m^{-2} for direct irradiance (Ohmura et al., 1998). Only measurements passing the quality check procedures described by Roesch et al. (2011) has been considered here.

Table 1. Geographical coordinates of the two BSRN stations.

Station	Latitude (positive North, ISO 19115)	Longitude (positive East, ISO 19115)	Elevation a.s.l. (m)
Carpentras	44.083	5.059	99
Payerne	46.815	6.944	491

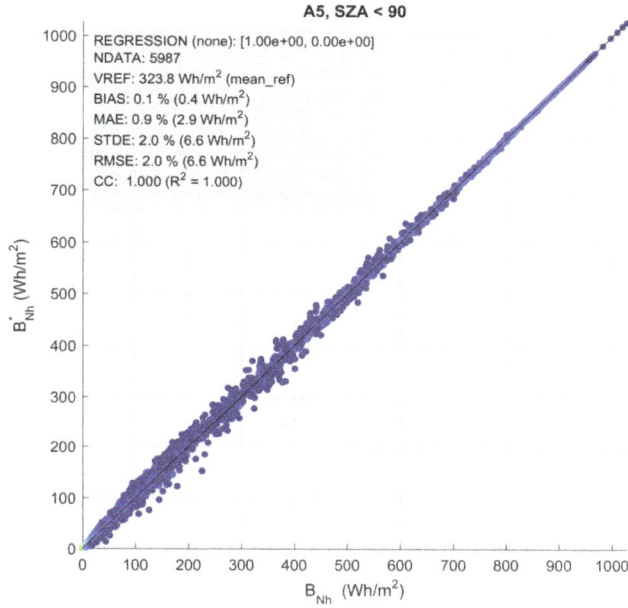

Figure 1. Correlogram between B_{Nh} (horizontal axis) and B_{Nh}^{*} computed with $\theta_{S}^{\mathrm{eff},A5}$ (vertical axis) for all solar zenith angles (SZA) for Carpentras.

For each station, every 1 min, the actual θ_{S} was accurately computed by the means of the SG2 algorithm (Blanc and Wald, 2012). The actual direct irradiance on horizontal surface B_{N} can be computed using Eq. (1). Then, hourly measurements are simulated by aggregating B_{N} and B over 1 h, yielding hourly irradiations B_{Nh} and B_{h}. Only hours with no missing nor invalid data have been selected. Given these hourly time-series, $\theta_{S}^{\mathrm{eff}}$ is computed with the six proposed practices. Then, using Eq. (2), an estimated time series B_{Nh}^{*} is computed from the actual B_{h} time series:

$$B_{Nh}^{*} = \frac{B_{h}}{\cos\theta_{S}^{\mathrm{eff}}}. \tag{9}$$

Finally, the actual B_{Nh} and estimated B_{Nh}^{*} time series are compared. The deviations: $B_{Nh}^{*} - B_{Nh}$ are computed and then summarized by the bias, root mean square error and correlation coefficient. The smaller the discrepancies, the more accurate the practice.

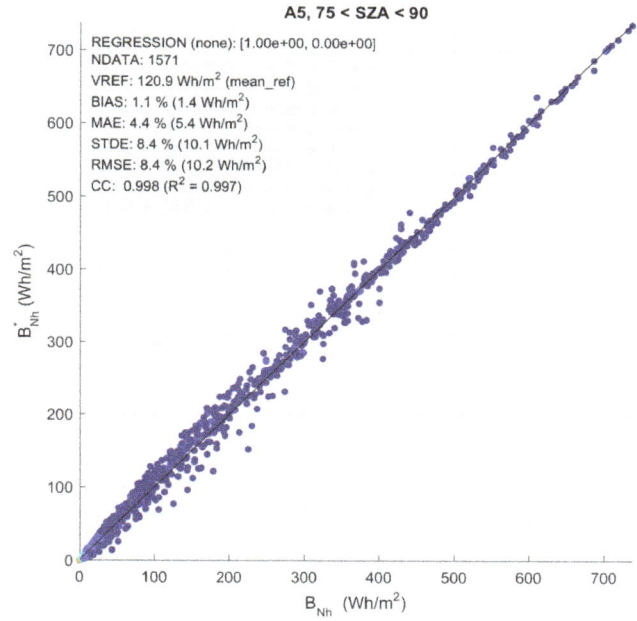

Figure 2. Correlogram between B_{Nh} (horizontal axis) and B_{Nh}^{*} computed with $\theta_{S}^{\mathrm{eff},A5}$ (vertical axis) for the subset "low sun" (SZA: solar zenith angle) for Carpentras.

4 Results

A subset of the data, called "low sun", has been created to better study the cases of sun low above horizon, i.e. $\theta_{S} > 75°$. Tables 2 and 3 report the results at Carpentras and Payerne for daylight time (all angles) and for the subset "low sun" for the six different practices. Correlograms between B_{Nh} (horizontal axis) and B_{Nh}^{*} (vertical axis) at Carpentras and Payerne for practice A5 are shown in Figs. 1–4.

One observes that the error depends on the range of θ_{S}. For large θ_{S}, errors are much greater than for smaller θ_{S}. One may also observe that errors are far from being negligible for most practices.

Errors are the greatest for practices A0, A1 and A2. The bias ranges from 5 to 6 Wh m^{-2} for all θ_{S}, and from 16 to 21 Wh m^{-2} for $\theta_{S} > 75°$. In the latter case, it means a relative bias of 16–17 % which is quite large. The RMSE ranges from 16 to 25 Wh m^{-2} for all θ_{S}, and from 31 to 48 Wh m^{-2} for $\theta_{S} > 75°$ – relative values are 30–40 %. Correlation coefficients are very large as a whole. The minima are observed for large θ_{S} and are greater than 0.974.

Better results are attained for practices A3 and A4. The bias is 4–5 Wh m^{-2} for all θ_{S}, and 13–15 Wh m^{-2} for $\theta_{S} > 75°$. It corresponds to respectively 1 and 12–13 % in relative values. The RMSE is 13–15 Wh m^{-2} (relative RMSE is 4–5 %) for all θ_{S} and ranges from 24 to 28 Wh m^{-2} for $\theta_{S} > 75°$ – relative RMSE is 23–24 %. Correlation coefficients are very large as a whole. The minima are observed for large θ_{S} and are greater than 0.986.

Table 2. Performance of each practice for Carpentras for all angles and the subset "low sun". Relative values are computed relative to the mean value of B_{Nh}. Best results are in bold.

	Bias (Wh m^{-2})		RMSE (Wh m^{-2})		Correlation coefficient	
	All angles	Low sun	All angles	Low sun	All angles	Low sun
A0	6 (2 %)	21 (17 %)	25 (8 %)	48 (39 %)	0.997	0.974
A1	6 (2 %)	21 (17 %)	25 (8 %)	48 (40 %)	0.997	0.974
A2	6 (2 %)	19 (16 %)	20 (6 %)	38 (31 %)	0.998	0.986
A3	5 (1 %)	15 (12 %)	15 (5 %)	28 (23 %)	0.999	0.993
A4	5 (1 %)	15 (12 %)	15 (5 %)	28 (23 %)	0.999	0.993
A5	**0 (0 %)**	**1 (1 %)**	**7 (2 %)**	**10 (8 %)**	**1.000**	**0.998**

Table 3. Performance of each practice for Payerne for all angles and the subset "low sun". Relative values are computed relative to the mean value of B_{Nh}. Best results are in bold.

	Bias (Wh m^{-2})		RMSE (Wh m^{-2})		Correlation coefficient	
	All angles	Low sun	All angles	Low sun	All angles	Low sun
A0	5 (2 %)	17 (17 %)	20 (6 %)	38 (37 %)	0.998	0.980
A1	5 (2 %)	17 (17 %)	20 (6 %)	38 (37 %)	0.998	0.981
A2	5 (2 %)	16 (16 %)	16 (5 %)	31 (30 %)	0.999	0.989
A3	4 (1 %)	13 (13 %)	13 (4 %)	24 (24 %)	0.999	0.994
A4	4 (1 %)	13 (13 %)	13 (4 %)	24 (24 %)	0.999	0.994
A5	**1 (0 %)**	**3 (3 %)**	**6 (2 %)**	**9 (9 %)**	**1.000**	**0.998**

The best results are attained for practice A5. The bias is very small: 0 Wh m^{-2} or close to, for all θ_S and 1–3 Wh m^{-2} for $\theta_S > 75°$. The RMSE is 6–7 Wh m^{-2} – relative RMSE is 2 % – for all SZA and 9–10 Wh m^{-2} for $\theta_S > 75°$ – relative RMSE is 8–9 %. Correlation coefficients are very large as a whole and greater than 0.998. One may observe in Figs. 1 to 4 that the points are well aligned along the $y = x$ line with a very small scattering.

The performances decrease with large θ_S. It should be noted that the decrease is much less pronounced with practice A5 than with the others. For example, the RMSE for A5 increases from 7 to 10 Wh m^{-2} for Carpentras, and from 6 to 9 Wh m^{-2} for Payerne, while it doubles for the other practices, e.g. from 25 to 48 Wh m^{-2} for Carpentras and A0.

The azimuth of the sun Ψ_S is defined as the angle between the projection of the direction of the sun on the horizontal plane and a reference direction. The ISO convention is to count Ψ_S clockwise from North where its value is 0. Thus, it is $\frac{\pi}{2}$ for East, π for South and $\frac{3\pi}{2}$ for West. The effective solar azimuth Ψ_S^{eff} may be computed with the following equations (ESRA, 2000):

$$\Psi_S^{eff} = \pi - \cos^{-1}\left[\frac{\left(\sin\Phi\cos\theta_S^{eff} - \sin\delta\right)}{\left(\cos\Phi\sin\theta_S^{eff}\right)}\right] \text{ before noon}$$

$$\Psi_S^{eff} = \pi + \cos^{-1}\left[\frac{\left(\sin\Phi\cos\theta_S^{eff} - \sin\delta\right)}{\left(\cos\Phi\sin\theta_S^{eff}\right)}\right] \text{ afternoon} \quad (10)$$

where Φ is the latitude and δ the declination angle.

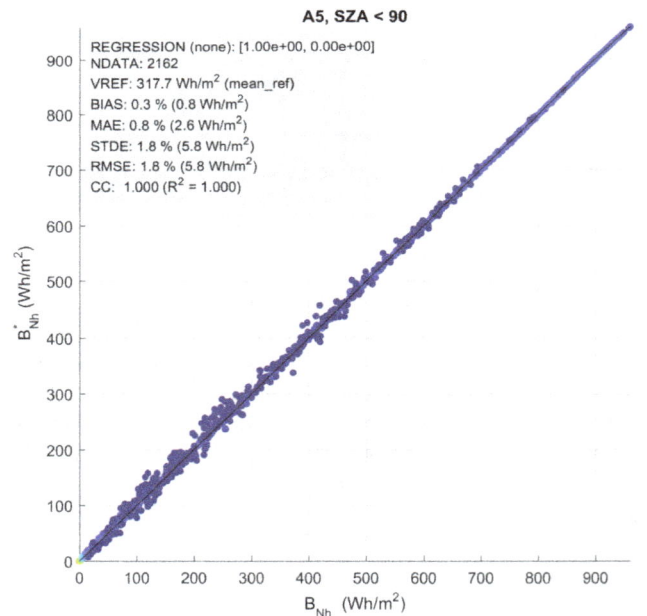

Figure 3. Correlogram between B_{Nh} (horizontal axis) and B_{Nh}^* computed with $\theta_S^{eff,A5}$ (vertical axis) for all solar zenith angles (SZA) for Payerne.

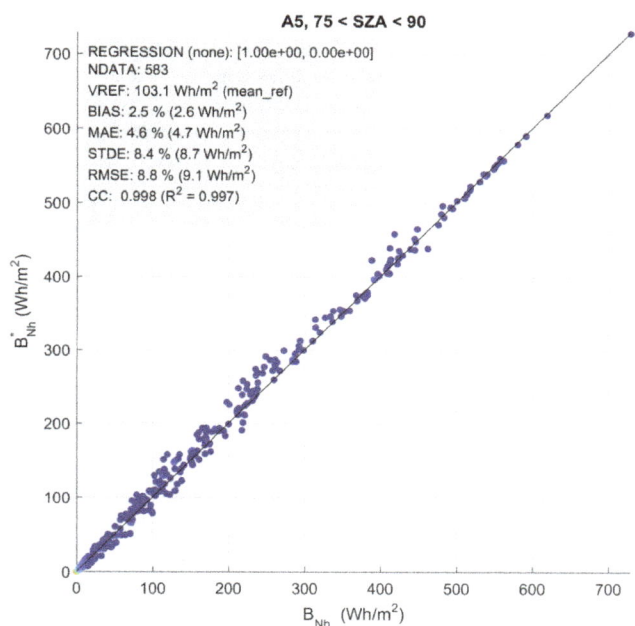

Figure 4. Correlogram between B_{Nh} (horizontal axis) and B_{Nh}^* computed with $\theta_S^{eff,A5}$ (vertical axis) for the subset "low sun" (SZA: solar zenith angle) for Payerne.

5 Conclusions

Several practices may be used to compute the effective solar angles. Though dealing with a limited number of cases, this study has shown that errors are far from being negligible for most practices. One must be careful in selecting the practice.

The practice A5 produces the best results by far with no bias and small RMSE. It is followed by A4 and A3. The worst ones are A0, A1 and A2. For all methods, performances show a tendency to decrease during sunrise and sunset hours. The errors may double, except for A5 which shows little degradation in performance for large θ_S.

Practically, how to implement these practices? Practices A0 to A4 needs a library to compute θ_S every 1 min. Among several solutions, such as the Python code PyEphem available at http://rhodesmill.org/pyephem/, one may use the equations in ESRA (2000) which form the Solar Geometry 1 library (SG1) or the more accurate SG2 library (Blanc and Wald, 2012); both are available at http://www.oie.mines-paristech.fr/Valorisation/Outils/Solar-Geometry/. If one does not possess software for computing this angle, the web site SoDa Service for professionals in solar radiation (www.soda-pro.com) offer free-of-charge efficient services that implement SG2 and deliver time series of solar zenith and azimuth angles and declination angle.

Implementing practice A4 requests in addition the computation of the hourly irradiation at the top of atmosphere on horizontal and normal-to-sun surfaces. The above mentioned tools, including the SoDa Service, may be used in that purpose.

Implementing practice A5 may be costly if one does not possess software for the estimation of hourly – or better resolution – direct irradiation on both normal and horizontal surfaces under cloud-free skies. The ESRA clear-sky model (Rigollier et al., 2004) is easy to implement (code available at http://www.oie.mines-paristech.fr/Valorisation/Outils/Clear-Sky-Library/) but several others are, too. If one does not possess a clear-sky model, the SoDa Service offer free-of-charge a service that implements the McClear model and that delivers in one click time series of hourly irradiation on horizontal and normal-to-sun surfaces for any place in the world.

Acknowledgements. The authors thank the two reviewers whose comments help in improving this text. The research leading to these results has been undertaken within the Task #46 of the Solar Heating and Cooling Implementing Agreement of the International Energy Agency, and has been partly funded by the French Agency ADEME, research grant no. 1105C0028. The authors thank the operators of the Carpentras and Payerne stations for their valuable measurements and the Alfred Wegener Institute for hosting the BSRN website from which data may be downloaded. The McClear service in the SoDa Service is made available to anyone by the MINES ParisTech and Transvalor within the Copernicus Atmosphere Monitoring Service implemented by the ECMWF on behalf of the European Commission.

References

Blanc, P. and Wald, L.: The SG2 algorithm for a fast and accurate computation of the position of the Sun, Solar Energy, 86, 3072–3083, doi:10.1016/j.solener.2012.07.018, 2012.

ESRA – European Solar Radiation Atlas: Fourth edition, includ. CD-ROM, edited by: Scharmer, K., Greif, J., Scientific advisors: Dogniaux, R., Page, J. K., Authors: Wald, L., Albuisson, M., Czeplak, G., Bourges, B., Aguiar, R., Lund, H., Joukoff, A., Terzenbach, U., Beyer, H. G., and Borisenko, E. P., published for the Commission of the European Communities by Presses de l'Ecole, Ecole des Mines de Paris, Paris, France, 2000.

Hall, I., Prairie, R., Anderson, H., and Boes, E.: Generation of Typical Meteorological Years for 26 SOLMET stations, SAND78-1601, Sandia National Laboratories, Albuquerque, 1978.

Kalogirou, S.: Generation of Typical Meteorological Year (TMY-2) for Nicosia, Cyprus, Renewable Energy, 28, 2317–2334, doi:10.1016/S0960-1481(03)00131-9, 2003.

Korany, M., Boraiy, M., Eissa, Y., Aoun, Y., Abdel Wahab, M. M., Alfaro, S. C., Blanc, P., El-Metwally, M., Ghedira, H., Hungershoefer, K., and Wald, L.: A database of multi-year (2004–2010) quality-assured surface solar hourly irradiation measurements for the Egyptian territory, Earth Syst. Sci. Data Discuss., 8, 737–758, doi:10.5194/essdd-8-737-2015, 2015.

Lefèvre, M., Oumbe, A., Blanc, P., Espinar, B., Gschwind, B., Qu, Z., Wald, L., Schroedter-Homscheidt, M., Hoyer-Klick, C., Arola, A., Benedetti, A., Kaiser, J. W., and Morcrette, J.-J.: Mc-Clear: a new model estimating downwelling solar radiation at ground level in clear-sky conditions, Atmos. Meas. Tech., 6, 2403–2418, doi:10.5194/amt-6-2403-2013, 2013.

Ohmura, A., Gilgen, H., Hegner, H., Mueller, G., Wild, M., Dutton, E. G., Forgan, B., Froelich, C., Philipona, R., Heimo, A., Koenig-Langlo, G., McArthur, B., Pinker, R., Whitlock, C. H., and Dehne, K.: Baseline Surface Radiation Network (BSRN/WCRP): New precision radiometry for climate research, B. Am. Meteorol. Soc., 79, 2115–2136, doi:10.1175/1520-0477(1998)079<2115:BSRNBW>2.0.CO;2, 1998.

Rigollier, C., Lefèvre, M., and Wald, L.: The method Heliosat-2 for deriving shortwave solar radiation from satellite images, Solar Energy, 77, 159–169, doi:10.1016/j.solener.2004.04.017, 2004.

Roesch, A., Wild, M., Ohmura, A., Dutton, E. G., Long, C. N., and Zhang, T.: Assessment of BSRN radiation records for the computation of monthly means, Atmos. Meas. Tech., 4, 339–354, doi:10.5194/amt-4-339-2011, 2011.

27

Spatial patterns of European droughts under a moderate emission scenario

J. Spinoni[1], G. Naumann[2], and J. Vogt[3]

[1]European Commission, Joint Research Centre, Institute for the Protection and Security of the Citizen, Ispra, Italy
[2]CONICET, National Scientific Technical and Research Council, Buenos Aires, Argentina
[3]European Commission, Joint Research Centre, Institute for the Environment and Sustainability, Ispra, Italy

Correspondence to: J. Spinoni (jonathan.spinoni@gmail.com)

Abstract. Meteorological drought is generally defined as a prolonged deficiency of precipitation and is considered one of the most relevant natural hazards as the related impacts can involve many different sectors. In this study, we investigated the spatial patterns of European droughts for the periods 1981–2010, 2041–2070, and 2071–2100, focusing on the projections under a moderate emissions scenario. To do that, we used the outputs of the KNMI-RACMO2 model, which belongs to the A1B family and whose spatial resolution is $0.25° \times 0.25°$. By means of monthly precipitation and potential evapotranspiration (PET), we computed the Standardized Precipitation Index (SPI) and the Standardized Precipitation Evapotranspiration Index (SPEI) at the 12-month accumulation scale. Thereafter, we separately obtained drought frequency, duration, severity, and intensity for the whole of Europe, excluding Iceland. According to both indicators, the spatial drought patterns are projected to follow what recently characterized Europe: southern Europe, who experienced many severe drought events in the last decades, is likely to be involved by longer, more frequent, severe, and intense droughts in the near future (2041–2070) and even more in the far future (2071–2100). This tendency is more evident using the SPEI, which also depends on temperature and consequently reflects the expected warming that will be highest for the Mediterranean area in Europe. On the other side, less severe and fewer drought events are likely to occur in northern Europe. This tendency is more evident using the SPI, because the precipitation increase is projected to outbalance the temperature (and PET) rise in particular in Scandinavia. Regarding the mid-latitudes, the SPEI-based analyses point at more frequent drought events, while the SPI-based ones point at less frequent events in these regions.

1 Introduction

Drought is a slowly developing natural hazard which can affect large areas and populations, can propagate through the full hydrological cycle, and may have both immediate consequences as well as long-term economic and environmental impacts (Vogt et al., 2011; Vogt and Somma, 2000). Drought is a temporary condition that can result in irreversible damages to ecosystems and can take place in almost all climates, not only in areas prone to land degradation (Winslow et al., 2011).

Due to its complex evolution and onset, there is no single definition of drought, and the scientific literature usually distinguishes between meteorological, agricultural, hydrological, ground-water, streamflow, and socioeconomic droughts (Mishra and Singh, 2010). In this study we refer to meteorological drought because we used meteorological variables – precipitation and mean temperature – to compute the drought indicators at the 12-month accumulation scale. However, in the literature, the 3-month accumulation scale is sometimes used for meteorological drought, while the 12-month scale is used for hydrological drought (Mishra and Singh, 2010, 2011). Meteorological drought is usually defined as a deficit in precipitation over a defined period compared with climatological normal values, but drought conditions can also be triggered by high temperatures, low relative humidity, strong and desiccating winds, etc. (Zampieri et al., 2009). The impacts of meteorological and hydrological drought events involve a wide variety of sectors, e.g. agriculture (Ciais et al., 2005), soil (Hirschi et al., 2011), ecology (McDaniels et al., 1997), and energy (Hightower and Pierce, 2008).

At the global level, recent climate change seems to have caused only little increase in drought (Dai, 2011; Sheffield et al., 2012; Spinoni et al., 2014; Trenberth et al., 2014). However, at the regional level some areas experienced a remarkable increase in drought frequency and severity over the last decades, for example in the Carpathian region (Spinoni et al., 2013), southern Europe (van der Schrier et al., 2013; Beguería et al., 2014; Spinoni et al., 2015a), and in particular the Mediterranean region (Briffa et al., 2009; Hoerling et al., 2012). According to the latest report of the Intergovernmental Panel on Climate Change (IPCC, 2014), global warming will generally drive the dry regions to drier conditions and the wet ones to wetter conditions. This can be applied to Europe as a whole, as southern Europe is projected to get drier and northern Europe to get wetter (Spinoni et al., 2015a). Though there is indeed a general tendency towards a drier world (Sherwood and Fu, 2014), at the global level the paradigm *the dry gets drier, and the wet gets wetter* should be handled with care (Greve et al., 2014).

This study has been conducted in the framework of the GAP-PESETA project, the follow-up of the Joint Research Centre's PESETA II project, whose main goal was to gain insights into the patterns of climate change impacts in Europe and was focused on many topics (agriculture, energy, floods, forest fires, etc.) but provided only preliminary information regarding drought, i.e. about streamflow drought only (Ciscar et al., 2014). We dealt with spatial drought patterns from 1950 to 2100, investigating in particular the evolution of drought events under a moderate emissions scenario (family A1B, see IPCC, 2000) in the near future (period: 2041–2070) and in the far future (2071–2100) compared with the recent past (1981–2010). The examined area is Europe, including European Russia, but excluding Iceland, Greenland, the Azores, the Canary Islands, and Madeira. We computed two drought indicators, the Standardized Precipitation Index (SPI; McKee et al., 1993) and the Standardized Precipitation Evapotranspiration Index (SPEI; Vicente-Serrano et al., 2010), at the monthly scale for every grid point (spatial resolution: $0.25° \times 0.25°$) belonging to the examined area. Such indicators have been used to calculate the frequency, duration, severity, and intensity of drought events on the same grid.

This study has three main goals: firstly, it aims at providing new insights about future droughts in Europe at a spatial resolution higher than most of the existing studies, which usually analyze data at $0.5° \times 0.5°$ or $1° \times 1°$ resolution, or focus on smaller regions (see, e.g. Blenkinsop and Fowler, 2007); secondly, it focuses on drought events, differently than most of the publications, which usually fo-

cus on drought-related climate indicators (see, e.g. Heinrich and Gobiet, 2012); thirdly, it aims at evaluating whether and where a potential evapotranspiration (PET) increase will be the leading factor for drought and, on the other side, whether and where the drought trends will be driven by precipitation. In this study, we did not consider potential human-induced changes in future drought patterns, because meteorological droughts do not directly depend on human activities, as other types of droughts do, e.g. the hydrological drought (Wanders and Yada, 2015).

After the introduction, Sect. 2 deals with data and methods: we describe the data inputs and the quality checks, we motivate the choice of the indicators and their computation algorithms, and we define the drought variables. Section 3 deals with results and discussions: we present figures about the change of frequency, duration, severity, and intensity of drought events in Europe between the near (and far) future and the recent past. Section 3 also focuses on the debate about which will be the most important meteorological driver for future droughts in Europe. Section 4 briefly summarizes the main findings and anticipates possible future developments.

2 Data and methods

2.1 Scenario model, input variables, and study region

The PESETA II project based its projections on the outputs of the version 2.1 of the KNMI regional atmospheric climate model RACMO (van Meijgaard et al., 2008), consequently we chose such model for our analyses. The RACMO2 is provided by the Royal Meteorological Institute of the Netherlands, its driving global circulation model (GCM) is the ECHAM5-RT3, and it belongs to the A1B scenario family. The special report on the emissions scenarios published by the IPCC in 2000 (IPCC, 2000) divides them into four main groups: A1, A2, B1, and B2. The A1 family describes a future world of very rapid economic growth, global population that peaks in the 2050s and declines thereafter, and the rapid introduction of new and more efficient technologies. This category is further subdivided into categories which correspond to different directions of technological change in the energy system: fossil intensive energy sources (A1FI), nonfossil energy sources (A1T), and balance across all sources (A1B). The A1B scenario foresees moderate but increasing emissions of carbon dioxide (CO_2) and methane (CH_4) until the 2050s, followed by a slow decrease. Nitrous oxide (N_2O) emissions are projected to be stable through the whole of the century, and the sulfur dioxide (SO_2) emissions are projected to increase until the 2030s and thereafter decrease. In the latest assessment report (the IPCC AR5, 2014), it was discussed how the various scenarios are able to represent what effectively happened in the last decades and it turned out that the A1 class effectively pictures the actual global change, though it seems to slightly overestimate the emissions and the rise of temperature.

From the RACMO2 model, we directly obtained monthly precipitation, and we derived PET using the FAO-56 Penman–Monteith algorithm (Allen et al., 1998), from January 1950 to December 2100 (spatial resolution of $0.25° \times 0.25°$). The selected area is Europe, but we excluded the small atlantic islands (Madera, the Azores, and the Canary Islands) and Iceland, for a few grid points there showed suspect precipitation data in the 2000s according to the quality-checks and the homogenization tests performed with the Multiple Analysis of Series for Homogenization software (MASH, version 3.03, Szentimrey, 1999).

The choice of the algorithm to derive PET is a key point regarding the computation of drought variables (van der Schrier et al., 2011; Trenberth et al., 2014). We chose the Penman–Monteith equation because the outputs of the RACMO2 model include the input variables needed, i.e. temperature, relative humidity, solar radiation, and wind speed. Other common formulations are based on mean temperature (Thornthwaite; Thornthwaite, 1948) or on minimum and maximum temperature (Hargreaves and Samani, 1982; Allen et al., 1998). Recently, contradictory opinions emerged about the validity of such simplified formulations: Sheffield et al. (2012) pointed at the overestimation of drought trends based on temperature-only-based PET estimations, while Dai (2011) and van der Schrier et al. (2011) affirmed that the choice of the PET formulation does not remarkably affect drought indicators such as the PDSI (Palmer, 1965) and the sc-PDSI (Wells et al., 2004) in Europe. However, keeping in mind that we used *potential* and not *actual* evapotranspiration (AET) and that this could bias the results in soil moisture stressed areas (Brutsaert and Parlange, 1998), we discarded the temperature-only-based PET models as they may influence too much the projections of drought events due to the projected temperature rise in the A1B scenario, especially in the far future.

2.2 Drought indicators, events, and derived variables

The monthly precipitation and PET series – from 1950 to 2100 – have been used to compute the Standardized Precipitation Index (SPI: McKee et al., 1993), which depends on precipitation, and the Standardized Precipitation Evapotranspiration Index (SPEI, Vicente-Serrano et al., 2010), which depends on the difference between precipitation and PET. We chose the SPI for it is probably the most applied drought indicator at the European level (e.g. Lloyd-Hughes and Saunders et al., 2002), but considering that the scenarios describe a warming future (IPCC, 2014), we assumed that it is important to consider also one indicator that directly or indirectly depends on temperature; thus we included the SPEI that is nowadays frequently applied in Europe (e.g. Vicente-Serrano et al., 2014) in our analyses.

We computed the SPI and the SPEI at the 12-month scale, and we fitted the cumulated precipitation with the Gamma distribution (Thom, 1958) for the SPI and the cumulated difference between precipitation and PET with the log-logistic distribution (Shoukri et al., 1988) for the SPEI, following the approaches suggested by the authors who first presented such indicators, being also aware that different distributions could best fit the SPI and the SPEI in Europe, depending on the local climate features (Stagge et al., 2014, 2015). Dealing with standardized indicators, the length of the record used (Wu et al., 2005) and the choice of the baseline period (Guttman, 1999) are key issues: we used all the available data in the period 1950–2100 to fit the distributions. If a shorter period (e.g. 30-year interval) is chosen as the baseline for an indicator computed over a 151-year period and the selected period is characterized by frequent and severe drought events, this will influence the computation of the indicator and the other periods are bound to be characterized by less frequent and severe drought events. Because we were not interested in the absolute number of drought events (and their duration and severity) in a certain period, but we aimed at studying whether the drought events will be longer, more severe, frequent, or intense in the future compared to the recent past, the entire baseline ensures more robust comparisons between different periods.

This study focuses on the drought events, not on the indicators, and we separately analyzed them according to the SPI and the SPEI. Using the monthly series of the indicators at a grid point level, we followed Guerrero-Salazar and Yevjevich (1975) and McKee et al. (1993) to define the events. In this study, a drought event takes place every time the indicator falls below -1 for at least 2 consecutive months, and it ends when the indicator rises above 0. For a given period, the drought frequency is therefore the number of drought events. The duration of a drought event refers to the number of months of the event, its severity to the integral area of the event (the sum of the indicator values below zero, in absolute values, during the occurrence of the event), and its intensity to the ratio between severity and duration. We computed the frequency and the average duration, severity, and intensity of the drought events over 30-year periods: recent past (1981–2010), near future (2041–2070), and far future (2071–2100). The results are presented as comparisons between recent past and, respectively, near (Fig. 1) and far future (Fig. 2).

3 Results and Discussion

3.1 Drought patterns in the near future (2041–2070)

The drought patterns of the near future (2041–2070) have been evaluated vs. the reference period 1981–2010. The frequency (DF; number of events), duration (DD; months), severity (DS; score), and intensity (DI; score) of the events shown in Fig. 1 are expressed per decade.

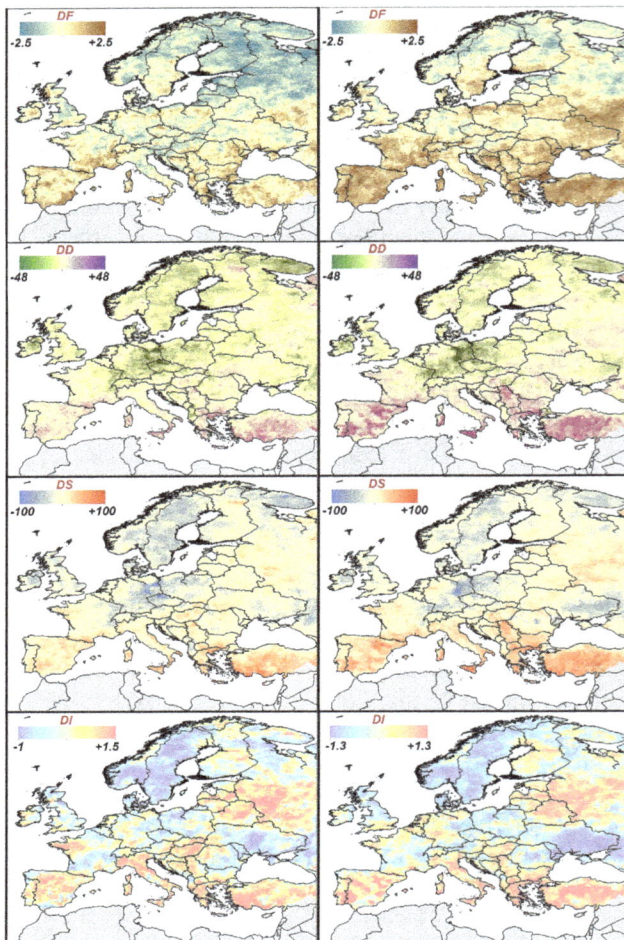

Figure 1. Comparisons between near future and recent past for the frequency (DF), duration (DD), severity (DS), and intensity (DI) of drought events, according to the SPI-12 (left panel) and the SPEI-12 (right panel). All the values are expressed per decade.

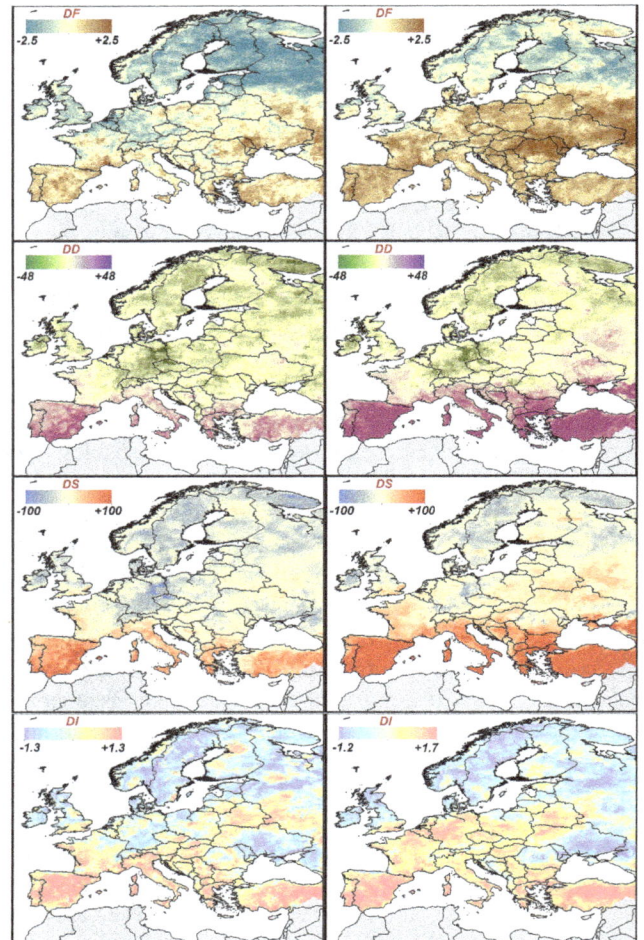

Figure 2. Comparisons between far future and recent past for the frequency (DF), duration (DD), severity (DS), and intensity (DI) of drought events, according to the SPI-12 (left panel) and the SPEI-12 (right panel). All the values are expressed per decade.

According to the SPI, the drought events are projected to be more frequent in the near future (2041–2070) than in the recent past (1981–2010) in the Mediterranean region, in the Balkans, along the Black Sea coast and in general in southern Europe. On the contrary, Scandinavia (but not southern Sweden) and the Baltic republics are projected to be involved by fewer drought events. According to the SPEI, the gradient between northern and southern Europe is remarkable and also the European mid-latitudes are projected to be involved by more frequent droughts (up to 2.3 more events per decade).

Regarding the duration and according to the SPI, the drought events are projected to be longer in the near future than in the recent past in Spain, the French Riviera, Sicily and Sardinia, Greece, and Turkey, and shorter in Northern Ireland, central Europe, and Scandinavia. According to the SPEI, they will be longer in the entire Mediterranean region and the Balkans, and shorter in the same regions highlighted by the SPI. According to both the indicators, the

drought events are projected to follow, on average, the same spatial patterns regarding the severity and the duration. Instead, the spatial patterns of the drought intensity are less homogeneous. The average intensity is projected to be higher in central Spain, north-eastern France, northern Italy, the Carpathian region, the Balkans, Belarus, Greece, and Turkey. On the contrary, it is projected to be lower especially in Scandinavia and Ukraine.

In general, it seems that the PET increase (included in the SPEI) will result in more frequent drought events in the near future in southern Europe, but it will not be the main cause for longer or more severe events everywhere in southern Europe, as they are similarly projected by both the SPI and the SPEI. On the contrary, the precipitation increase will outbalance the PET (and temperature) increase in northern Europe and is projected to cause less frequent events there. Central Europe shows contradictory tendencies: less (SPI) or more

(SPEI) frequent and intense events depending on the indicator, shorter and less severe events for both indicators.

A more detailed description of the varying importance of precipitation deficit vs. temperature extremes in a changing climate goes beyond the scopes of this paper. However, the authors more thoroughly discussed on that in Spinoni et al. (2015a, b) and the reader can also find more details on other recent publications, such as Paulo et al. (2012); Vicente-Serrano et al. (2014); Kingston et al. (2015).

3.2 Drought patterns in the far future (2071–2100)

We repeated the analyses discussed in Sect. 3.1, this time for the far future (2071–2100) compared with the recent past (1981–2010), see Fig. 2. According to the SPI, southern Europe is projected to be involved in more frequent droughts in the far future than in the recent past, and this higher frequency will be more evident according to the SPEI. Northern Europe will be hit by fewer droughts for both the indicators, while central Europe shows different patterns: no particular changes for the SPI, excluding a frequency decrease for Germany vs. more drought events reaching up to latitudes corresponding to Denmark for the SPEI. To summarize, it seems that the PET increase will be the driver for drought frequency increase in the far future as well as in the near future for most of Europe, excluding Scandinavia and northern Russia.

From Fig. 2 we can notice that, according to the SPI, the Mediterranean region – and southern Europe in general – are expected to be hit by longer, more severe, and intense droughts in the far future. Such increasing tendencies are even more pronounced for duration and severity in Spain, Greece, and Turkey, according to the SPEI. Instead, the decrease in northern European regions is lower in absolute values. If we compare the far future with the near future (not shown), the drought duration and severity increase in the far future compared to the near future is relevant only in Spain according to the SPI, while it is relevant in every part of Europe below 45° N of Latitude according to the SPEI. Once again, it seems that the global warming tendency will cause the increase of drought length and severity at the end of the century, even under a moderate emissions scenario.

The intensity of the drought events is projected to increase for both the indicators in the far future in southern Europe and Belarus, and to decrease in Irealand, Scotland, Scandinavia, Ukraine, and Russia. Germany again shows contradictory patterns: according to the SPI, fewer, shorter, less severe, and less intense drought events will occur, while according to the SPEI the droughts will be more frequent and intense, but also shorter.

We underline that the results discussed in this chapter and in the previous one have been derived from indicators based on 12-month accumulation, i.e. a whole year, consequently the seasonal effects are lost as well as the seasonality of precipitation and evapotranspiration. To overcome such a shortage of data, we plan performing new analyses based on

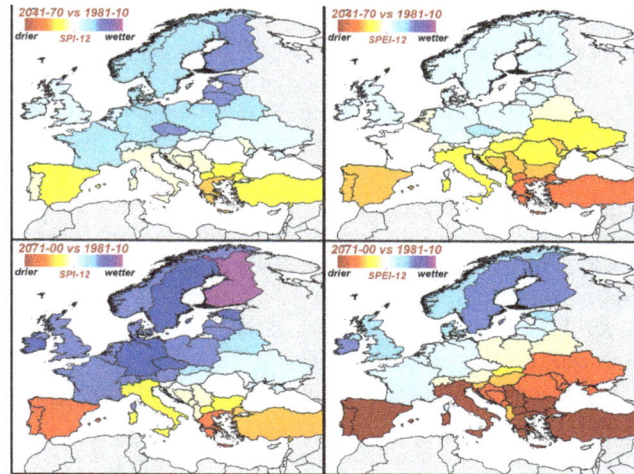

Figure 3. Country tendencies towards drier or wetter near and far future, according to the SPI-12 (left panel) and the SPEI-12 (right panel).

shorter accumulation periods (SPI-3 and SPEI-3), in order to study the seasonal spatial drought patterns which may be different from annual patterns in some parts of Europe, both during the current and future climate.

3.3 Projected drought patterns at a country scale

To provide a general picture of the future European drought tendencies, we analyzed the complete monthly records of the indicators from 1950 to 2100. We averaged them at a country level, and we compared their values for the three periods already investigated in the previous chapters. Because this part is based on country averages and European Russia is not a country, we excluded it from the analyses.

Figure 3 shows that the southern European countries are projected to be drier in the near future than in the recent past. In particular, the SPEI indicator projects drier conditions than the SPI indicators for the Mediterranean countries, the Balkans, and Turkey. According to the SPI, the central and northern European countries are projected to be wetter in the near future, in particular the Baltic republics and Finland. In general, in the far future a more extreme shift towards drier conditions for southern and eastern Europe can be seen, in particular according to the SPEI, and a more extreme shift towards a wetter central and northern Europe can be seen, in particular according to the SPI. In the last decades of the current century such shifts are projected to be extreme also by the SPI, especially regarding the shifts towards wetter conditions, but also towards drier conditions for the Iberian Peninsula and Greece.

These findings reinforce the general idea of climate change in Europe: from the 1950s to 2010s, southern Europe faced a drying tendency and northern Europe a tendency toward wetness (Lloyd-Hughes and Saunders, 2002; IPCC, 2014; Briffa et al., 2009; Hannaford et al., 2011; Hoerling et al., 2012; Spinoni et al., 2015a). Such spatial patterns are projected to continue – and increase in magnitude – until the end of the century, as also discussed by, e.g. Burke et al. (2006), Blenkinsop and Fowler (2007), and Warren et al. (2009) using different drought indicators and moderate emission scenarios. In particular, compared to our findings, Warren et al. (2009) presented similar results (drought variables tend to increase in southern and eastern Europe, and decrease in northern Europe) by using SPI-5 and SPI-12 projected in 2050–2098 under different emission scenarios; the only remarkable differences involve Romania, where a more pronounced tendency towards longer drought events was found by Warren et al. (2009).

4 Conclusions

In this study we investigated the tendency of drought events in Europe until 2100 under a moderate emissions scenario, the KNMI-RACMO2. To do that, we based our analyses on two drought indicators, the SPI and the SPEI, computed at a 12-month scale. Regarding the drought events, we defined four quantities (frequency, duration, severity, and intensity), and we compared the recent past (1981–2010) vs. the near (2041–2070) and the far future (2071–2100). The spatial drought patterns have been analyzed both at $0.25° \times 0.25°$ and at the country scale.

We looked for the answers to a few questions: are the drought events likely to become more frequent in the future? Will they be longer? Will they be more severe and intense? Southern Europe, that already experienced a drying trend in the second part of the 20th century (see, e.g. IPCC, 2014), is projected to be affected by more frequent, severe, intense, and longer drought events in the near future and even more in the far future. The drying trend is driven by the combination of PET increase and precipitation decrease. Oppositely, northern Europe, the area that experienced a trend toward wetness from the 1970s onwards (IPCC, 2014), is projected to be involved by fewer and less intense droughts, mainly due to the projected precipitation increase.

The results of this study should be considered as a preliminary step towards more detailed analyses regarding the projections of European drought events. Many different aspects could be introduced to refine the methodologies and the outputs. Among them we plan to study seasonal droughts, adding new indicators, using other models belonging to the A1B (see Meehl et al., 2005, for a summary) and different scenario families, performing tests to compute the statistical error intervals of the climate projections, and coupling the drought information with other climate extremes.

Acknowledgements. We would like to thank the two anonymous referees for their precious suggestions which helped improving the manuscript.

References

Allen, R. G., Pereira, L. S., Raes, D., and Smith, M.: Crop evapotranspiration-Guidelines for computing crop water requirements, FAO Irrigation and drainage paper, 56, FAO, Rome, 15 pp., 1998.

Beguería, S., Vicente-Serrano, S. M., Reig, F., and Latorre, B.: Standardized precipitation evapotranspiration index (SPEI) revisited: parameter fitting, evapotranspiration models, tools, datasets and drought monitoring, Int. J. Climatol., 34, 3001–3023, 2014.

Blenkinsop, S. and Fowler, H. J.: Changes in European drought characteristics projected by the PRUDENCE regional climate models, Int. J. Climatol., 27, 1595–1610, 2007.

Briffa, K. R., van Der Schrier, G., and Jones, P. D.: Wet and dry summers in Europe since 1750: evidence of increasing drought, Int. J. Climatol., 29, 1894–1905, 2009.

Brutsaert, W. and Parlange, M. B.: Hydrologic cycle explains the evaporation paradox, Natur, 396, 30–30, 1998.

Burke, E. J., Brown, S. J., and Christidis, N.: Modeling the recent evolution of global drought and projections for the twenty-first century with the Hadley Centre climate model, J. Hydrometeorol., 7, 1113–1125, 2006.

Ciais, P., Reichstein, M., Viovy, N., Granier, A., Ogée, J., Allard, V., Aubinet, M., Buchmann, N., Bernhofer, C., Carrara, A., Chevallier, F., De Noblet, N., Friend, A. D., Friedlingestein, P., Grünwald, T., Heinesch, B., Keronen, P., Knohl, A., Krinner, G., Loustau, D., Manca, G., Matteucci, G., Miglietta, F., Ourcival, J. M., Papale, D., Pilegaard, K., Rambal, S., Seufert, G., Soussana, J. F., Sanz, M. J., Schulze, E. D., Vesala, T., and Valentini, R.: Europewide reduction in primary productivity caused by the heat and drought in 2003, Nature, 437, 529–533, 2005.

Ciscar, J. C., Feyen, L., Soria, A., Lavalle, C., Raes, F., Perry, M., Nemry, F., Demirel, H., Rozsai, M., Dosio, A., Donatelli, M., Srivastava, A., Fumagalli, D., Niemeyer, S., Shrestha, S., Ciaian, P., Himics, M., Van Doorslaer, B., Barrios, S., Ibáñez, N., Forzieri, G., Rojas, R., Bianchi, A., Dowling, P., Camia, A., Libertà, G., San Miguel, J., de Rigo, D., Caudullo, G., Barredo, J.-I., Paci, D., Pycroft, J., Saveyn, B., Van Regemorter, D., Revesz, T., Vandyck, T., Vrontisi, Z., Baranzelli, C., Vandecasteele, I., Batista e Silva, F., and Ibarreta, D.: Climate Impacts in Europe. The JRC Peseta II Project, JRC-IPTS Working Paper JRC8701, Institute for Prospective and Technological Studies, Joint Research Centre, Ispra, Italy, 2014.

Dai, A.: Drought under global warming: a review, Wiley Interdisciplinary Reviews: Climate Change, 2, 45–65, 2011.

Greve, P., Orlowsky, B., Mueller, B., Sheffield, J., Reichstein, M., and Seneviratne, S. I.: Global assessment of trends in wetting and drying over land, Nat. Geosci., 7, 716–721, 2014.

Guerrero-Salazar, P. and Yevjevich, V.: Analysis of drought characteristics by the theory of runs, Fort Collins, Colorado: Colorado State University, Hydrology Papers, 80, 1–44, 1975.

Guttman, N. B.: Accepting the standardized precipitation index: a calculation algorithm, J. Am. Water Resour. As., 35, 311–322, 1999.

Hannaford, J., Lloyd-Hughes, B., Keef, C., Parry, S., and Prudhomme, C.: Examining the large-scale spatial coherence of European drought using regional indicators of precipitation and streamflow deficit, Hydrol. Process., 25, 1146–1162, 2011.

Hargreaves, G. H. and Samani, Z. A.: Estimating potential evapotranspiration, J. Irr. Drain. Div.-ASCE, 108, 225–230, 1982.

Heinrich, G. and Gobiet, A: The future of dry and wet spells in Europe: A comprehensive study based on the ENSEMBLES regional climate models, Int. J. Climatol., 32, 1951–1970, 2012.

Hightower, M. and Pierce, S. A.: The energy challenge, Nature, 452, 285–286, 2008.

Hirschi, M., Seneviratne, S. I., Alexandrov, V., Boberg, F., Boroneant, C., Christensen, O. B., Formayer, H., Orlowsky, B., and Stepanek, P.: Observational evidence for soil-moisture impact on hot extremes in southeastern Europe, Nat. Geosci., 4, 17–21, 2011.

Hoerling, M., Eischeid, J., Perlwitz, J., Quan, X., Zhang, T., and Pegion, P.: On the increased frequency of Mediterranean drought, J. Climate, 25, 2146–2161, 2012.

Kingston, D. G., Stagge, J. H., Tallaksen, L. M., and Hannah, D. M.: European-Scale Drought: Understanding Connections between Atmospheric Circulation and Meteorological Drought Indices, J. Climate, 28, 505–516, 2015.

IPCC: IPCC Special Report on Emissions Scenarios. A Special Report of IPCC WG III, Cambridge University Press, Cambridge UK and New York, NY, 599 pp., 2000.

IPCC: Summary for policymakers, in: Climate Change 2014: Impacts, Adaptation, and Vulnerability. Part A: Global and Sectoral Aspects. Contribution of Working Group II to the Fifth Assessment Report of the Intergovernmental Panel on Climate Change, edited by: Field, C. B., Barros, V. R., Dokken, D. J., Mach, K. J., Mastrandrea, M. D., Bilir, T. E., Chatterjee, M., Ebi, K. L., Estrada, Y. O., Genova, R. C., Girma, B., Kissel, E. S., Levy, A. N., MacCracken, S., Mastrandrea, P. R., and White, L. L., Cambridge University Press, Cambridge, United Kingdom and New York, NY, USA, 1–32, 2014.

Lloyd-Hughes, B. and Saunders, M. A.: A drought climatology for Europe, Int. J. Climatol., 22, 1571–1592, 2002.

McDaniels, T. L., Axelrod, L. J., Cavanagh, N. S., and Slovic, P.: Perception of ecological risk to water environments, Risk Analysis, 17, 341–352, 1997.

McKee, T. B., Doeskin, N. J., and Kleist, J.: The relationship of drought frequency and duration to time scales, in: Proceedings of the 8th Conference on Applied Climatology, American Meteorological Society, Boston, MA, 179–184, 1993.

Meehl, G. A., Arblaster, J. M., and Tebaldi, C.: Understanding future patterns of increased precipitation intensity in climate model simulations, Geophys. Res. Lett., 32, L18719, doi:10.1029/2005GL023680, 2005.

Mishra, A. K. and Singh, V. P.: A review of drought concepts, J. Hydrol., 391, 202–216, 2010.

Mishra, A. K. and Singh, V. P.: Drought modeling – A review, J. Hydrol., 403, 157–175, 2011.

Palmer, W. C.: Meteorological drought, U.S. Weather Bureau Research Paper, 45, 58 pp., 1965.

Paulo, A. A., Rosa, R. D., and Pereira, L. S.: Climate trends and behaviour of drought indices based on precipitation and evapotranspiration in Portugal, Nat. Hazards Earth Syst. Sci., 12, 1481–1491, doi:10.5194/nhess-12-1481-2012, 2012.

Sheffield, J., Wood, E. F., and Roderick, M. L.: Little change in global drought over the past 60 years, Nature, 491, 435–438, 2012.

Sherwood, S. and Fu, Q.: A Drier Future?, Science, 343, 737–739, 2014.

Shoukri, M. M., Mian, I. U. H., and Tracy, D. S.: Sampling properties of estimators of the log-logistic distribution with application to Canadian precipitation data, Can. J. Stat., 16, 223–236, 1988.

Spinoni, J., Antofie, T., Barbosa, P., Bihari, Z., Lakatos, M., Szalai, S., Szentimrey, T., and Vogt, J.: An overview of drought events in the Carpathian Region in 1961–2010, Adv. Sci. Res., 10, 21–32, doi:10.5194/asr-10-21-2013, 2013.

Spinoni, J., Naumann, G., Carrao, H., Barbosa, P., and Vogt, J. V.: World drought frequency, duration, and severity for 1951–2010, Int. J. Climat., 34, 2792–2804, 2014.

Spinoni, J., Naumann, G., Vogt, J., and Barbosa, P.: European drought climatologies and trends based on a multi-indicator approach, Global Planet. Change, 127, 50–57, 2015a.

Spinoni, J., Naumann, G., Vogt, J. V., and Barbosa, P.: The biggest drought events in Europe from 1950 to 2012, Journal of Hydrology: Regional Studies, 3, 509–524, doi:10.1016/j.ejrh.2015.01.001, 2015b.

Stagge, J. H., Tallaksen, L. M., Xu, C. Y., and van Lanen, H. A. J.: Standardized precipitation-evapotranspiration index (SPEI): Sensitivity to potential evapotranspiration model and parameters, Proceedings of FRIEND-Water 2014, IAHS Red Book No. 363, Montpellier, France, 367–373, 2014.

Stagge, J. H., Tallaksen, L. M., Gudmundsson, L., van Loon, A. F., and Stahl, K.: Candidate distributions for climatological drought indices (SPI and SPEI), Int. J. Climatol., doi:10.1002/joc.4267, in press, 2015.

Szentimrey, T.: Multiple analysis of series for homogenization (MASH), Proceedings of the Second Seminar for Homogenization of Surface Climatological Data, Budapest, Hungary, WMO & WCDMP No. 41, 27–46, 1999.

Thom, H. C.: A note on the gamma distribution, Mon. Weather Rev., 86, 117–122, 1958.

Thornthwaite, C. W.: An approach toward a rational classification of climate, Geogr. Rev., 38, 55–94, 1948.

Trenberth, K. E., Dai, A., van der Schrier, G., Jones, P. D., Barichivich, J., Briffa, K. R., and Sheffield, J.: Global warming and changes in drought, Nature Climate Change, 4, 17–22, 2014.

van der Schrier, G., Jones, P. D., and Briffa, K. R.: The sensitivity of the PDSI to the Thornthwaite and Penman-Monteith parameterizations for potential evapotranspiration, J. Geophys. Res.-Atmos., 116, doi:10.1029/2010JD015001, 2011.

van der Schrier, G., Barichivich, J., Briffa, K. R., and Jones, P. D.: A scPDSI-based global dataset of dry and wet spells for 1901–2009, J. Geophyis. Res., 118, 4025–4048, 2013.

van Meijgaard, E., van Ulft, L. H., van de Berg, W. J., Bosveld, F. C., van den Hurk, B. J. J. M., Lenderink, G., and Siebesma, A. P.: The KNMI regional atmospheric climate model RACMO version 2.1, Technical Report TR-302, Koninklijk Nederlands Meteorologisch Instituut, De Bilt, The Netherlands, 50 pp., 2008.

Vicente-Serrano, S. M., Beguería, S., and López-Moreno, J. I.: A multiscalar drought index sensitive to global warming: the standardized precipitation evapotranspiration index, J. Climate, 23, 1696–1718, 2010.

Vicente-Serrano, S. M., Lopez-Moreno, J. I., Beguería, S., Lorenzo-Lacruz, J., Sanchez-Lorenzo, A., García-Ruiz, J. M., Azorin-Molina, C., Moran-Tejeda, E., Revuelto, J., Trigo, R., Coelho, F., and Espejo, F.: Evidence of increasing drought severity caused by temperature rise in southern Europe, Environ. Res. Lett., 9, 044001, doi:10.1088/1748-9326/9/4/044001, 2014.

Vicente-Serrano, S. M., van der Schrier, G., Beguería, S., Azorin-Molina, C., and Lopez-Moreno, J. I.: Contribution of precipitation and reference evapotranspiration to drought indices under different climates, J. Hydrol., 526, 42–54, 2015.

Vogt, J. and Somma, F. (Eds.): Drought and drought mitigation in Europe, Kluwer, Dordrecht, 325 pp., 2000.

Vogt, J. V., Safriel, U., Von Maltitz, G., Sokona, Y., Zougmore, R., Bastin, G., and Hill, J.: Monitoring and assessment of land degradation and desertification: Towards new conceptual and integrated approaches, Land Degrad. Dev., 22, 150–165, 2011.

Wanders, N. and Yada, Y.: Human and climate impacts on the 21th century hydrological drought, J. Hydrol., 526, 208–220, 2015.

Warren, R., Yu, R., Osborn, T., and de la Nava Santos, S.: Future European drought regimes under mitigated and un-mitigated climate change, IOP Conf. Ser.: Earth Environ. Sci., 6, 292012, doi:10.1088/1755-1307/6/29/292012, 2009.

Wells, N., Goddard, S., and Hayes, M. J.: A self-calibrating Palmer drought severity index, J. Climate, 17, 2335–2351, 2004.

Winslow, M. D., Vogt, J. V., Thomas, R. J., Sommer, S., Martius, C., and Akhtar-Schuster, M.: Science for improving the monitoring and assessment of dryland degradation, Land Degrad. Dev., 22, 145–149, 2011.

Wu, H., Hayes, M. J., Wilhite, D. A., and Svoboda, M. D.: The effect of the length of record on the standardized precipitation index calculation, Int. J. Climatol., 25, 505–520, 2005.

Zampieri, M., D'Andrea, F., Vautard, R., Ciais, P., de Noblet-Ducoudré, N., and Yiou, P.: Hot European summers and the role of soil moisture in the propagation of Mediterranean drought, J. Climate, 22, 4747–4758, 2009.

28

Twenty-first century wave climate projections for Ireland and surface winds in the North Atlantic Ocean

Sarah Gallagher[1], **Emily Gleeson**[1], **Roxana Tiron**[2], **Ray McGrath**[1], **and Frédéric Dias**[3]

[1]Research, Environment and Applications Division, Met Éireann, Glasnevin Hill, Glasnevin, Dublin 9, Ireland
[2]OpenHydro Ltd., Greenore, Co. Louth, Ireland
[3]UCD School of Mathematics and Statistics, University College Dublin, Belfield, Dublin 4, Ireland

Correspondence to: Sarah Gallagher (sarah.gallagher@met.ie)

Abstract. Ireland has a highly energetic wave and wind climate, and is therefore uniquely placed in terms of its ocean renewable energy resource. The socio-economic importance of the marine resource to Ireland makes it critical to quantify how the wave and wind climate may change in the future due to global climate change. Projected changes in winds, ocean waves and the frequency and severity of extreme weather events should be carefully assessed for long-term marine and coastal planning. We derived an ensemble of future wave climate projections for Ireland using the EC-Earth global climate model and the WAVEWATCH III® wave model, by comparing the future 30-year period 2070–2099 to the period 1980–2009 for the RCP4.5 and the RCP8.5 forcing scenarios. This dataset is currently the highest resolution wave projection dataset available for Ireland. The EC-Earth ensemble predicts decreases in mean (up to 2 % for RCP4.5 and up to 3.5 % for RCP8.5) 10 m wind speeds over the North Atlantic Ocean (5–75° N, 0–80° W) by the end of the century, which will consequently affect swell generation for the Irish wave climate. The WAVEWATCH III® model predicts an overall decrease in annual and seasonal mean significant wave heights around Ireland, with the largest decreases in summer (up to 15 %) and winter (up to 10 %) for RCP8.5. Projected decreases in mean significant wave heights for spring and autumn were found to be small for both forcing scenarios (less than 5 %), with no significant decrease found for RCP4.5 off the west coast in those seasons.

1 Introduction

Due to its location in the North Atlantic Ocean, Ireland is advantageously placed in Europe from a renewable wind and wave energy perspective. The west coast consists of an energetic and variable wind and wave climate which consequently has a large potential for renewable energy extraction (Gallagher et al., 2016a), and is a promising location for the deployment of future Wave Energy Converters (WECs). Ireland is also regularly in the path of energetic swells and mid-latitude cyclones, a concern for public safety in the coastal and marine environment, and the viability of any potential WECs deployment. The prevailing wind directions are from the south and west with annual mean surface (10 m) wind speeds in the offshore of 5–10 m s^{-1}. Annual mean significant wave heights (SWH) vary from 1–2 m in the wind-sea dominated Irish Sea, to 3–4 m off the swell dominated Atlantic coast of Ireland. In winter mean SWHs are over 5 m off the west coast (Gallagher et al., 2014). Although not part of this study, the careful assessment of any potential WECs deployment locations would also benefit from the consideration of currents, in addition to the wind and wave climate, and in particular wave-current interactions (WCI), which can vary wave power estimation (for example, Barbariol et al., 2013). In the case of Ireland, although generally not large around the coast, with the exception of some areas off the east and north (SEI, 2004), currents could play an important role in the accurate estimation of available wave power.

Quantifying how the wave climate of Ireland may change in the future is crucial for the long-term planning of large-scale deployments of marine renewable energy installations, as well as other marine activities. The most recent climate projections in IPCC AR5 (Intergovernmental Panel on Climate Change Fifth Assessment Report, Collins et al., 2013) for the North Atlantic wind climate under CMIP5 (Coupled Model Intercomparison Project 5, Taylor et al., 2012), project an overall increase in mean sea-level pressure (MSLP) during winter and summer, fitting with slacker winds. The CMIP5 is a unified framework for co-ordinating global climate models (Taylor et al., 2012). It defines a series of experiments both multi-century and decadal under defined greenhouse gas (GHG) emission scenarios or Representative Concentration Pathways (RCP) (Moss et al., 2010). Various scenarios for changing GHG emissions have been defined: low (RCP2.6), medium-low (RCP4.5), medium (RCP6.0) and high (RCP8.5) emission scenarios. In general average wind changes over the North Atlantic Ocean by the end of the century have been found to be small and negative under climate warming, with the large natural variability of the region dominant (Collins et al., 2013).

Uncertainties in some aspects of the projections of future wind climate regimes, most notably in future storm track positioning, results in low confidence in wave climate projections for the North Atlantic region (Church et al., 2013; Woolf and Wolf, 2013). Studies used to examine the future wave climate have tended to consist of small numbers of ensembles with varying methodologies (Hemer et al., 2013a). However for example, studies such as those by the Coordinated Ocean Wave Climate Project (COWCLIP) (Hemer et al., 2013b) predict a reduction in mean annual, spring and summer SWH in the Northeast Atlantic by the end of the 21st century, consistent with the study presented in this paper. Global projections using CMIP5 models under the RCP4.5 and RCP8.5 emission scenarios (Dobrynin et al., 2012) and a 20-member ensemble of statistically modelled projections derived from CMIP5 data (Wang et al., 2014) also show a decrease in mean annual SWH for Atlantic mid-latitudes.

In summary, there is still large uncertainty in wave climate projections for the end of the twenty-first century for the North Atlantic region in general, and regionally for Ireland. There has not been any high resolution wave projection studies focused on the country. This study focuses specifically on Irish coastal waters with the finest resolution grid around Ireland varying in grid spacing from 15 km at the grid boundaries to 1 km in the nearshore. Using two CMIP5 GHG emission scenarios, RCP4.5 (medium-low) and RCP8.5 (high), the projected changes in surface wind speed over the North Atlantic for the end of the century were examined. This ensured that we could explore changes in the swell generating regions of the North Atlantic Ocean, which influence the Irish wave climate. The projected wave climate changes for Ireland were also examined, focusing on the seasonal changes in mean SWH. The projected spring and autumn changes for both RCP4.5 and RCP8.5 are investigated, whereas Gallagher et al. (2016b) focussed only on the summer and winter projections.

The paper is organised as follows: Section 2 provides details about the EC-Earth and WAVEWATCH III® models used in this study followed by a summary of the validation

carried out in Gallagher et al. (2016b). In Sect. 3 the results are presented and discussed. In Sect. 4 we conclude the findings of this study.

2 Model details and validation

2.1 EC-Earth

The EC-Earth global climate model (version 2.3) used in this study consisted of an atmosphere-land surface module coupled to an ocean-sea ice module (Hazeleger et al., 2012). Its atmospheric component was based on cycle 31r1 of the European Centre for Medium-Range Weather Forecasts (ECMWF) Integrated Forecasting System and has a spectral resolution of T159L62 (1.125° or ~125 km grid spacing with 62 model levels). The Nucleus for European Modelling of the Ocean (version 2; Madec, 2008) was used for the oceanic component of the model (resolution 1° or ~110 km grid spacing with 42 vertical levels). The Louvain-la-Neuve Sea ice Model (version 2; Fichefet and Morales Maqueda, 1997) was used for the sea-ice module of the model. The OASIS (Ocean Atmosphere Sea Ice Soil) coupler (version 3; Valcke, 2006) was used for the two-way coupling between the atmosphere-land surface and ocean-sea ice components.

Three of the 14 EC-Earth ensemble members were available for use in this analysis and are representative of the spread of the ensemble. These ensemble members cover the range of interannual variability of the winds, although the spread of the EC-Earth ensemble mean annual wind speeds was not found to be large. Each member consists of an historical simulation and two future simulations (RCP4.5 and RCP8.5) and are denoted *meiX*, *me4X* and *me8X* where $X = 1, 2, 3$ and denotes the ensemble number.

2.2 WAVEWATCH III®

The WAVEWATCH III® (Tolman and the WAVEWATCH® III Development Group, 2014) model was used for the wave model simulations. We implemented three nested grids using a confined local area grid for Irish waters and two larger regional grids for the Northeast and North Atlantic Ocean, respectively. These grids are shown in Fig. 1. The North and Northeast Atlantic grids (a) and (b) in Fig. 1 were two-way nested. The finest resolution grid was run afterwards using the wave spectra output from the second grid to force it at its boundaries. Grid three, (c) in Fig. 1, was constructed using an unstructured grid formulation (Roland, 2008). Input and dissipation terms were formulated as in Ardhuin et al. (2010), tuned for the ECMWF global winds. For each of the three wave model grids shown in Fig. 1, simulations were driven by EC-Earth winds and sea ice fields from three historical, three RCP4.5 and three RCP8.5 ensemble members, respectively. Finally, one simulation was driven by ERA-Interim data to check the quality of the EC-Earth data in terms of capturing the present wind and wave climate.

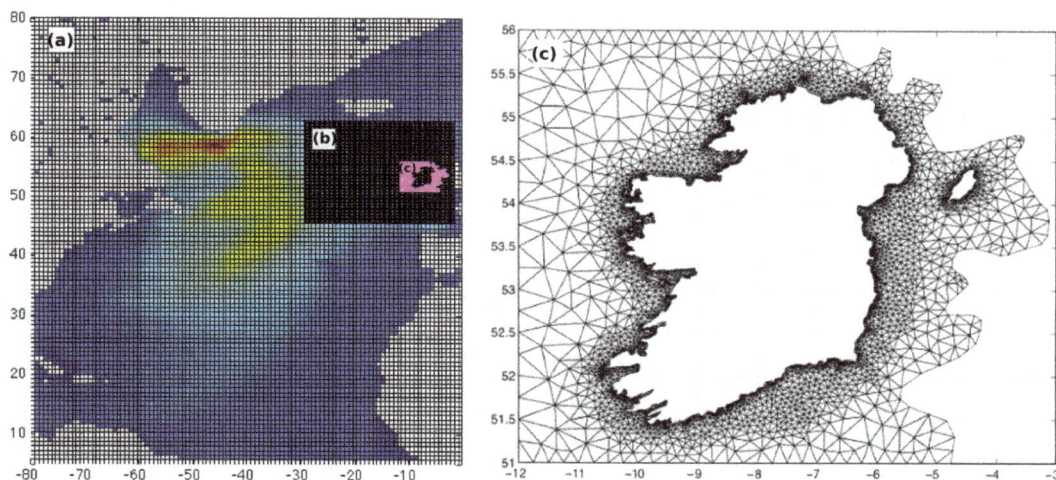

Figure 1. Left panel: the three wave model grids as described in Sect. 2.2. (**a**) The largest resolution North Atlantic grid has a $0.75° \times 0.75°$ resolution. (**b**) The grid for the Northeast Atlantic has a $0.25° \times 0.25°$ resolution. (**c**) The unstructured grid around Ireland has a resolution ranging from 15 km offshore to 1 km in the nearshore. Right panel: wave model unstructured grid used for the finest resolution domain around Ireland (**c**). This grid has 4473 nodes and the resolution varies from 15 km offshore to 1–2 km in the nearshore.

2.3 Validation

The validation carried out for this study is described in detail in Gallagher et al. (2016b), and a short summary is provided here. Overall, there is reasonable agreement (mostly within $\pm 10\%$) between the 1981–2009 wind (means and percentiles) output from the EC-Earth historical ensemble and the ERA-Interim dataset, seasonally averaged over the North Atlantic basin. Wave buoy observations from the Irish Marine Weather Buoy Network, maintained by Met Éireann and the Marine Institute, were compared to the ERA-Interim driven wave model output (on the same grid as the wave projection simulations) and good agreement was found between the time-series with correlation coefficients of 0.93 or greater at each of the buoy locations examined. Seasonal differences of less than ± 5–10% were found for the ERA-Interim driven wave model versus the historical EC-Earth driven wave model SWH (annual and seasonal), for the Atlantic and Celtic Sea regions. This was consistent across each individual ensemble member for the mean and 95th percentile of SWH.

3 Results and discussion

In this section we present the changes in the seasonal mean 10 m winds over the North Atlantic, and SWH off Irish Atlantic coasts, for the period 2070–2099 relative to 1980–2009. Although results for the Irish Sea are shown (off the east coast), they are excluded from our analysis. This is because the EC-Earth model grid has a spacing (~ 125 km) of comparable size to the Irish Sea and the winds are considered too coarse in resolution to accurately drive a wave model for this region.

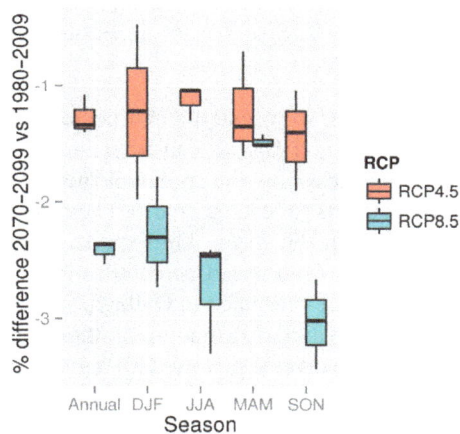

Figure 2. EC-Earth projected changes (%) in the mean 10 m wind speed over the North Atlantic Ocean for the period 2070–2099 relative to 1980–2009. Results for both the RCP4.5 (red) and RCP8.5 (blue) future climate scenarios are shown. The whiskers on the boxes show the spread of the results across the ensemble members (three RCP4.5 and three RCP8.5 members).

The projected changes (%) in the EC-Earth mean 10 m wind speed averaged over the North Atlantic Ocean were estimated annually and for each season: winter (December, January, February; DJF hereafter); spring (March, April, May; MAM hereafter); summer (June, July August; JJA hereafter); and autumn (September, October, November; SON hereafter), for the period 2070–2099 (future) relative to 1980–2009 (historical). Results for both the RCP4.5 and RCP8.5 future climate scenarios are shown in Fig. 2. Similar to other CMIP5 experiments for the North Atlantic region

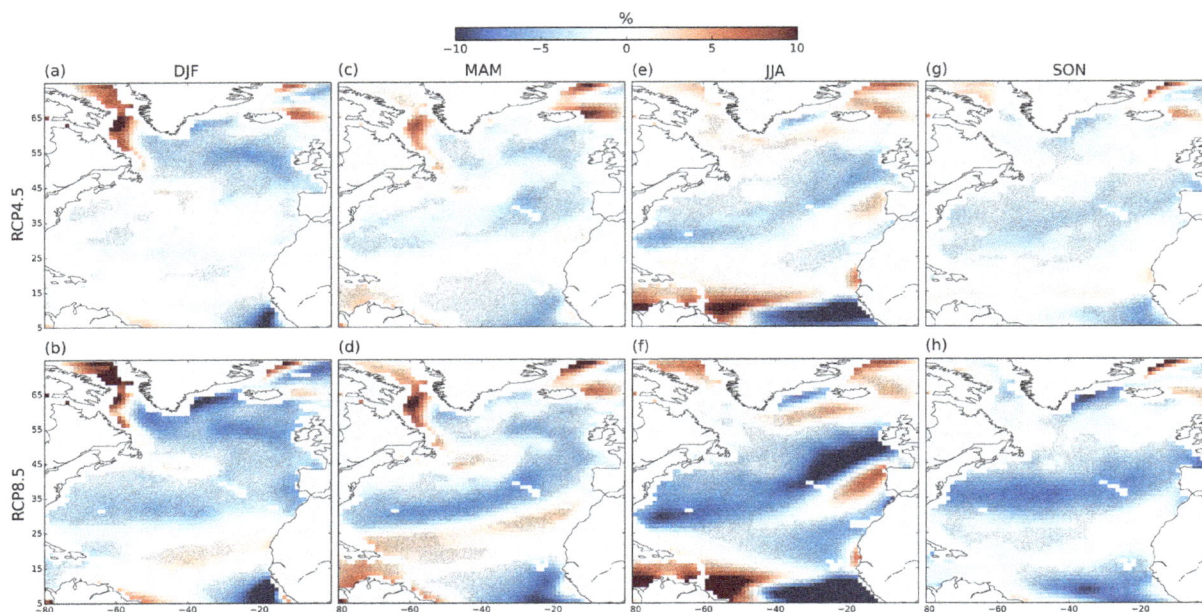

Figure 3. EC-Earth projected changes (%) in 10 m wind speed over the North Atlantic Ocean for the period 2070–2099 relative to 1980–2009. Ensemble mean for RCP4.5: **(a)** DJF **(c)** MAM **(e)** JJA **(g)** SON. Ensemble mean for RCP8.5: **(b)** DJF **(d)** MAM **(f)** JJA **(h)** SON. Hatching covers areas where the changes are greater than twice the inter-ensemble standard deviation of the past period 1980–2009.

(Collins et al., 2013), we found a small decrease North Atlantic winds in DJF and JJA, as can be seen in Fig. 2. The whiskers on the boxes show the spread of the results across the ensemble members for each scenario. There are decreases of between 1–3 % in the mean wind speed. Projected decreases in all of the wind speed percentiles (averaged over the North Atlantic) are presented in Gallagher et al. (2016b).

Examining the spatial distribution of these changes in Fig. 3 reveals a more diverse picture with regions of increasing and decreasing surface winds. This figure shows areal plots of the changes in the seasonal ensemble mean 10 m winds (ensembles of three members for both RCP4.5 and RCP8.5 scenarios) for the future relative to the historical period. Regions of relative decrease (negative changes, blue) can be seen in the swell generating areas for Ireland in the Northeast Atlantic for all seasons, for both scenarios, with larger decreases in the mean winds under the RCP8.5 scenario. This is particularly true for RCP8.5 DJF and JJA seasonal means where the decreases are more negatively pronounced (darker blue) and there are larger areas of statistical significance based on the following metric: where the change in the shown parameter exceeds twice the inter-ensemble standard deviation of the historical period 1981–2009, hatching is applied to the areal plot. This is also reflected in the more robust decreases found in SWH in Fig. 4 for these seasons. It should be noted that an ensemble of three members is small and a larger ensemble would be preferable in order to estimate a robust inter-ensemble spread (the hatching). However, the computational resources available to run the wave

projections only enabled a limited set of wave model simulations to be run (ten 30-year simulations for each of the three nested wave model grids: an ERA-Interim hindcast and nine EC-Earth driven model runs as described in Sect. 2.1).

The changes in the ensemble mean SWH (three members for each RCP scenario) for the future relative to the historical period are presented in Fig. 4. The ensemble mean annual SWH in Fig. 4b and c shows a reduction of 5–10 % off the Atlantic coastline under the RCP4.5 and RCP8.5 scenarios, with very little variation between the individual ensemble members (not shown). For DJF, as can be seen in Fig. 4e and f, the mean SWH decreases by over 5 % under RCP4.5 and by up to 10 % for RCP8.5 off the west coast. In JJA, there are mean decreases of up to 15 % off the south coast under RCP8.5 (Fig. 4l). The reduction in mean SWH for RCP4.5 in JJA is lower (Fig. 4k), at just over 5 % off the south coast. Nevertheless, this reduction is still significant using the inter-model standard deviation test. The projections in MAM and SON show smaller, less robust decreases of between 1–4 % in SWH under both scenarios (Fig. 4h–i and 4n–o, respectively). Off the west coast, the decrease in SWH under the RCP4.5 scenario for these seasons was not significant. These projected changes in SWH may consequently reduce the amount of wave power available for exploitation for any future deployments of WECs. However, as was shown in Gallagher et al. (2016a), Ireland possesses a large and energetic wave energy resource, particularly off the west coast, and this reduction is expected to have a minimal impact on

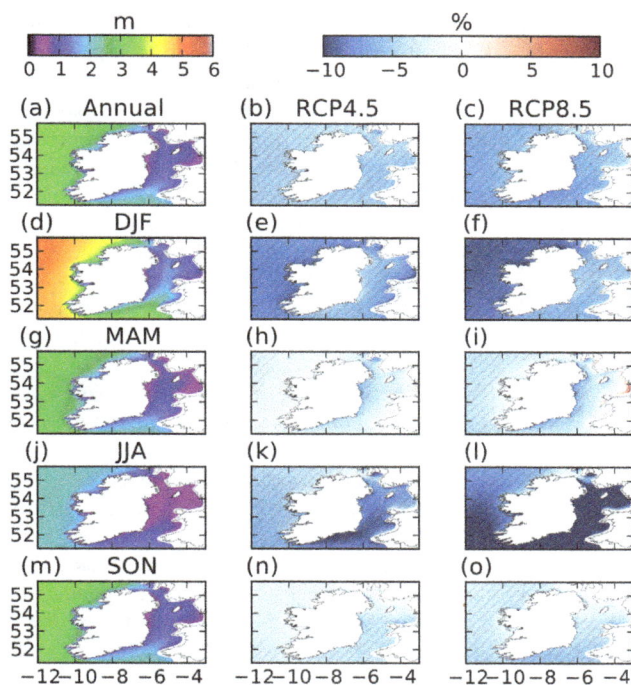

Figure 4. Ensemble mean SWH (m): (**a**) annual (**d**) DJF (**g**) MAM (**j**) JJA and (**m**) SON for the historical period (1980–2009). Projected changes (%) in annual ensemble mean SWH for the period 2070–2099 relative to 1980–2009 for ensemble mean RCP4.5: (**b**) annual (**e**) DJF (**h**) MAM (**k**) JJA (**n**) SON and ensemble mean RCP8.5: (**c**) annual (**f**) DJF (**i**) MAM (**l**) JJA and (**o**) SON. Hatching denotes areas where the magnitude of the ensemble mean change exceeds twice the inter-model standard deviation.

the overall potential for ocean renewable energy extraction in Ireland.

The geographical track of mid-latitude North Atlantic depressions crossing the region around Ireland (50–57° N, 3–13° W) for the future period relative to the historical was also examined in Gallagher et al. (2016b) and a summary of the findings are included here. Using a heuristic algorithm, low pressure systems were tracked using constraints on speed, direction of movement and the deepening rate. Based on the tracking, the number of depressions were counted for each ensemble member in the RCP8.5 scenario. A decrease in the frequency of depressions crossing this region was found for all depression counts examined (MSLP core minima ranging from 940 to 990 hPa). We found no evidence that the historical and future data samples of depressions were from differing distributions using a two-sample Kolmogorov–Smirnov test. It should also be noted that substantial uncertainty exists in the ability of CMIP5 models to project changes for Northern Hemisphere winter storm tracks in the North Atlantic Ocean (Church et al., 2013; Zappa et al., 2013).

4 Conclusions

Using the EC-Earth global climate model and the WAVE-WATCH III® wave model we estimated how climate change might affect the wave climate around Ireland and the North Atlantic for the future period 2070–2099 relative to 1980–2009. We used three EC-Earth ensemble members, where each member consisted of an historical simulation and two future realisations for the RCP4.5 and RCP8.5 emission scenarios, respectively. EC-Earth MSLP fields were also used to examine how the number of depressions crossing Irish waters is projected to change towards the end of the century.

The EC-Earth ensemble projections show an average decrease in mean surface wind speeds over the North Atlantic Ocean for all seasons, greater under RCP8.5 than RCP4.5 which results in an average decrease in mean SWH for Ireland annually and for all seasons. The largest decreases were found off the south coast of Ireland in JJA (15 %) and off the west coast in DJF (10 %) for the RCP8.5 emissions scenario. The projected changes in mean SWH for MAM and SON are small and less robust than for DJF and JJA, and should be treated with caution due to the large natural variability in the wave climate of Ireland. No significant changes were found for MAM and SON off the west coast for the RCP4.5 scenario. The small number of ensemble members reduces the robustness of the estimated projections but nevertheless this is the first set of high resolution wave projections for Ireland.

Future work includes a plan to run a much larger multi-model ensemble, in order to improve the estimates of uncertainties. In addition, higher resolution driving data, such as that from the next suite of global climate simulations under CMIP6 might improve the representation of storm tracks and extremes in the wind and wave climate of the region. Any careful assessment of the future wave climate of Ireland should also consider both winds and WCI, and ideally be carried out using coupled atmosphere-wave-oceanic numerical models.

Author contributions. E. Gleeson performed and analysed the EC-Earth simulations. R. Tiron and S. Gallagher designed and performed the WAVEWATCH III® simulations. S. Gallagher analysed the WAVEWATCH III® simulations. R. McGrath developed the storm tracking algorithm code and implemented it on the EC-Earth dataset. S. Gallagher and E. Gleeson analysed the storm tracking algorithm results. S. Gallagher and E. Gleeson prepared the manuscript with contributions from F. Dias.

Acknowledgements. The study was funded by Science Foundation Ireland (SFI) under the research project "High-end computational modelling for wave energy systems" (10/IN.1/I2996). The numerical simulations were performed on the Fionn cluster at the Irish Centre for High-end Computing (ICHEC) and at the Swiss National Computing Centre under the PRACE-2IP project (FP7 RI-283493) "Nearshore wave climate analysis of the west coast of

Ireland". We also wish to thank the editor and two reviewers for their suggestions and comments.

References

Ardhuin, F., Rogers, E., Babanin, A., Filipot, J.-F., Magne, R., Roland, A., van der Westhuysen, A., Queffeulou, P., Lefevre, J.-M., Aouf, L., and Collard, F.: Semi-empirical dissipation source functions for wind-wave models: Part I, definition, calibration and validation, J. Phys. Oceanogr., 40, 1917–1941, 2010.

Barbariol, F., Benetazzo, A., Carniel, S., and Sclave, M.: Improving the assessment of wave energy resources by means of coupled wave-ocean numerical modeling, Renew. Energ., 60, 462–471, doi:10.1016/j.renene.2013.05.043, 2013.

Church, J., Clark, P., Cazenave, A., Gregory, J., Jevrejeva, S., Levermann, A., Merrifield, M., Milne, G., Nerem, R., Nunn, P., Payne, A., Pfeffer, W., Stammer, D., and Unnikrishnan, A.: Sea Level Change, in: Climate Change 2013: The Physical Science Basis. Contribution of Working Group I to the Fifth Assessment Report of the Intergovernmental Panel on Climate Change, edited by: Stocker, T. F., Qin, D., Plattner, G.-K., Tignor, M., Allen, S. K., Boschung, J., Nauels, A., Xia, Y., Bex, V., and Midgley, P. M., Cambridge University Press, Cambridge, United Kingdom and New York, NY, USA, 2013.

Collins, M., Knutti, R., Arblaster, J., Dufresne, J.-L., Fichefet, T., Friedlingstein, P., Gao, X., Gutowski, W. J., Johns, T., Krinner, G., Shongwe, M., Tebaldi, C., Weaver, A. J., and Wehner, M.: Long-term Climate Change: Projections, Commitments and Irreversibility, in: Climate Change 2013: The Physical Science Basis. Contribution of Working Group I to the Fifth Assessment Report of the Intergovernmental Panel on Climate Change, edited by: Stocker, T. F., Qin, D., Plattner, G.-K., Tignor, M., Allen, S. K., Boschung, J., Nauels, A., Xia, Y., Bex, V., and Midgley, P. M., Cambridge University Press, Cambridge, United Kingdom and New York, NY, USA, 2013.

Dobrynin, M., Murawsky, J., and Yang, S.: Evolution of the global wind wave climate in CMIP5 experiments, Geophys. Res. Lett., 39, L18606, doi:10.1029/2012GL052843, 2012.

Fichefet, T. and Morales Maqueda, M. A.: Sensitivity of a global sea ice model to the treatment of ice thermodynamics and dynamics, J. Geophys. Res., 102, 12609–12646, 1997.

Gallagher, S., Tiron, R., and Dias, F.: A long-term nearshore wave hindcast for Ireland: Atlantic and Irish Sea coasts, 1979–2012, Ocean Dynam., 64, 1163–1180, doi:10.1007/s10236-014-0728-3, 2014.

Gallagher, S., Tiron, R., Whelan, E., Gleeson, E., Dias, F., and McGrath, R.: The nearshore wind and wave energy potential of Ireland: a high resolution assessment of availability and accessibility, Renew. Energ., 88, 494–516, doi:10.1016/j.renene.2015.11.010, 2016a.

Gallagher, S., Gleeson, E., Tiron, R., Dias, F., and McGrath, R.: Wave Climate Projections for Ireland for the end of the Twenty-First Century including analysis of EC-Earth winds over the North Atlantic Ocean, Int. J. Climatol., doi:10.1002/joc.4656, in press, 2016b.

Hazeleger, W., Wang, X., Severijns, C., Ştefănescu, S., Bintanja, R., Sterl, A., Wyser, K., Dutra, E., Baldasano, J. M., Bintanja, R., Bougeault, P., Caballero, R., Ekman, A. M. L., Christensen, J. H., van den Hurk, B., Jimenez, P., Jones, C., Kållberg, P., Koenigk, T., McGrath, R., Miranda, P., Van Noije, T., Palmer, T., Parodi, J. A., Schmith, T., Selten, F., Storelvmo, T., Sterl, A., Tapamo, H., Vancoppenolle, M., Viterbo, P., and Willén, U.: EC-Earth V2. 2: description and validation of a new seamless earth system prediction model, Clim. Dynam., 39, 2611–2629, 2012.

Hemer, M. A., Katzfey, J., and Trenham, C. E.: Global dynamical projections of surface ocean wave climate for a future high greenhouse gas emission scenario, Ocean Model, 70, 221–245, 2013a.

Hemer, M. A., Fan, Y., Mori, N., Semedo, A., and Wang, X. L.: Projected changes in wave climate from a multi-model ensemble, Nat. Clim. Change, 3, 471–476, 2013b.

Madec, G.: NEMO ocean engine. Note du Pole de modélisation, No. 27, Institut Pierre-Simon Laplace (IPSL), France, ISSN: 1288-1619, 2008.

Moss, R. H., Edmonds, J. A., Hibbard, K. A., Manning, M. R., Rose, S. K., van Vuuren, D. P., Carter, T. R., Emori, S., Kainuma, M., Kram, T., Meehl, G. A., Mitchell, J. F. B., Nakicenovic, N., Riahi, K., Smith, S. J., Stouffer, R. J., Thomson, A. M., Weyant, J. P., and Wilbanks, T. J.: The next generation of scenarios for climate change research and assessment, Nature, 463, 747–756, 2010.

Roland, A.: Development of WWM (Wind Wave Model) II: Spectral wave modelling on unstructured meshes, PhD thesis, Institute of Hydraulics and Water Resource Engineering, Technical University Darmstadt, Germany, 2008.

Sustainable Energy Ireland (SEI): Tidal & Current Energy Resources in Ireland, http://www.seai.ie/Publications/Renewables_Publications_/Ocean/Tidal_Current_Energy_Resources_in_Ireland_Report.pdf (last access: 14 April 2016), Technical report, Sustainable Energy Ireland, 2004.

Taylor, K. E., Stouffer, R. J., and Meehl, G. A.: An Overview of CMIP5 and the experiment design, B. Am. Meteorol. Soc., 93, 485–498, doi:10.1175/BAMS-D-11-00094.1, 2012.

Tolman, H. and the WAVEWATCH® III Development Group: User manual and system documentation of WAVEWATCH III version 4.18, Technical Report 316, NOAA/NWS/NCEP/MMAB, 2014.

Valcke, S.: OASIS3 User Guide (prism 2-5), CERFACS Technical Report TR/CMGC/06/73, CERFACS, Toulouse, France, 60 pp., 2006.

Wang, X. L., Feng, Y., and Swail, V. R.: Changes in global ocean wave heights as projected using multimodel CMIP5 simulations, Geophys. Res. Lett., 41, 1026–1034, doi:10.1002/2013GL058650, 2014.

Woolf, D. and Wolf, J.: Impacts of climate change on storms and waves, MCCIP Science Review 2013, 20–26, doi:10.14465/2013.arc03.020-026, 2013.

Zappa, G., Shaffrey, L. C., Hodges, K. I., Samson, P. G., and Stephenson, D. B.: A multimodel assessment of future projections of North Atlantic and European extratropical cyclones in the CMIP5 climate models, J. Climate, 26, 5846–5862, doi:10.1175/JCLI-D-12-00573.1, 2013.

29

Intense air-sea exchange and heavy rainfall: impact of the northern Adriatic SST

P. Stocchi and S. Davolio

Institute of Atmospheric Sciences and Climate, Italian National Research Council, CNR-ISAC,
Bologna, 40129, Italy

Correspondence to: P. Stocchi (p.stocchi@isac.cnr.it)

Abstract. Over the northern Adriatic basin, intense air-sea interactions are often associated with heavy precipitation over the mountainous areas surrounding the basin. In this study, a high-resolution mesoscale model is employed to simulate three severe weather events and to evaluate the effect of the sea surface temperature on the intensity and location of heavy rainfall. The sensitivity tests show that the impact of SST varies among the events and it mainly involves the modification of the PBL characteristics and thus the flow dynamics and its interaction with the orography.

1 Introduction

Northeastern Italy (NEI) is often affected by heavy rainfall events. Sometimes they produce a daily accumulation that can reach values as high as 40 % of the mean annual amount, even in less than 12 h, thus leading to severe flash floods, damages and human casualties (Barbi et al., 2012). Heavy precipitation over the Alps is directly (e.g., by orographic uplift) or indirectly (e.g., by orographic cyclogenesis) related to the influence of mountain ranges on atmospheric motions. Heavy precipitations over NEI are often associated with Sirocco or Bora winds (Manzato, 2007; Davolio et al., 2015, 2016), thus involving intense air-sea interactions. In this situation, the Adriatic Sea acts as a source of moisture and heat for the atmosphere and contributes to destabilize the air mass in the boundary layer (PBL), which is then transported toward the mountains where convective instability is released. As pointed out by Dorman et al. (2007) and Pullen et al. (2007) turbulent surface heat fluxes and sea surface temperature (SST) variations are important parameters that characterize intense air-sea exchanges.

The relationship between SST and precipitation is well recognized in the tropics, where ocean conditions drive the atmosphere and higher SSTs are generally accompanied by increased convection and precipitation (Trenberth and Shea, 2005). Toy and Johnson (2014) highlighted that even mesoscale SST fronts influence the PBL stability, resulting

in an enhancement of horizontal convergence and precipitation. At mid-latitudes, however, the impact of the SST on atmospheric phenomena is still debatable and different studies have addressed the possible role of the SST during severe weather events (Miglietta et al., 2011; Pastor et al., 2015).

In the last decade, the importance of Mediterranean SST representation in meteorological models has been investigated for heavy rainfall events. In particular, Lebeaupin et al. (2006) showed that an averaged variation of SST ($\pm 1.5\,°\mathrm{C}$) impacted the intensity and localization of rainfall, while high-resolution SST analyses did not produce relevant improvements in precipitation forecasts. Other studies (e.g. Berthou et al., 2014) based on coupled models simulations confirmed that intense rainfalls can be sensitive to the evolution of SST during the event.

Numerical weather prediction (NWP) models usually keep SST fixed at its initial value or allow just slow changes according to surface fluxes. This SST representation is generally unrealistic even for short-range forecasts, especially in small and shallow basins like the Adriatic Sea (Davolio et al., 2015; Ricchi et al., 2016). This framework motivated the present study aimed at investigating the impact of the northern Adriatic SST on the intensity and location of the rainfall, identifying the relevant physical mechanisms involved. We focused on representative severe events that occurred over NEI, in different months of late Sum-

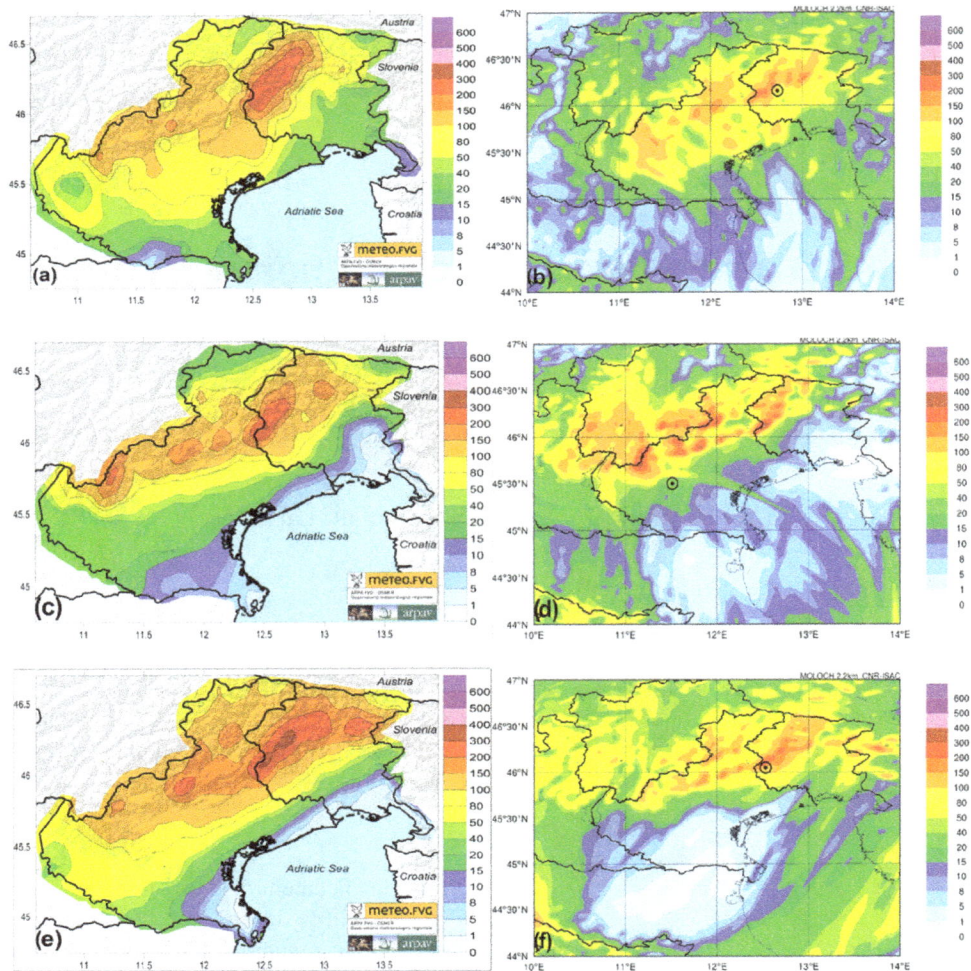

Figure 1. 24 h accumulated precipitation (mm). Observations (**a, c, e**) and MOLOCH control simulations (**b, d, f**) at: 00:00 UTC, 16 September 2006 (Chievolis case – **a**, **b**); 00:00 UTC, 2 November 2010 (Vicenza case – **c**, **d**); 00:00 UTC, 12 November 2012 (Piancavallo case – **e, f**). Only a portion of the MOLOCH integration domain is shown. Borders of Veneto region (west) and FVG region (east) are shown. Observed precipitation is obtained by interpolation of data provided by the dense network (about 260 rain-gauges) managed by the FVG regional meteorological agency (OSMER – ARPA FVG). Black circles in (**b**), (**d**) and (**f**) indicate the location of Chievolis, Vicenza and Piancavallo, respectively.

mer/Autumn, which present different rainfall characteristics (orographic, stratiform and convective) and intense Sirocco wind. A high-resolution NWP system was used to simulate the heavy precipitation events and to perform sensitivity experiments. This study represents a first modelling analysis of the SST effect in this area, and a preliminary step toward a full coupling between atmospheric and ocean models foreseen in the framework of the Italian flagship project RIT-MARE (http://www.ritmare.it).

2   Overview of the events

The three analysed cases represent typical rainfall events affecting eastern Alps. Only a short description is provided in the following. Associated with an approaching upper-level

trough, deepening in the Mediterranean and slowly moving eastward, winds blew from southwest in the middle troposphere while close to the surface, over the Adriatic Sea, Sirocco wind from southeast impinged on the Alpine barrier. The advection of warm and moist air favoured intense orographic precipitation, with embedded convective activity, due to thermodynamic profiles becoming progressively more unstable in the course of the event. The three selected cases are: 14–16 September 2006 (Chievolis), 30 October–2 November 2010 (Vicenza) and 10–12 November 2012 (Piancavallo). For the last two cases, a more detailed description can be found in Davolio et al. (2016). Chievolis event occurred in the Friuli Venezia Giulia (FVG) region in late summer and was characterized by convective precipitation exceeding 300 mm in 24 h. The other two cases (Piancav-

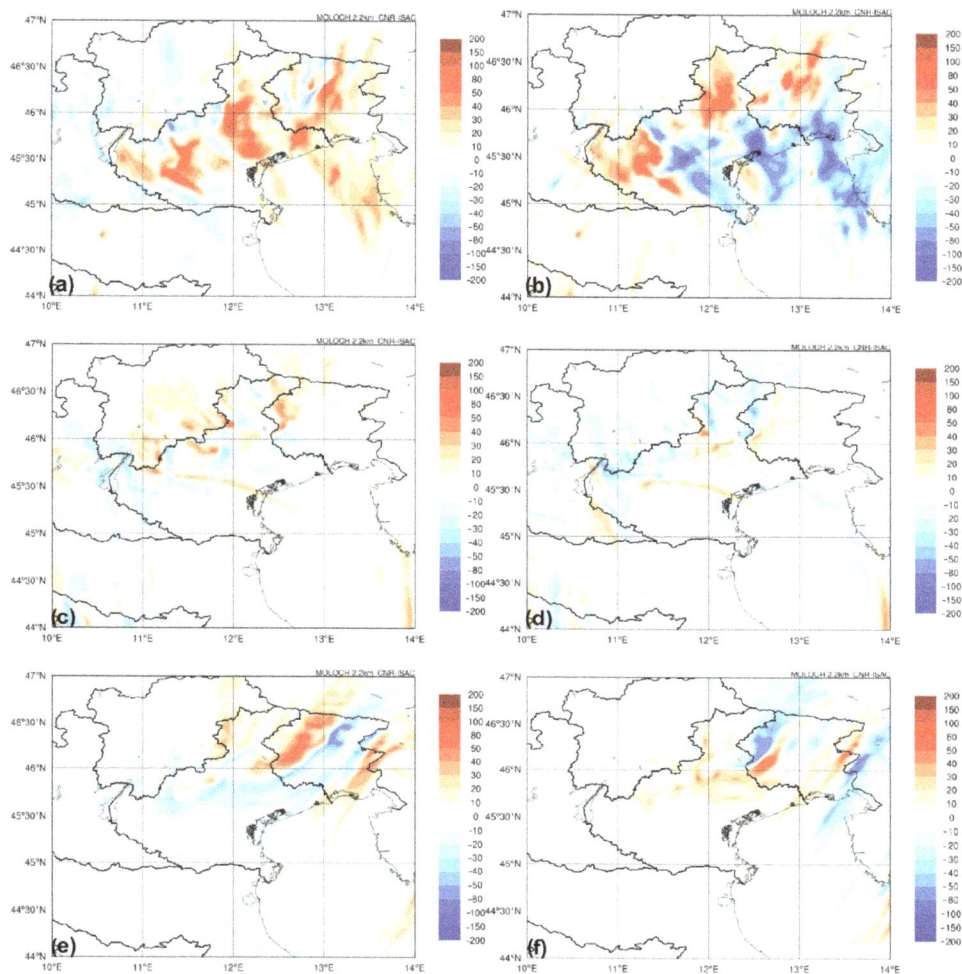

Figure 2. Difference of 24 h accumulated precipitation (mm) fields between CNTRL and SST2M (**a, c, e**) or SST2P (**b, d, f**) at: 00:00 UTC, 16 September 2006 (Chievolis case **a, b**); 00:00 UTC, 2 November 2010 (Vicenza case **c, d**); 00:00 UTC, 12 November 2012 (Piancavallo case **e, f**).

allo and Vicenza), occurred in autumn, were responsible for floods and affected both FVG and Veneto regions with orographic rainfall up to 250 and 300 mm in 24 h, respectively (Fig. 1).

3 NWP system setup

The NWP system is based on BOLAM hydrostatic and MOLOCH non-hydrostatic models, both developed at CNR-ISAC (Davolio et al., 2015). In the present study, BOLAM runs over an European domain, with horizontal resolution of about 11 km and 50 vertical levels. MOLOCH (its whole integration domain is shown in Fig. 4) is nested in BOLAM. Its integration, covering the Adriatic basin, is initialized with a 3 h BOLAM forecast in order to avoid a sudden change in the grid resolution from the global to the 2.2 km MOLOCH grid-spacing (50 vertical level), based on pure interpolation. The

model chain is initialized at least 12 h before the onset of intense rainfall, in order to allow the PBL to adjust to the modified SST. Specifically, BOLAM is initialized at 12:00 UTC on 14 September 2006 (Cheivolis) and 10 October 2012 (Piancavallo), and at 18:00 UTC on 30 October 2010 (Vicenza). BOLAM-MOLOCH control simulation (CNTRL) is driven by IFS-ECMWF forecasts, thus the SST field is provided through the OSTIA analyses (Donlon et al., 2012). In order to analyse SST effects on forecast precipitation, for each event the analysed SST field is modified by increasing (SST2P) or decreasing (SST2M) its value over the Adriatic Sea by 2 °C. This value seems reasonable considering the uncertainties in different SST analyses (Lebeaupin et al., 2006; Davolio et al., 2015).

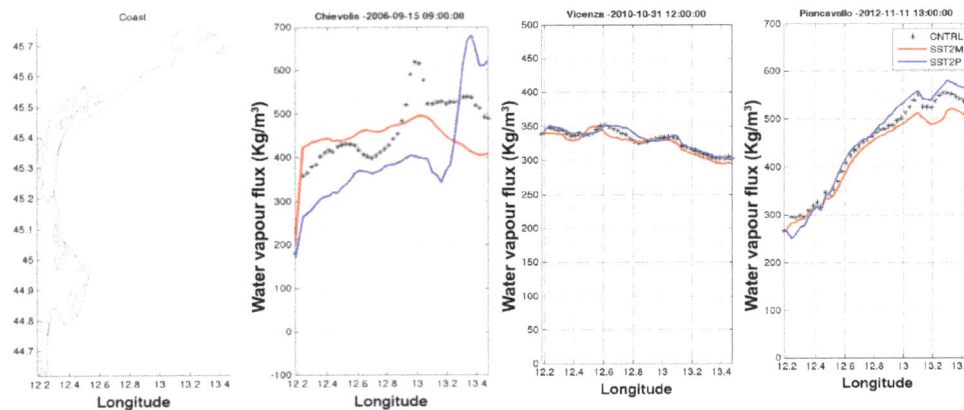

Figure 3. Vertically integrated water vapour fluxes across the Northern Adriatic coastline (shown in the left part of the panel) computed at: 09:00 UTC, 15 September 2006 (Chievolis); 12:00 UTC, 31 October 2010 (Vicenza); 13:00 UTC, 11 November 2012 (Piancavallo). Coloured lines indicate the results of MOLOCH experiments: CNTRL (black stars), SST2M (red) and SST2P (blue).

4   Results

4.1   Representation of the intense precipitation events

In all the analysed cases, the precipitation distribution is strongly correlated with the Alpine orography as shown by the observation (Fig. 1a, c, e). In particular for the Chievolis and Piancavallo events, high values of precipitation are recorded along the Pre-Alpine slopes in FVG region, with maxima exceeding 300 mm on the locations corresponding to Chievolis and Piancavallo. For the Vicenza case, precipitation maxima are distributed along the pre-Alps from Veneto to FVG, with two maxima exceeding $200\,mm\,24\,h^{-1}$.

The reference simulations (CNTRL) succeed in reproducing the precipitation fields in terms of intensity and distribution with only a slightly tendency to underestimate the maxima (Fig. 1b, d, f). Therefore, these simulations can be suitably used as a reference to perform sensitivity studies, aimed at understanding the role of SST during these events.

4.2   Impact of SST on precipitation

Figure 2 shows the difference of daily-accumulated precipitation between the CNTRL simulation and the simulations with modified SST (SST2P and SST2M) for each case study. Red colour indicates the areas where CNTRL run is more rainy. While for Vicenza event the precipitation differences due to SST changes can be explained in terms of rather small displacements of the rainfall location, for the Chievolis and Piancavallo events the impact of SST is larger. In this case, SST2M is generally drier than the CNTRL, while in SST2P rainfall is shifted closer to the coast and towards the eastern part of the domain. For Vicenza event the impact of SST is very limited. For Piancavallo event, a colder (warmer) SST is associated with upstream (downstream) displacement of the intense precipitation area with respect to the orography. Moreover, in SST2M intense rainfall is displaced over the

plain and affects a large portion of the Po Valley. This behaviour can be explained in terms of interaction between the low-level southerly flow and the orography, as detailed in the following. However a more detailed analysis of the interaction between flow and orography in terms of the change of the Froude number with SST is ongoing.

4.3   Surface fluxes, water vapour and wind

In order to better understand which processes involved in an intense rain event are sensitive to the SST, heat and moisture surface fluxes as well as the low-level wind field and water vapour transport have been analysed.

The temporal evolution (not shown) of surface sensible and latent heat fluxes averaged over the northern Adriatic Sea for the three experiments shows a similar behaviour for the three events: an increase (decrease) of averaged fluxes in response to a warmer (colder) SST. Also the values of the fluxes are similar, ranging from about $200\,W\,m^{-2}$ for SST2P to $50\,W\,m^{-2}$ for SST2M, with CNTR in between. Maxima are associated with the most intense phase of Sirocco wind.

The evolution of vertically integrated water vapour fluxes across the northern Adriatic coastline has been also analysed in order to evaluate the amount of water vapour moving northward from the sea towards the Alps. It is worth noting that this does not provide information about the sources of moisture. For this purpose, the development of a diagnostic tool, aimed at providing water vapour budget in the atmosphere over the Adriatic Sea, is ongoing and it will be applied in a future work. Figure 3 provides an overall view of the spatial evolution of the moisture inflow associated with the southeasterly Sirocco wind, up to a prescribed altitude of 3000 m. The results refer to the time of strongest Sirocco and provide a comparison among the three experiments (CNTRL, SST2P and SST2M).

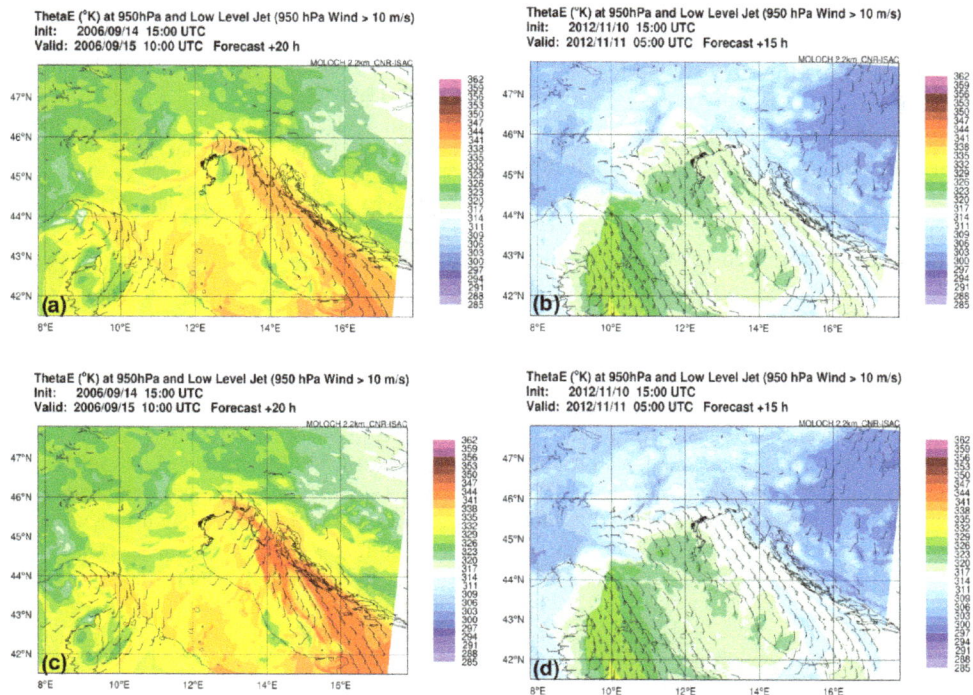

Figure 4. 950 hPa Equivalent potential temperature and wind barb (wind speed exceeding $10 \, \text{m s}^{-1}$) for Chievolis (**a, c**) and Piancavallo (**b, d**) events. CNTRL simulations (**a, b**) are compared with SST2P for Chievolis (**c**) and SST2M for Piancavallo (**d**).

Contrary to the averaged surface fluxes, increasing SST does not lead systematically to an increase of inward water vapour transportation, along the northern Adriatic coast, which seems instead dependent on the single events. Vicenza case presents more constant water vapour fluxes during the event with values around $350 \, \text{kg m}^{-3}$ very weakly sensitive to the SST. For Chievolis event, a SST increase produces an evident shift from west to east of intense water vapour fluxes. For Piancavallo event, an increase of SST results in an increase of water vapour fluxes, although quite limited. These behaviours are consistent with the results obtained for the precipitation fields (Fig. 2) where the largest differences with respect to the CNTRL simulation are found for Chievolis and Piancavallo events.

The analysis of heat surface fluxes and water vapour fluxes across the coast suggests that changes in SST do not impact directly on the precipitation through a substantial modification of the amount of moisture impinging on the Alps and feeding the precipitation. Instead, SST seems to play an important but indirect role in determining the location of rainfall, influencing the PBL characteristics. In fact, the changes in moisture fluxes (Fig. 3), and consequently in precipitation (Fig. 2) can be ascribed to different low-level wind fields in the sensitivity experiments, as shown in Fig. 4.

In the Chievolis event, increasing SST (Fig. 4c) produces a weakening of the southeasterly Sirocco wind and more intense southwesterly winds descending from the Apennines

and entering the northern Adriatic basin. Different characteristics of the PBL over the sea seem responsible for different superposition of these two flows, thus determining a southwestward shift of the moisture advection and of the precipitation (Fig. 2b). Also for the Piancavallo case (Fig. 4b, d), the SST variation impacts the PBL characteristics and thus changes the flow regime across the Alpine barrier. A colder SST (Fig. 4d) enhances the stability and thus the blocking of the low-level flow impinging on the orography. Therefore, flow-around is favoured (with respect to flow-over), producing an evident westward deflection of the wind and a shift of the precipitation upstream and over the Po Valley (Fig. 2e). For Vicenza events (not shown), no relevant differences appear.

5 Conclusions

Three intense rain events were simulated using a high-resolution model to evaluate the effect of the SST on heavy rainfall in the NEI Alps. Different SST fields have been imposed as low-level boundary conditions over the Adriatic Sea and these preliminary results show that the impact of SST on precipitation varies among different events.

A warmer SST increases the surface heat fluxes over the sea, but does not necessary affect the vertical integrated water vapour flux across the coast (i.e. water vapour available for the precipitations on the Alps), which is probably modulated mainly by large-scale/mesoscale circulation. The response of heavy precipitation to a SST change is complex: SST affects the PBL characteristics and thus the flow dynamics and its interaction with orography.

This study can be considered as a first step toward a more detailed investigation of the effect of the air-sea interaction in this area. In particular a more detailed evaluation of the water balance in the atmosphere is already ongoing together with further sensitivity simulations using high-resolution satellite SST analyses or SST field from an ocean model to evaluate the impact of small-scale SST features.

Acknowledgements. This work was supported by the Italian flagship project RITMARE. This work represents a contribution to the HyMeX international program. The authors thank also Andrea Cicogna, Agostino Manzato and Arturo Pucillo (OSMER – ARPA FVG) for fruitful discussions and for providing observed rainfall maps.

References

Barbi, A., Monai, M., Racca, R., and Rossa, A. M.: Recurring features of extreme autumnall rainfall events on the Veneto coastal area, Nat. Hazards Earth Syst. Sci., 12, 2463–2477, doi:10.5194/nhess-12-2463-2012, 2012.

Berthou, S., Mailler, S., Drobinski, P., Arsouze, T., Bastin, S., Béranger, K., and Lebeaupin Brossier, C.: Sensitivity of an intense rain event between an atmosphere-only and atmosphere-ocean regional coupled model: 19 September 1996, Q. J. Roy. Meteor. Soc., 141, 258–271, doi:10.1002/qj.2355, 2014.

Davolio, S., Stocchi, P., Carniel, S., Benetazzo, A., Bohm, E., Ravaioli, M., Riminucci, F., and Li X.: Exceptional Bora outbreak in winter 2012: Validation and analysis of high-resolution atmospheric model simulations in the northern Adriatic area, Dynam. Atmos. Oceans, 71, 1–20, 2015.

Davolio, S., Volonté, A., Manzato, A., Pucillo, A., Cicogna, A., and Ferrario, M. E.: Mechanisms producing different precipitation patterns over North-Eastern Italy: insights from HyMeX-SOP1 and previous events, Q. J. Roy. Meteor. Soc., doi:10.1002/qj.2731, 2016.

Donlon, C. J., Martin, M., Stark, J., Roberts-Jones, J., Fiedler, E., and Wimmer, W.: The operational sea surface temperature and sea ice analysis (OSTIA) system, Remote Sens. Environ., 116, 140–158, 2012.

Dorman, C. E., Carniel, S., Cavaleri, L., Sclavo, M., Chiggiato, J., Doyle, J., Haack., T., Pullen, J., Grbec, B., Vilibić, I., Janeković, I., Lee, C., Malačič, V., Orlić, M., Paschini, E., Russo, A., and Signell, R. P.: February 2003 marine atmospheric conditions and the Bora over the northern Adriatic, J. Geophys. Res., 112, C03S03, doi:10.1029/2005JC003134, 2007.

Lebeaupin, C., Ducrocq, V., and Giordani, H.: Sensitivity of torrential rain events to the sea surface temperature based on high-resolution numerical forecasts, J. Geophys. Res., 111, D12110, doi:10.1029/2005JD006541, 2006.

Manzato, A.: The 6 h climatology of thunderstorms and rainfalls in the Friuli Venezia Giulia plain, Atmos. Res., 83, 336–348, 2007.

Miglietta, M. M., Moscatello, A., Conte, D., Mannarini, G., Lacorata, G., and Rotunno, R.: Numerical analysis of a Mediterranean "hurricane" over south-eastern Italy: Sensitivity experiments to sea surface temperature, Atmos. Res., 101, 412–426, 2011.

Pastor, F., Valiente, J. A., and Estrela, M. J.: Sea surface temperature and torrential rains in the Valencia region: modelling the role of recharge areas, Nat. Hazards Earth Syst. Sci., 15, 1677–1693, doi:10.5194/nhess-15-1677-2015, 2015..

Pullen, J., Doyle, J. D., Haack, T., Dorman, C. D., Signell, R. P., and Lee, C. M.: Bora event variability and the role of air-sea feedback, J. Geophys. Res., 112, C03S18, doi:10.1029/2006JC003726, 2007.

Ricchi, A., Miglietta M . M., Falco, P. P., Benetazzo, A., Bonaldo, D., Bergamasco, A., Sclavo, M., and Carniel S.: On the use of coupled ocean-atmosphere-wave model during an extreme cold air outbreak over the Adriatic Sea, Atmos. Res., 172–173, 48–65, 2016.

Toy, M. D. and Johnson, R. H.: The influence of an SST front on a heavy rainfall event over coastal Taiwan during TiMREX, J. Atmos. Sci., 71, 3223–3249, 2014.

Trenberth, K. E. and Shea, D. J.: Relationships between precipitation and surface temperature, Geophys. Res. Lett., 32, 1–4, 2005.

30

The VIADUC project: innovation in climate adaptation through service design

L. Corre[1], **P. Dandin**[2], **D. L'Hôte**[3], **and F. Besson**[1]

[1]Direction de la Climatologie, Météo-France, Toulouse, France
[2]Centre national de recherches météorologiques (CNRM), Météo-France, Toulouse, France
[3]Strate Ecole de Design, Sèvres, France

Correspondence to: L. Corre (lola.corre@meteo.fr)

Abstract. From the French National Adaptation to Climate Change Plan, the "Drias, les futurs du climat" service has been developed to provide easy access to French regional climate projections. This is a major step for the implementation of French Climate Services. The usefulness of this service for the end-users and decision makers involved with adaptation planning at a local scale is investigated.

As such, the VIADUC project is: to evaluate and enhance Drias, as well as to imagine future development in support of adaptation. Climate scientists work together with end-users and a service designer. The designer's role is to propose an innovative approach based on the interaction between scientists and citizens. The chosen end-users are three Natural Regional Parks located in the South West of France. The latter parks are administrative entities which gather municipalities having a common natural and cultural heritage. They are also rural areas in which specific economic activities take place, and therefore are concerned and involved in both protecting their environment and setting-up sustainable economic development.

The first year of the project has been dedicated to investigation including the questioning of relevant representatives. Three key local economic sectors have been selected: i.e. forestry, pastoral farming and building activities. Working groups were composed of technicians, administrative and maintenance staff, policy makers and climate researchers. The sectors' needs for climate information have been assessed. The lessons learned led to actions which are presented hereinafter.

1 Introduction

"Drias, les futurs du climat" service (Lémond et al., 2011) was launched in 2012, as a response of the French scientific community to society's need for climatic information, within the frame of the National Adaptation to Climate Change Plan. The service is mainly composed of a national data website and the associated support. French regional climate projections are easily and freely available. Drias includes general information about the state of the climate system and output from climate modelling. The site also offers users the opportunity to raise question on data access or interpretation, through a dedicated hotline. Lastest advances developed in 2014 include the availability of climate impact data (e.g. water resource, among others), and climate indices based on recent scenarios for socio-economic development (the Representative Concentration Pathways scenarios, used in the 5th IPCC report, 2013). The Volume 4 of "The climate of France for the 21st Century" report entitled "Regionalized scenarios 2014 edition" (Ouzeau et al., 2014) refers to the same climate simulations. Among other results, they show significant rises of averaged temperatures, number of days of summer heat waves, and drought events. Interestingly enough, this update of the reference report has been linked with the Drias service. This will enhance the coherency of various services delivered by the scientific community, which is a key request made by users.

The Drias service has been accompanied by a multidisciplinary users committee. The users feedbacks were very positive and highlighted how the website filled the gap between research climate laboratories and society. Yet the committee was mainly composed of advanced users, such as scientists coming from various disciplines (i.e. environmentalists, agronomists, hydrologists, geographers, economists, sociologists, and so forth), and consulting companies involved in environment and policy management, or supporting adaptation administrative entities. Confrontations with non specialist users tell a somewhat different story. The site has been described as a complex tool. In particular, end-users underline their inability to draw any useful conclusion from the huge amount of data available. Moreover, the information delivered to a wide audience seemed highly virtual, and did not meet their principle requirements. This brought up the following question: how can we imagine a service to better support the end-users at large to improve relevant actions for climate adaptation?

2 Objectives and partners

The Viaduc project has been set up for 3 years. It is to evaluate and improve the already existing Drias climate service, as well as to imagine future developments by having the scientific community better involved. To address this issue, a team built on the know-how from the Meteorological service (climatologists and climate scientists) and end-users with a service designer was put together. The role of the climate scientists was to present to the users the existing Météo-France deliverables that could meet their needs, and with the designer, to hear their feedback and to develop prototypes that could ultimately feed the production suite.

The chosen end-users are three Natural Regional Parks located in the South West France. Such parks are administrative entities which gather together municipalities defined by a common natural and cultural heritage. They are also rural areas in which economic activities do take place, and therefore are concerned and involved in both protecting their environment and setting up sustainable economic development. Sometimes referred to as adaptation practioners (Lorenz et al., 2015), this category of users is currently under-study, despite being recognized as very much exposed to climate action challenges (Porter et al., 2014).

Between scientists and end-users, the innovative point of the project was to involve a service designer as an interpreter. The service design methodology can be described as the activity of planning and organizing people, infrastructure, communication and material components of the service. The aims were to improve the quality of a service by focusing on the users' needs (e.g. Erlhoff and Marshall, 2008; Stickdorn and Schneider, 2010). Stating that "people have good reason for doing what they do", design considered users' behaviour as the significant consequence of the way they think. Observing and analysing these behaviours allowed pointing-out what does and what doesn't make sense for users. This qualitative diagnostic constituted the basis for the design process

A USER BASED STRATEGY

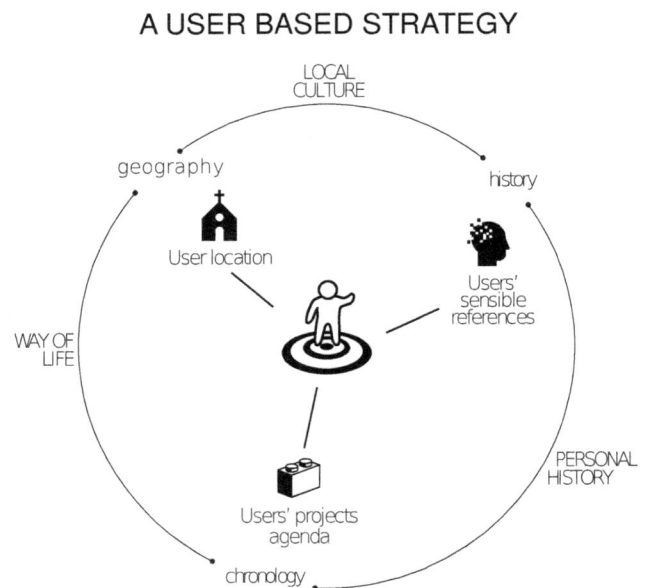

Figure 1. The service design methodology: a user based strategy.

(Fig. 1). Once understood what users considered as meaningful, it became possible to propose and develop solutions that will naturally fit their routines. Within the Viaduc project, the designer was looking for a consensus between the scientific possibilities and constraints and the users' requirements, taking their expectations and references into account. His role thus implied to monitor and ease the dialogue between users and scientists. When the project began, the designer had no theoretical knowledge of climate science.

3 Undertaken actions

For the first year of the project, it was decided to work primarily with the task officers in charge of climate action in the Parks. The aim was to understand their plans, abilities, and needs. Progressively, the existing climate adaptation actions were modified and improved: e.g. tailormade exchanges and communications for the local elective representatives, dedicated training for State agents acting locally, and technical working groups with focus on forestry, pastoral farming, and building. Each of these activities corresponds to a priority economical sector in one of the Parks. As an example, in the Haut-Languedoc Park, actions were underway regarding forestry. Working groups gathered experts and scientists from the main French forest and agronomy institutes, local forest technicians and owners, and various actors involved in this economic sector, ranging from regional or local State to companies representatives.

The work progressed with various sectors widely depending on the advancement of the needs identified at the beginning of the project. Thus, foresters had a very precise need of climate data to cross-reference with their own dataset on tree evolution (i.e. growth, reddening, among others). The climate scientists simply provided the data, helped to use them and discussed the associated uncertainties. Based on this collaboration, the Institute for Forest Development (IDF) produced a climatic forest atlas as a tool for forest management in the context of climate change (Institut pour le développement forestier/Centre national de la propriété forestière, 2014). In the fields of building and pastoral farming, the identification of needs took much longer. It appeared that the main feedback consisted in a strong need of training for them, expressed by the technical agents who appeared to be the key actors for echoing the climate message.

Beyond supporting actions locally, key messages and lessons which can help the climate research community to improve climate services were delivered. Three important notions have emerged. The first is on the local scale: end-users' involvement appeared to be inversely proportional to their spatial distance from climate change manifestations. The second is on the sector's characteristic time scale: the end-users' awareness appeared to be proportional to the time scales associated with their activity. As tree growth and climate change time scales are both longer than a few decades, foresters are very much exposed to climate change impacts and are already looking for adaptation solutions. On the contrary, pastoral farmers adjust their activity to weather conditions on a daily basis. Not being familiar with long prospects, they were far less responsive to climate scientists warnings. The third important notion concerned impact parameters: to be understood, climate projection need to be translated into sectorial and economical impact parameters. For example, communicating in terms of vegetation sensitivity to summer forest fires, and associated costs, might be more efficient than presenting expected changes through evapotranspiration.

4 The designer point-of-view

After one year, the designer delivered a critical report. On the positive side, he highlighted the scientists' transparency and attention to details. On the less positive side, he pointed-out the need to facilitate the appropriation and understanding of the climate projections by users. In his opinion, projections are a research tool made and used by researchers, and most of the time not suitable for non-specialists. Projections mention a changing climate in a too distant future (average time over 30 years) and a too global place (spatial average over the globe, or a whole continent). Moreover, the ensemble of futures that the projections highlighted is more frustrating than satisfying, raising questions instead of providing answers. Projections are depersonalized and decontextualized. Projections are intangible.

As a result of these observations, the designer made some recommendations. First, to make the representation of the scientific results more easily understandable, he suggested

to keep to communication standards (style guide). Moreover, figures had to be straightforward and tailored for each category of users. These two points have been already been identified as critical issues to be adressed by the scientific community in order to deliver understandable messages. For example, Kelleher and Wagener (2011) provided ten guidelines for effective data visualisation in scientific publications. Difficulties associated with tailored communication have been analysed by Lorenz et al. (2015). They pointed-out a gap between users comprehension and preferences, as well as a lack of within-users-group homogeneity.

The designer also insisted on the need for narrative process (storytelling) to accompany scientific figures. The complexity should be introduced progressively, using animations and allowing users to enrich themselves the content. Last but not least, the designer insisted on the necessity of providing a meaningful context. In order to arouse interest, projections need to meet with a personal project (at a given time and space, for a given user). Efficient communication on climate change depended on this timing. The designer suggested taking advantage of windows of opportunity associated with long-term projects, for which people will be concerned with the effects of climate change. He also shed some light on weaknesses of the current service, only delivering future scenarios without any possibility for the users to link this virtual future with known and experienced references. Many clues were lgiven to the scientists in order to help them improve their understanding of the expected products and services they may already deliver for the needs of adaptation.

5 In practice

5.1 Soil moisture in Aveyron: using storytelling

Let's examine a first attempt to apply the concepts presented above, based on the evolution of the water resources in Aveyron, a French department located near Toulouse, in the Midi-Pyrénées region. We began with the operational product delivered by Météo-France to water resource managers (Fig. 2). From the latter, this product is very useful, since it allows to place the actual situation in a climatological context (see Dandin et al., 2012). On the contrary, for a wider audience – the task officers involved in the Viaduc project and more than 40 State agents met during trainings – and from the designer's point of view, it was not usable. First attempts were made to improve the products, always bearing in mind that a massive production, able to cover the entire country, must be feasible. Figure 3 shows how animations and storytelling can be used to deliver a clear and striking message of the same scientific material.

This example has been challenged in front of various end-users (i.e. local electives, farmers, among others), and the level of interest was in general very positive. Indeed, its strength consists in anchoring the future in past and present climate references. Past and present situations are stamped in

Soil moisture index - Aveyron (department 12)

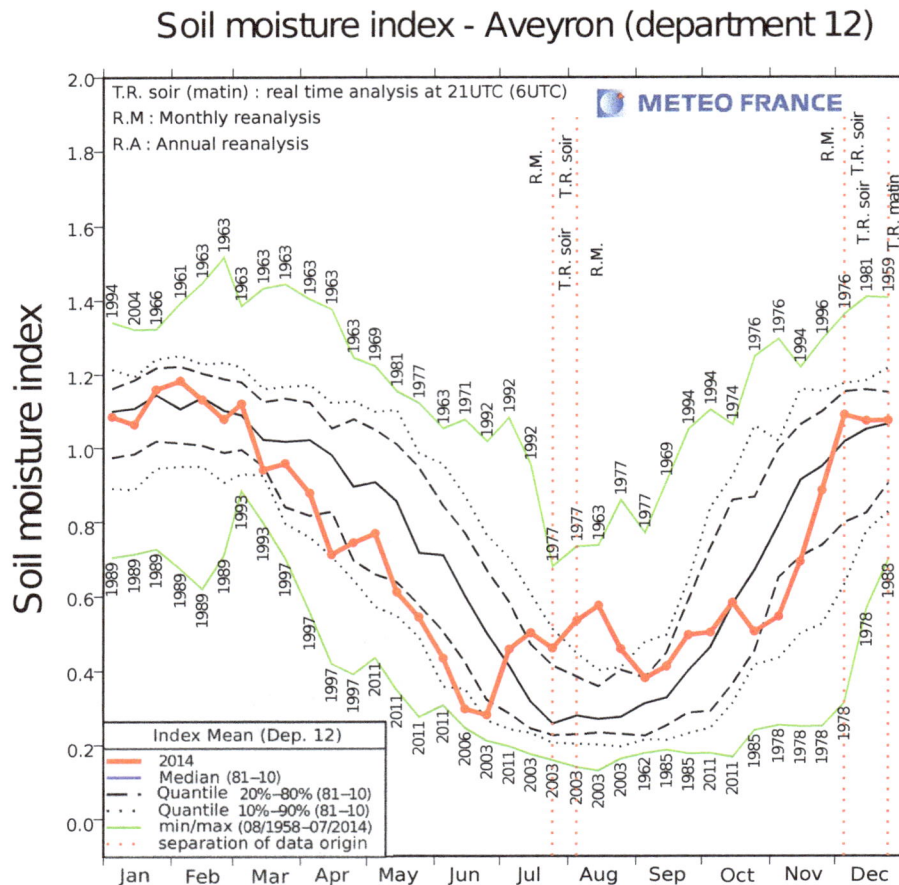

Figure 2. Time sketch showing the evolution of the Soil Moisture Index for a French department (red) in 2014, with climatological reference.

the collective memory and useful for future predictions. Another advantage is that it constitutes a feasible attempt based on an already existing production system. Yet, further developments remain to be done in order to take into account the projections' uncertainties (from multi-models and multi-scenarios).

5.2 Temperature in Clermont-Ferrand: from daily variability to climate change

Climate change science is based on the detection of long-term changes in mean conditions of climate variables, such as the mean temperature (Hegerl et al., 2007). As they are superimposed with large natural weather-related variability, people in general do not perceive such changes. Several studies suggest that the lack of public concern for climate change is linked with the lack of perception and personal experience of the ongoing changes (Spence et al., 2011; Weber, 2010). When temperature may vary by more than 10 °C from one day to another at a given place, why should people worry of a long-term 2 °C global change? Figure 4 is a very preliminary attempt to link the sensible day-to-day weather-related variability with the long-term ongoing climate change. Here

the focus is on the summer minimal temperature from a given local weather station (Clermont-Ferrand), between 1960 and 2009. Perceived daily temperature values are represented together with the longer time averages (from monthly to seasonal and decadal values).

6 Conclusions

The Viaduc project is a thought-experiment which aims at improving the existing climate services. A dual approach is used, based on supporting actions locally on the one hand, and identifying feasible productions on the other hand. The service design methodology helps sheding light on key messages. Firstly, there is a need to provide local climate information, not only about the future, but also about the past and present. This confirms the strong need for high quality, dense (close) and easily available observations. Secondly, there is a need to communicate in terms of impact. This underlines the need to enhance collaboration with impact communities. Thirdly, the appropriation and understanding of the climate results must be facilitated. This last point is both crucial and difficult, as tailoring the results to each user category pre-

Soil moisture in Aveyron, year 1959

Soil moisture in Aveyron, Observed records 1959-2010 (min, max)

Soil moisture in Aveyron, Climatology 1981-2010

Soil moisture in Aveyron, year 2014

Soil moisture in Aveyron, End 21st century climatology

Understanding how soil moisture varies according to seasons

Understanding soil moisture records (min max)

Understanding soil moisture mean Spotting major record: 2003 heat wave (long term memory activation)

Spotting ongoing situation (short term memory activation)

Comparing projected climatology to to major record

past

further into mathematical abstraction

present

future

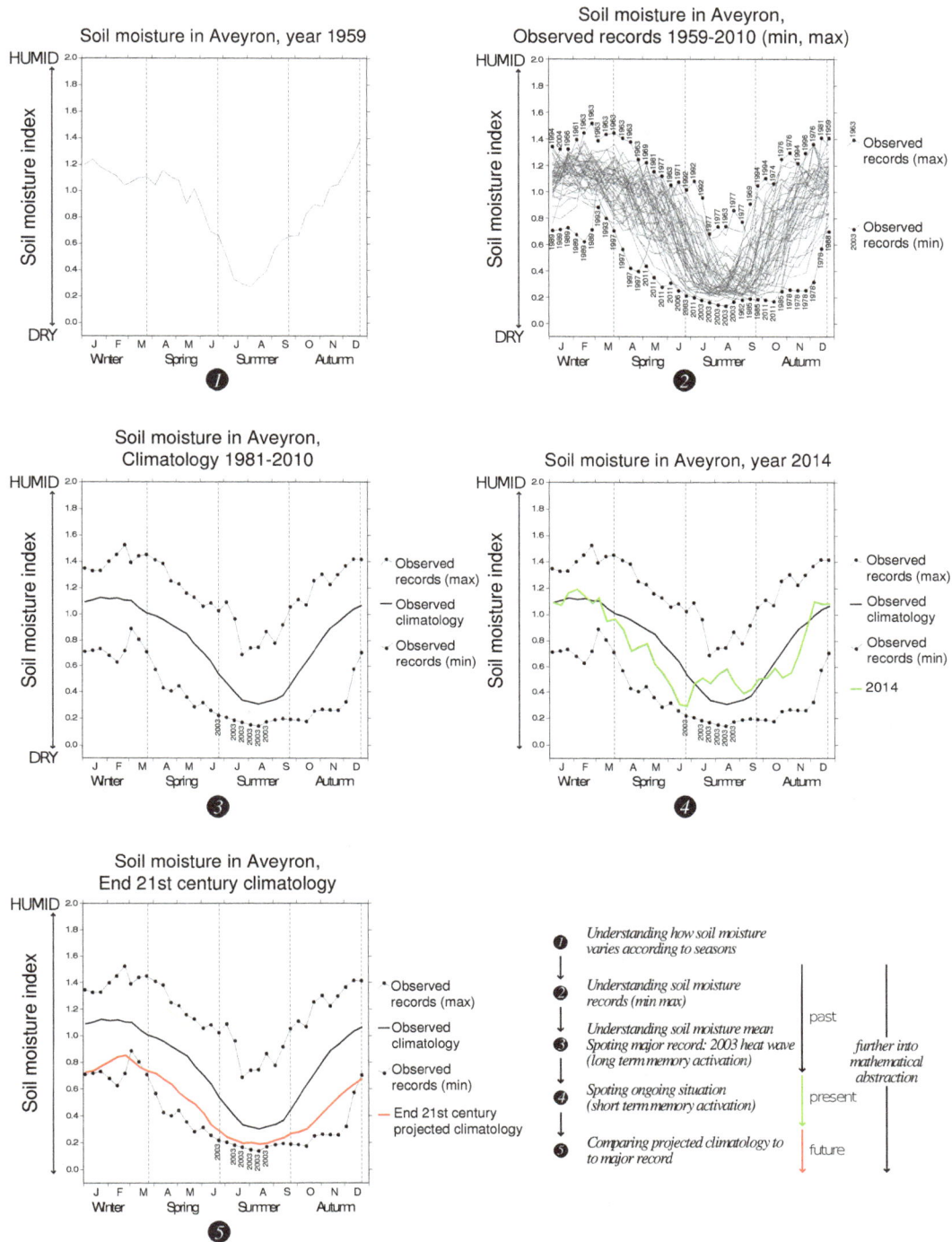

Figure 3. Attempt to ease the understanding of the product presented in Fig. 1, by using animations and storytelling.

vents generalization. Comments guiding the users step-by-step to help them depict the information and understand the message can be easily provided when direct contact is established between producers and users. However, challenges remain (e.g. reports or website products).

Future work will consist in converting the designer theoretical recommendations into feasible productions and prototypes with a view to enrich the Drias services. Many key messages are hidden behind what has been exchanged with end-users during the first year of the project. They bring some sparkle and life to our procedures and old habits. Be-

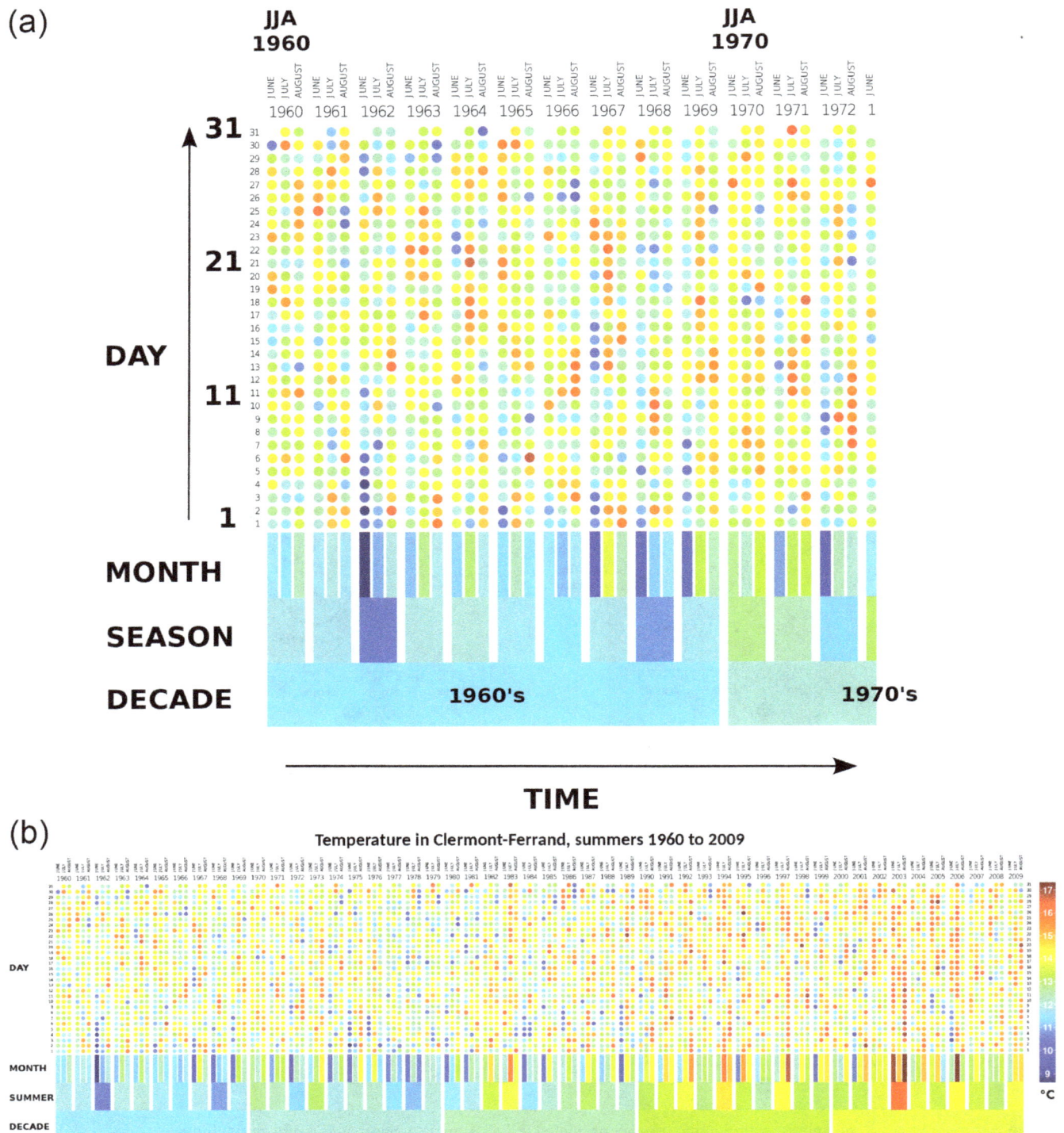

Figure 4. Summer minimal temperature in Clermont-Ferrand. Data are from a new high quality reference dataset (Gibelin et al., 2014) composed of homogenized monthly series and quality controlled daily series. **(a)** Guidelines on how to read the plot. Colors correspond to temperature values. The upper part of the plot (dots) shows daily temperatures. Each column corresponds to a distinct month (June, July or August). Dots along the vertical show temperatures from day 1 to the last day of the month. The lower part of the plot (rectangles) shows monthly, seasonal and decadal mean temperatures. Each column of dots is averaged and the corresponding monthly mean is shown by an underlying rectangle. Similarly, 3 consecutive monthly means are averaged and the corresponding summer mean is drawn below. Finally, the summer decadal mean is represented using the same color code at the baseline. The time goes from left to right. **(b)** Summer minimal temperature evolution in Clermont-Ferrand, from 1960 to 2009.

ing able to transform some of the emerging ideas into real improvements of products and services delivered by the scientific community, to support climate adaptation is a continuing challenge.

Acknowledgements. This work was supported by the MEDDE GICC programme. The authors thank Yves M. Tourre from LDEO of Columbia University for insightful comments.

References

Dandin, P., Besson, F., Blanchard, M., Céron, J.-P., Franchistéguy, L., Rousset-Regimbeau, F., Soubeyroux, J.-M., Baillon, M., Vidal, J.-P., Singla, S., Martin, E., and Habets, F.: Partnerships on water resource management in France, in: Climate exchange, 2nd Edn., World Meteorological Organization, Geneva, Switzerland, 76–78, 2012.

Erlhoff, M. and Marshall, T. (Eds.): Design Dictionary – Perspectives on design terminology, Birkhaüser Verlag AG, Basel, Germany, 2008.

Gibelin, A. L., Dubuisson, B., Corre, L., Deaux, N., Jourdain, S., Laval, L., Piquemal, J. M., Mestre, O., Dennetière, D., Desmidt, S., and Tamburini, A.: Evolution de la température en France depuis les années 1950: Constitution d'un nouveau jeu de séries homogénéisées de référence, La Météorologie, 87, 45–53, 2014.

Hegerl, G. C., Zwiers, F. W., Braconnot, P., Gillett, N. P., Luo, Y., Marengo Orsini, J. A., Nicholls, N., Penner, J. E., and Stott, P. A.: Understanding and Attributing Climate Change, in: Climate Change 2007: The Physical Science Basis. Contribution of Working Group I to the Fourth Assessment Report of the Intergovernmental Panel on Climate Change, edited by: Solomon, S., Qin, D., Manning, M., Chen, Z., Marquis, M., Averyt, K. B., Tignor, M., and Miller, H. L., Cambridge University Press, Cambridge, UK and New York, NY, USA, 2007.

Institut pour le développement forestier/Centre national de la propriété forestière (Ed.), 2014: Bioclimsol, un outil d'aide à la décision face au changement climatique [Special issue], Forêt entreprise, no. 218, Paris, France, 2014.

IPCC: Climate Change 2013: The Physical Science Basis, in: Contribution of Working Group I to the Fifth Assessment Report of the Intergovernmental Panel on Climate Change, edited by: Stocker, T. F., Qin, D., Plattner, G.-K., Tignor, M., Allen, S. K., Boschung, J., Nauels, A., Xia, Y., Bex, V., and Midgley, P. M., Cambridge University Press, Cambridge, UK and New York, NY, USA, 1535 pp., 2013.

Kelleher, C. and Wagener, T.: Ten guidelines for effective data visualization in scientific publications, Environ. Modell. Softw., 26, 822–827, doi:10.1016/j.envsoft.2010.12.006, 2011.

Lémond, J., Dandin, P., Planton, S., Vautard, R., Pagé, C., Déqué, M., Franchistéguy, L., Geindre, S., Kerdoncuff, M., Li, L., Moisselin, J.-M., Noël, T., and Tourre, Y.-M.: DRIAS: a step toward Climate Services in France, Adv. Sci. Res., 6, 179–186, doi:10.5194/asr-6-179-2011, 2011.

Lorenz, S., Dessai, S., Forster, P., and Paavola, J.: Tailoring the visual communication of climate projections for local adaptation practitioners in Germany and the UK, SRI Papers 81, Sustainability Research Institute, School of Earth and Environment, The University of Leeds, Leeds, 2015.

Ouzeau, G., Déqué, M., Jouini, M., Planton, S., and Vautard, R.: Scénarios régionalisés – édition 2014 – pour la métropole et les régions d'outre-mer. Le climat de la France au XXIe siècle, Volume 4, rapport de la mission Jean Jouzel, août 2014, Ministère de l'Ecologie, du Développement Durable, des Transports et du Logement, Paris, France, 62 pp., 2014.

Porter, J., Demeritt, D., and Dessai, S.: The Right Stuff? Informing Adaptation to Climate Change in British Local Government, SRI Papers 76, Sustainability Research Institute, School of Earth and Environment, The University of Leeds, Leeds, 2014.

Spence, A., Poortinga, W., Butler, C., and Pidgeon, N. F.: Perceptions of climate change and willingness to save energy related to flood experience, Nat. Clim. Change, 1, 46–49, 2011.

Stickdorn, M. and Schneider, J.: This is service design thinking, BIS publishers, Amsterdam, the Netherlands, 2010.

Weber, E. U.: What shapes perceptions of climate change?, Wiley Interdisciplinary Reviews: Climate Change, 1, 332–342, 2010.

31

Flux measurements by the NRC Twin Otter atmospheric research aircraft: 1987–2011

Raymond L. Desjardins[1], Devon E. Worth[1], J. Ian MacPherson[2], Matthew Bastian[2], and Ramesh Srinivasan[2]

[1]Science and Technology Branch, Agriculture and Agri-Food Canada, Ottawa, Ontario, Canada
[2]National Research Council of Canada, Aerospace, Flight Research Laboratory, Ottawa, Ontario, Canada

Correspondence to: Raymond L. Desjardins (ray.desjardins@agr.gc.ca)

Abstract. Over the past 30 years, the Canadian Twin Otter research group has operated an aircraft platform for the study of atmospheric greenhouse gas fluxes (carbon dioxide, ozone, nitrous oxide and methane) and energy exchange (latent and sensible heat) over a wide range of terrestrial ecosystems in North America. Some of the acquired data from these projects have now been archived at the Flight Research Laboratory and Agriculture and Agri-Food Canada. The dataset, which contains the measurements obtained in eight projects from 1987 to 2011 are now publicly available. All these projects were carried out in order to improve our understanding of the biophysical controls acting on land-surface atmosphere fluxes. Some of the projects also attempted to quantify the impacts of agroecosystems on the environment. To provide information on the data available, we briefly describe each project and some of the key findings by referring to previously published relevant work. As new flux analysis techniques are being developed, we are confident that much additional information can be extracted from this unique data set.

1 Introduction

For over three decades, the National Research Council of Canada (NRC) Twin Otter atmospheric research aircraft has been involved in projects associated with the exchange of energy and trace gases between the biosphere and the atmosphere (MacPherson and Bastian, 2003; Desjardins et al., 1982). These studies have focussed on quantifying atmosphere-biosphere interactions as well as quantifying greenhouse gas (GHG) emissions from agroecosystems. The aircraft has been flown in a number of national and international projects involving a large number of collaborators from university and government research organizations (Pederson et al., 1995; Sellers et al., 1992, 1997; Stewart et al., 1998; Kustas et al., 2005; Jackson et al., 1999). Most of the data from these flux projects have been reorganized to a common format and stored in digital form (MacPherson and Bastian, 2003). The purpose of this publication is to provide information on the data that are available and briefly mention some of the interesting results that have been obtained in order to promote further use of these data. In the near future,

these data will be transferred to the Government of Canada open data portal (www.open.canada.ca), where they will be downloadable by the public, free of charge. The use of the data will be subject to the Open Government License. We encourage researchers who are interested in analyzing aircraft-based flux data to contact the authors.

2 Aircraft-based flux measurements

The NRC Twin Otter atmospheric research aircraft (Fig. 1) has been instrumented to accurately measure the three components of atmospheric motion and the vertical fluxes of heat, momentum, water vapour, carbon dioxide (CO_2), ozone (O_3), methane (CH_4) and nitrous oxide (N_2O), as well as other supporting meteorological and spectral data (Desjardins and MacPherson, 1991; Desjardins et al., 2000). Since one of the aircraft's primary research roles is in the field of land/atmosphere interactions, a number of remote sensing instruments were used to document surface conditions during flux-measuring campaigns. For several research

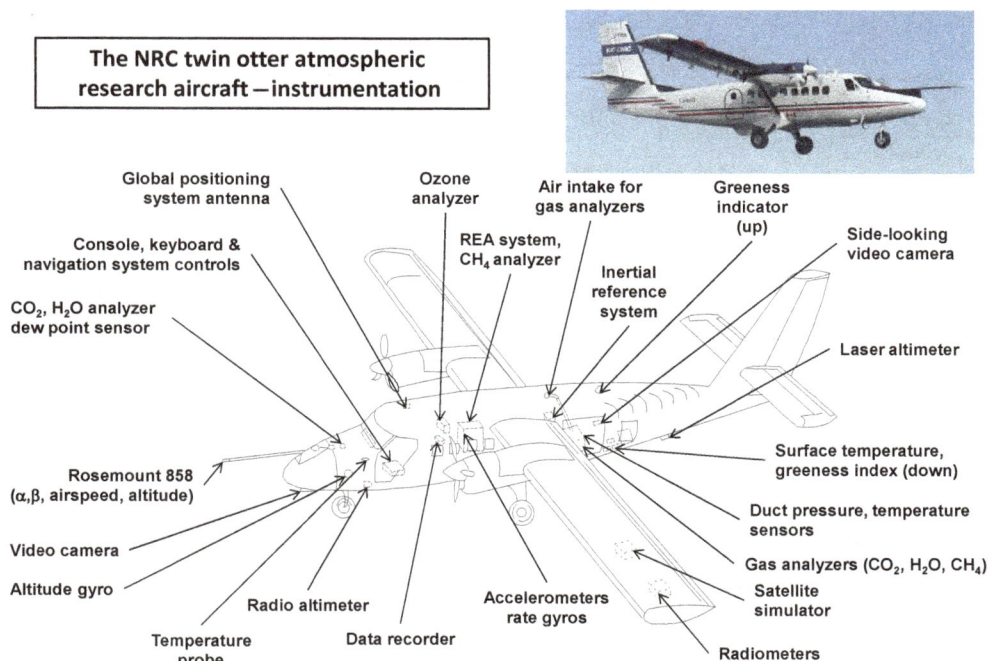

Figure 1. Instrumentation on-board the NRC Twin Otter, 1987–2011. Actual instrumentation varied slightly by measurement campaign.

projects, the fluxes of trace gases with very low atmospheric concentrations (e.g. N_2O, CH_4), were measured using the Relaxed Eddy Accumulation (REA) technique (Pattey et al., 2006). Details on the aircraft, the instrumentation and various sampling strategies are presented in Desjardins et al. (2000).

3 Summary of measurements campaigns and their significant findings

The Twin Otter Flux Data Archive presently available contains data from eight flux projects flown from 1987 to 2011 (Fig. 2 and Table 1). A short description of each project with a sample of some of the interesting findings follows.

3.1 The First ISLSCP Field Experiment (FIFE)

The First ISLSCP (International Satellite Land Surface Climatology Project) Field Experiment, carried out in Kansas, was designed to obtain satellite and ground observations for the development and validation of models relating surface fluxes to spectral data. An overview of FIFE was published by Sellers et al. (1992). The Twin Otter participated in four intensive field campaigns. Three campaigns took place in 1987 and one in 1989. Grid flights were shown to be an excellent way to document the spatial differences in mass and energy exchange associated with the FIFE site (Desjardins et al., 1992). Budget analysis of the boundary layer using grid flight and double-stack patterns (Betts et al., 1990, 1992) revealed an underestimation of the sensible and latent heat

Figure 2. Location of measurement campaigns flown by the NRC Twin Otter, 1987–2011.

Table 1. Summary of NRC Twin Otter measurement campaigns, 1987–2011.

#	Campaign	Location	Year	Dates	Landcover types	Project flights	Project flight hours	Measured fluxes[a]	Flight patterns (Typical altitude, m a.g.l.)
1	FIFE	Central Kansas, USA	1987	26 Jun–16 Oct	Mixed agriculture; grassland	42	103	UW, H, LE, CO_2	Line (variable), stack (various 90–750), L pattern (90, 180, 270), grid (90), sounding (near surface – max. 2500), regional (150)
1	FIFE	Central Kansas, USA	1989	27 Jul–12 August	Mixed agriculture; grassland	16	69	UW, H, LE, CO_2	Line (variable), stack (various 90–1300), L pattern (50, 180, 270), grid (90), sounding (near surface – max. 2500), regional (150)
2	CODE	San Joaquin Valley, USA	1991	10 Jul–5 August	Mixed agriculture (cotton, grapes, orchards); grassland; urban	24	62	UW, H, LE, CO_2, O_3	Line (30, 60), stack (30, 60, 90), grid (30), sounding (near surface – max. 2500), regional (150), city (150)
3	BOREAS	Southern and Northern Study Areas of the BOREAS project, Saskatchewan and Manitoba, Canada	1994	25 May–19 Sep	Boreal forest; mixed forest; mixed agriculture (wheat, canola, pasture); lakes	57	154	UW, H, LE, CO_2, O_3	Grid (30), Line (40), stack (40, 100, 170), sounding (near surface – max. 2900), regional (40)
3	BOREAS	Southern and Northern Study Areas of the BOREAS project, Saskatchewan and Manitoba, Canada	1996	9 Jul–8 August	Boreal forest; mixed forest; lakes	27	64	UW, H, LE, CO_2, O_3	Grid (30), Line (40), stack (various 20–300), sounding (near surface – max. 2900), regional (40)
4	SGP97	Oklahoma, USA	1997	18 Jun–17 Jul	Mixed agriculture (pasture, wheat, other crops); tallgrass prairie	27	83	UW, H, LE, CO_2, O_3	Line (40), stack (40, 100, 170), sounding (near surface – max. 2900), regional (40)
5	MAGS	Mackenzie River delta, Northwest Territories, Canada	1999	21 May–14 Jul	Boreal forest; river delta	24	70	UW, H, LE, CO_2, O_3	Line (65), stack (30, 60, 90), grid (65), regional (65), sounding (near surface – max. 2200)
6	Trace Gas – N_2O	Southeastern Ontario, Canada	2000	21 Mar–10 Apr	Mixed agricultural (corn, soybean); mixed forest	6	10	$UW, H, LE, CO_2, O_3, N_2O$ (REA)	Line (65)
6	Trace Gas – N_2O	Southeastern Ontario, Canada	2001	19 Mar–27 Apr	Mixed agricultural (corn, soybean); mixed forest	14	20	$UW, H, LE, CO_2, O_3, N_2O$ (REA)	Line (65)
6	Trace Gas – N_2O	Southern Saskatchewan, Canada	2002	2–17 Apr	Mixed agriculture (mixed grains, canola, pulses, forage/pasture)	13	20	$UW, H, LE, CO_2, O_3, N_2O$ (REA)	Line (60)
7	SMACEX	Walnut Creek Watershed, Iowa, USA	2002	15 Jun–6 Jul	Mixed agriculture (corn, soybean)	16	46	UW, H, LE, CO_2, O_3	Line (40), sounding (near surface – max. 2000)
6	Trace Gas – N_2O	Southeastern Ontario, Canada	2003	26 Mar–8 Oct	Mixed agricultural (corn, soybean); mixed forest	31	49	$UW, H, LE, CO_2, O_3, N_2O$ (REA)	Line (65)
6	Trace Gas – N_2O	Southeastern Ontario, Canada	2004	29 Mar–4 Jun	Mixed agricultural (corn, soybean); mixed forest	21	30	$UW, H, LE, CO_2, O_3, N_2O$ (REA)	Line (55)
8	Trace Gas – CH_4	Southeastern Ontario, Canada	2011	8 Apr–10 May	Mixed agricultural (corn, soybean); mixed forest; urban; wetland	10	15	UW, H, LE, CO_2, CH_4	Line (170), stack (150, 200, 250), sounding (near surface – max. 2000)

[a] UW, momentum; H, sensible heat; LE, latent heat; CO_2, carbon dioxide; O_3, ozone; N_2O, nitrous oxide; CH_4, methane.

ranging from 20 to 30 % based on energy balance considerations. Some of the underestimation was attributed to the short length of the flight tracks and the high-pass filtering of the data, which was done to minimize the impact of non-stationarity on flux measurements. Both factors are known to contribute to the non-closure of the energy budget, which is an ongoing challenge warranting further research (Eder et al., 2014; Mauder et al., 2007).

3.2 The California Ozone Deposition Experiment (CODE)

In the summer of 1991 the San Joaquin Valley Air Pollution Study Agency organized a 4-week international field experiment to measure ozone concentrations and fluxes in the San Joaquin Valley of California (Pederson et al., 1995). The objective was to improve the understanding of the exchange processes occurring between the atmosphere and various types of vegetation. Ozone deposition velocities were shown to have a high correlation with Normalized Difference Vegetation Index (NDVI) for all the major crops. With knowledge of the mean ozone concentration, this relationship can then be used in conjunction with satellite data to estimate the amount of ozone absorbed by crops (Desjardins et al., 1999). Mahrt et al. (1994) used repeated runs over the same track for the partial isolation of the influence of surface heterogeneity and transient mesoscale motion. Such transfer can best be quantified using aircraft-based measurements. The versatility of aircraft-based flux measurements was demonstrated by showing the change in the O_3 concentration and flux upwind and downwind of a major highway (Guo et al., 1995) and by showing the magnitude of the CO_2 concentration and flux at an altitude of 150 m above the city of Fresno (Desjardins et al., 2000).

3.3 The Boreal Ecosystem Atmosphere Study (BOREAS)

The Boreal Ecosystem Atmosphere Study (BOREAS) was a large international field and analysis campaign organized by the NASA Goddard Space Flight Centre (Sellers et al., 1997). It was designed to improve the understanding of the interactions between the Boreal Forest biome and the atmosphere. The experiment focussed on two study areas in the boreal forest region of Canada, one north of Prince Albert, Saskatchewan, and the other near Thompson, Manitoba. Several well-instrumented towers measured fluxes and supporting meteorological and radiative data over the main forest types in the two study areas. The Twin Otter flew three flight campaigns in 1994 and one in 1996. Sun et al. (1997) showed that at night, lake induced atmospheric circulation resulted in venting of CO_2 over the lake. This meant that the CO_2 flux could not be accurately measured over the forest at night by tower-based systems in close proximity to the lake. Chen et al. (1999) used the combination of land cover information and tower-based flux measurements to obtain flux estimates that could be compared to those obtained by flying a grid pattern 15 km by 15 km. Mauder et al. (2007) used Wavelet analysis on airborne flux measurements to investigate thermally induced mesoscale circulations and turbulent organized structures. According to independent large eddy simulation studies, these are considered as a major contributor to the energy budget non-closure problem. Their results put in question the practice of correcting eddy covariance measurements of the sensible and the latent heat flux for the lack of energy budget closure according to the Bowen ratio and of correcting carbon dioxide fluxes according to the energy balance residual.

3.4 Southern Great Plains (SGP97)

The Southern Great Plains 1997 (SGP97) Hydrology Experiment was motivated by the widespread interest among hydrologists, soil scientists and meteorologists in the problems of estimating soil moisture and temperature states at the continental scale, and their coupling to the atmosphere (Jackson et al., 1999). The main objective was to develop algorithms using remotely sensed microwave data to measure soil moisture at scales expected from future satellite-based microwave systems. Use of both the airborne thermal and microwave imagery collected during SGP97 along with tower and aircraft flux measurements spawned the development and validation of remote sensing-based land surface schemes for water, energy and carbon flux estimation largely based on the two-source energy balance (TSEB) modelling framework (Norman et al., 1995). Kustas et al. (2006) used corresponding flux field outputs from the Atmosphere-Land Exchange Inverse (ALEXI) and the disaggregated ALEXI model (Anderson et al., 2011) in combination with tower and aircraft-based flux measurements to investigate flux-footprint relations (Schuepp et al., 1990). With techniques like Wavelet analysis, which can provide the full flux contribution over distances as short as 100 m, these data lend themselves to additional studies such as combining satellite and aircraft spectral data with aircraft-based flux measurements.

3.5 Mackenzie GEWEX Study (MAGS)

The Mackenzie GEWEX Study (MAGS) was part of the Global Energy and Water Cycle Experiment (GEWEX). MAGS was a comprehensive study of the hydrologic cycle and energy fluxes of the Mackenzie River Basin, which is the largest North American source of fresh water to the Arctic Ocean. The role of the NRC Twin Otter atmospheric research aircraft in MAGS was to provide measurements of surface-atmosphere exchange of sensible and latent heat and CO_2 at temporal and spatial scales suitable for model testing. Mauder et al. (2008) used these data to demonstrate that flux measurements from a low-flying aircraft could be used to produce 2-dimensional maps of the sensible and latent heat flux for this complex ecosystem. Recent work has demonstrated how aircraft-based flux data can be used in combination with satellite-based input data to develop environmental response functions (Metzger et al., 2013). These permit an order-of-magnitude increase in spatio-temporal resolution and extend the resulting flux maps.

3.6 Soil Moisture-Atmosphere Coupling Experiment (SMACEX)

The Soil Moisture-Atmosphere Coupling Experiment (SMACEX) was conducted over the period from 15 June to 6 July 2002 over the Walnut Creek Watershed near Ames, Iowa a corn and soybean production area (Kustas et al., 2005). The data collected in SMACEX represent a relatively wide range of atmospheric, soil moisture, and vegetation states for use in modeling studies. This unique dataset consists of tower and aircraft flux measurements of heat, water vapour, momentum, carbon dioxide and ozone. It was collected simultaneously with a wide range of satellite data in order to develop satellite-based algorithms for land surface flux estimation. These algorithms were evaluated against the network of flux towers, and should in the future be compared with the aircraft-flux and regional remote sensing-based models such as ALEXI. Finally the relatively dense network of flux towers located in the experimental domain of corn and soybean fields afforded a unique opportunity to compare aircraft and tower fluxes (Prueger et al., 2005). These results were preliminary and require a more rigorous suite of analyses. As the techniques for accurately estimating mass and energy fluxes using aircraft-based technology evolve, it is very likely that the accuracy of some of the land surface and remote sensing modeling algorithms will be improved. This should result in a better understanding of the atmospheric boundary layer and other non-local effects on flux tower measurements.

3.7 Trace gas – N_2O

From 2000 to 2004, the Twin Otter was flown in a series of flux-measuring studies to quantify N_2O emissions. The objective was to utilize the REA system on the Twin Otter to measure the flux of N_2O (Pattey et al., 2007). One project, which took place in Southern Saskatchewan, focussed on comparing the N_2O emissions for a 10 km by 10 km area obtained using a large number of chambers to aircraft flux measurements by flying a grid pattern over the site (Pattey et al., 2007). A second project was carried out in Eastern Ontario along two 20 km transects (Desjardins et al., 2010). The measurements covered two periods when the N_2O emissions from agricultural lands are at their greatest: the spring burst period associated with spring thaw and the period just after the application of N fertilizers in the spring. They showed that N_2O emissions varied substantially from day-to-day but were very similar for the two agricultural areas 30 km apart.

The day-to-day difference between the aircraft-based flux measurements and estimates by the Denitrification and Decomposition (DNDC) model (Li, 2000) confirmed that the 25 % estimate commonly assumed for the indirect N_2O emissions as part of the IPCC methodology is reasonable (Desjardins et al., 2010).

3.8 Trace gas – CH_4

In 2011, CH_4 flux measurements were collected over an agricultural region in eastern Canada in order to verify CH_4 inventory from agricultural sources. A fast response CH_4 analyzer and a REA system were flown along several 20 km transects at an altitude of 60 m. This study showed that similar methane fluxes can now be obtained using the REA and the eddy covariance techniques using the available instrumentation. It also demonstrated that in many agricultural regions, there can be other important sources of methane beside livestock such as wetlands and waste water treatment plants, that act as large and confounding sources of methane (Desjardins et al., 2016).

4   Conclusions

This publication documents the availability of archived data from Twin Otter flux projects flown from 1987 to 2011. The archive includes files of time series of a large selection of variables stored at either 16 Hz (1987 to 1994) or 32 Hz (1996 to 2011). Most of these projects were conducted in collaboration with scientists from universities and other research agencies. The archive represents high-quality data from over 4000 flux runs flown over a wide range of landscapes. It is a resource that could be used to extend the scientific knowledge and the considerable body of literature already generated by these projects. During these studies, three flux measurements were calculated: raw, detrended and high-pass filtered. This was done as an attempt to deal with conditions of non-stationarity. It now appears that time-frequency analyses such as Wavelets might be a more effective way to deal with this condition. We have started redoing some of the analyses but it is only one example of the new type of analysis that can be done with these data. The development of environmental response functions using the data collected during these projects could greatly improve the value of satellite data for assessing crop conditions. Clearly these large scale, interdisciplinary field studies have greatly advanced thermal-based energy balance modeling schemes for reliable field to regional scale energy balance modeling (Kustas and Anderson, 2009). Utilizing these flux-aircraft measurements should provide greater insight to the accuracy of the predicted flux fields generated using satellite-based algorithms.

Acknowledgements. Funding agencies, organizers of large scale experiments such as P. Sellers, F. Hall, J. Pederson, T. Jackson, W. Kustas, R. Stewart and all the collaborators and PhD students who have contributed to these projects are gratefully acknowledged. We also thank R. W. A. Hutjes, S. Metzger and one anonymous reviewer for constructive criticism to improve this manuscript.

References

Anderson, M. C., Kustas, W. P., Norman, J. M., Hain, C. R., Mecikalski, J. R., Schultz, L., González-Dugo, M. P., Cammalleri, C., d'Urso, G., Pimstein, A., and Gao, F.: Mapping daily evapotranspiration at field to continental scales using geostationary and polar orbiting satellite imagery, Hydrol. Earth Syst. Sci., 15, 223–239, doi:10.5194/hess-15-223-2011, 2011.

Betts, A. K., Desjardins, R. L., MacPherson, J. I., and Kelly, R. D.: Boundary-layer heat and moisture budgets from FIFE, Bound.-Lay. Meteorol., 50, 109–138, 1990.

Betts, A. K., Desjardins, R. L., and MacPherson, J. I.: Budget analysis of the boundary layer grid flights during FIFE 1987, J. Geophys. Res., 97, 18533–18546, 1992.

Chen, J. M., Leblanc, S. G., Cihlar, J., Desjardins, R. L., and MacPherson, J. I.: Extending aircraft- and tower-based CO$_2$ flux measurements to a boreal region using a Landsat thematic mapper land cover map, J. Geophys. Res., 104, 16859–16877, 1999.

Desjardins, R. L. and MacPherson, J. I.: Water vapour flux measurements from aircraft, in: Land Surface Evaporation: Measurement and Parameterization, edited by: Schmugge, T. and André, J. C., Springer-Verlag Ltd, New York, NY, USA, 245–260, 1991.

Desjardins, R. L., Brach, E. J., Alvo, P., and Schuepp, P. H.: Aircraft monitoring of surface carbon dioxide exchange, Science, 216, 733–735, 1982.

Desjardins, R. L., Schuepp, P. H., MacPherson, J. I., and Buckley, D. J.: Spatial and temporal variations of the fluxes of carbon dioxide and sensible and latent heat over the FIFE site, J. Geophys. Res., 97, 18467–18475, doi:10.1029/92JD01089, 1992.

Desjardins, R. L., MacPherson, J. I., and Schuepp, P. H.: Ground truthing for satellites using aircraft-based flux measurements, Proceedings of the International Symposium on Applied Agrometeorology and Agroclimatology, Volos, Greece, 24–26 April 1996, 53–59, 1999.

Desjardins, R. L., MacPherson, I., and Schuepp, P. H.: Aircraft-based flux sampling strategies, in: Encyclopedia of Analytical Chemistry, edited by: Meyers, R. A., John Wiley and Sons Ltd., Chichester, UK, 3573–3588, 2000.

Desjardins, R. L., Pattey, E., Smith, W. N., Worth, D., Grant, B. B., Srinivasan, R., MacPherson, I. J., and Mauder, M.: Multiscale estimates of N$_2$O emissions from agricultural lands, Agr. Forest Meteorol., 150, 817–824, doi:10.1016/j.agrformet.2009.09.001, 2010.

Desjardins, R. L., Worth, D. E., Srinivasan, R., Pattey, E., VanderZaag, A. C., Mauder, M., Metzger, S., Worthy, D., and Sweeney, C.: Verifying Inventory Estimates of Agricultural Methane Emissions at a Regional Scale Using Aircraft-Based Flux Measurements, in preparation, 2016.

Eder, F., De Roo, F., HKohnert, K., Desjardins, R. L., Schmid, H. P., and Mauder, M.: Evaluation of two energy balance closure parameterizations, Bound.-Lay. Meteorol., 151, 195–219, doi:10.1007/s10546-013-9904-0, 2014.

Guo, Y., Desjardins, R. L., MacPherson, J. I., and Schuepp, P. H.: A simple scheme for partitioning aircraft-measured ozone fluxes into surface-uptake and chemical transformation, Atmos. Environ., 29, 3199–3207, doi:10.1016/1352-2310(95)00088-G, 1995.

Jackson, T. J., Le Vine, D. M., Hsu, A. Y., Oldak, A., Starks, P. J., Swift, C. T., Isham, J. D., and Haken, M.: Soil moisture mapping at regional scales using microwave radiometry: The Southern Great Plains Hydrology Experiment, IEEE T. Geosci. Remote Sens., 37, 2136–2151, 1999.

Kustas, W. P. and Anderson, M. C.: Advances in thermal infrared remote sensing for land surface modeling, Agr. Forest Meteorol., 149, 2071–2081, 2009.

Kustas, W. P., Hatfield, J. L., and Prueger, J. H.: The Soil Moisture-Atmosphere Coupling Experiment (SMACEX): Background, Hydrometeorological Conditions, and Preliminary Findings, J. Hydrometeorol., 6, 791–804, 2005.

Kustas, W. P., Anderson, M. C., French, A. N., and Vickers, D.: Using a remote sensing field experiment to investigate flux-footprint relations and flux sampling distributions for tower and aircraft-based observations, Adv. Water Resour., 29, 355–368, doi:10.1016/j.advwatres.2005.05.003, 2006.

Li, C.: Modeling trace gas emissions from agricultural ecosystems, Nutr. Cycl. Agroecosys., 58, 259–276, 2000.

MacPherson, J. I. and Bastian, M.: Archive of NRC Twin Otter Data From the 1991–2003 Flux Projects, National Research Council Canada, Institute for Aerospace Research, Ottawa, Ontario, Canada, Report LTR-FR-204, 16 pp., 2003.

Mahrt, L., MacPherson, J. I., and Desjardins, R.: Observations of fluxes over heterogeneous surfaces, Bound.-Lay. Meteorol., 67, 345–367, 1994.

Mauder, M., Desjardins, R. L., and MacPherson, J. I.: Scale analysis of airborne flux measurements over heterogeneous terrain in a boreal ecosystem, J. Geophys. Res., 112, D13112, doi:10.1029/2006JD008133, 2007.

Mauder, M., Desjardins, R. L., and MacPherson, I. J.: Creating surface flux maps from airborne measurements: Application to the Mackenzie area GEWEX study MAGS 1999, Bound.-Lay. Meteorol., 129, 431–450, 2008.

Metzger, S., Junkermann, W., Mauder, M., Butterbach-Bahl, K., Trancón y Widemann, B., Neidl, F., Schäfer, K., Wieneke, S., Zheng, X. H., Schmid, H. P., and Foken, T.: Spatially explicit regionalization of airborne flux measurements using environmental response functions, Biogeosciences, 10, 2193–2217, doi:10.5194/bg-10-2193-2013, 2013.

Norman, J. M., Kustas, W. P., and Humes, K. S.: A two-source approach for estimating soil and vegetation energy fluxes in observations of directional radiometric surface temperature, Agr. Forest Meteorol., 77, 263–293, 1995.

Pattey, E., Edwards, G., Strachan, I. B., Desjardins, R. L., Kaharabata, S., and Wagner Riddle, C.: Towards standards for measuring greenhouse gas fluxes from agricultural fields using instrumented towers, Can. J. Soil Sci., 86, 373–400, 2006.

Pattey, E., Edwards, G. C., Desjardins, R. L., Pennock, D. J., Smith, W., Grant, B., and MacPherson, J. I.: Tools for quantifying N_2O emissions from agroecosystems, Agr. Forest Meteorol., 142, 103–119, 2007.

Pederson, J. R., Massman, W. J., Mahrt, L., Delany, A., Oncley, S., Hartog, G. D., Neumann, H. H., Mickle, R. E., Shaw, R. H., Paw U, K. T., Grantz, D. A., MacPherson, J. I., and Desjardins, R.: California ozone deposition experiment: Methods, results, and opportunities, Atmos. Environ., 29, 3115–3132, 1995.

Prueger, J. H., Hatfield, J. L., Kustas, W. P., Hipps, L. E., MacDonald, J. I., Neale, C. M. U., Eichinger, W. E., Cooper, D. I., and Parkin, T. B.: Tower and aircraft eddy covariance measurements of water vapor, energy and carbon dioxide fluxes during SMACEX, J. Hydrometeorol., 6, 954–960, 2005.

Schuepp, P. H., Leclerc, M. Y., MacPherson, J. I., and Desjardins, R. L.: Footprint prediction of scalar fluxes from analytical solutions of the diffusion equation, Bound.-Lay. Meteorol., 50, 355–373, 1990.

Sellers, P. J., Hall, F. G., Asrar, G., Strebel, D. E., and Murphy, R. E.: An overview of the First International Satellite Land Surface Climatology Project (ISLSCP) Field Experiment (FIFE), J. Geophys. Res., 97, 18345–18371, 1992.

Sellers, P. J., Hall, F. G., Kelly, R. D., Black, A., Baldocchi, D. D., Berry, J., Ryan, M., Ranson, K. J., Crill, P. M., Lettenmaier, D. P., Margolis, H., Cihlar, J., Newcomer, J., Fitzjarrald, D. R., Jarvis, P. G., Gower, S. T., Halliwell, D., Williams, D., Goodison, B., Wickland, D. E., and Guertin, F. E.: BOREAS in 1997: Experiment overview, scientific results, and future directions, J. Geophys. Res., 102, 28731–28769, 1997.

Stewart, R. E., Leighton, H. G., Marsh, P., Moore, G. W. K., Ritchie, H., Rouse, W. R., Soulis, E. D., Strong, G. S., Crawford, R. W., and Kochtubajda, B.: The Mackenzie GEWEX Study: The Water and Energy Cycles of a Major North American River Basisn, B. Am. Meteorol. Soc., 79, 2665–2683, 1998.

Sun, J., Lenschow, D. H., Mahrt, L., Crawford, T. L., Davis, K. J., Oncley, S. P., MacPherson, J. I., Wang, Q., Dobosy, R. J., and Desjardins, R. L.: Lake-induced atmospheric circulations during BOREAS, J. Geophys. Res., 102, 29155–29166, 1997.

32

Methodologies to characterize uncertainties in regional reanalyses

M. Borsche[1], A. K. Kaiser-Weiss[1], P. Undén[2], and F. Kaspar[1]

[1]Deutscher Wetterdienst, National Climate Monitoring, Frankfurter Str. 135, 63067 Offenbach, Germany
[2]Swedish Meteorological and Hydrological Institute, Folkborgsvägen 17, 601 76 Norrköping, Sweden

Correspondence to: M. Borsche (michael.borsche@dwd.de)

Abstract. When using climate data for various applications, users are confronted with the difficulty to assess the uncertainties of the data. For both in-situ and remote sensing data the issues of representativeness, homogeneity, and coverage have to be considered for the past, and their respective change over time has to be considered for any interpretation of trends. A synthesis of observations can be obtained by employing data assimilation with numerical weather prediction (NWP) models resulting in a meteorological reanalysis. Global reanalyses can be used as boundary conditions for regional reanalyses (RRAs), which run in a limited area (Europe in our case) with higher spatial and temporal resolution, and allow for assimilation of more regionally representative observations. With the spatially highly resolved RRAs, which exhibit smaller scale information, a more realistic representation of extreme events (e.g. of precipitation) compared to global reanalyses is aimed for. In this study, we discuss different methods for quantifying the uncertainty of the RRAs to answer the question to which extent the smaller scale information (or resulting statistics) provided by the RRAs can be relied on. Within the European Union's seventh Framework Programme (EU FP7) project Uncertainties in Ensembles of Regional Re-Analyses (UERRA) ensembles of RRAs (both multi-model and single model ensembles) are produced and their uncertainties are quantified. Here we explore the following methods for characterizing the uncertainties of the RRAs: (A) analyzing the feedback statistics of the assimilation systems, (B) validation against station measurements and (C) grids derived thereof, and (D) against gridded satellite data products. The RRA ensembles (E) provide the opportunity to derive ensemble scores like ensemble spread and other special probabilistic skill scores. Finally, user applications (F) are considered. The various methods are related to user questions they can help to answer.

1 Introduction

Atmospheric reanalyses produce complete and physically consistent data products aiming for a best estimate of the state of the Earth's atmosphere (Dee et al., 2014). In the European Union's seventh Framework Programme (EU FP7) project Uncertainties in Ensembles of Regional Re-Analyses (UERRA), various European meteorological regional reanalyses (RRAs) are developed. Users turn to the spatially highly resolved RRAs for smaller scale information and a more realistic representation of extreme events (e.g. of precipitation) compared to global reanalyses, and would like to derive trends.

The main objectives of the UERRA project are to produce a long-term (several decades) high-resolution climate quality ensemble of European RRAs of Essential Climate Variables (ECVs) and to estimate the associated uncertainties in these RRAs. The RRAs and the uncertainty estimates are to be made publicly available so a large community can benefit from the research. Within the UERRA project and its precursor project European Reanalysis and Observations for Monitoring (EURO4M), gridded data are produced, and data rescue and digitization efforts are undertaken (Brunet et al., 2013). Data rescue efforts are concentrated on filling gaps in data from 1950 onwards. More on the structure, participants, and status of the UERRA project can be found on its website http://www.uerra.eu.

Table 1. Description of the regional reanalysis planned in the UERRA project.

Feature	Met Office	SMHI	HErZ	Météo France
Boundary conditions (forcings)	6 hourly ERA-Interim fields	6 hourly ERA-Interim and/or ERA-40 fields	3 hourly ERA-Interim and/or ERA-20C fields	HARMONIE @ 11 km to > 5.5 km; ALADIN (Horányi et al., 1996) @ 5.5 km
Model and domain	Unified Model, CORDEX EU-11	HARMONIE, Lambert projection, CORDEX EU-11	COSMO, CORDEX EU-11	MESCAN (Soci et al., 2013), Lambert projection
Ensemble members	20	1 (2 for the period 2006 to 2010)	10 to 20	1 (4 to 6 for the period 2006 to 2010)
Deterministic DA method	Hybrid Ensemble-4D-Var	3D-Variational upper-air/OI (optimal interpolation) surface analysis	Nudging	OI surface re-analysis after a static or dynamical downscaling
Ensemble DA method	Ensemble of 4D-Vars	Ensemble of 3D-Vars	LETKF with ensemble nudging	
Time range	1978 to 2013	1961 to 2013	5 years	1961 to 2011
Observation input	various sources of surface, aircraft, upper air, satellite observations, and precipitation	Surface (pressure), SHIP, (SYNOP BUOY, DRIBU); aircraft (AIREP, AMDAR); Upper air (TEMP, PILOT)	Surface (SYNOP (pressure), SHIP, BUOY, DRIBU); aircraft (AIREP, AMDAR); upper air (TEMP, PILOT)	SYNOP (pressure), SHIP, BUOY, 24 h precipitation from rain gauge and T_{min}/T_{max} after pre-processing
Temporal resolution	6 h (analysis), hourly (forecast)	6 h	1 h	6 and 24 h for precipitation
Horizontal resolution	12 km control grid; analysis increments on 24 km; 24 km ensemble	11 km	12 km	5.5 km
Vertical resolution	70 levels from near surface to 80 km	65 levels	40 levels (20 m to 22 km)	only surface

Table 1 summarizes the RRAs which are planned to be produced within the UERRA project at the Met Office (MO), Exeter, UK (developed upon Renshaw, 2013); the Swedish Meteorological and Hydrological Institute (SMHI), Norrköping, Sweden (developed upon Dahlgren et al., 2014 and now Bubnova et al., 1995); DWD's (Deutscher Wetterdienst) Hans-Ertel Centre for Weather Research (HErZ; Simmer et al., 2015) University of Bonn, Germany (Bollmeyer et al., 2015); and Météo France, Toulouse, France (Häggmark et al., 2000; Jansson et al., 2007; Soci et al., 2011, 2013). The RRAs differ in the numerical weather prediction (NWP) model used, the data assimilation system applied, in the boundary conditions applied, and in the observations used for the assimilation. All deterministic RRA lateral boundary conditions are derived from the global ERA-Interim (Dee et al., 2011) and ERA-40 (Uppala et al., 2005) reanalysis whereas for the ensemble realizations it is planned to take lateral boundary conditions from ERA-20C (Poli et al., 2013) or ERA5, the successor of ERA-Interim, into account.

These three groups (MO, SMHI, HErZ) develop RRAs based on their operational NWP models, which are the Unified Model (Davies et al., 2006), the HARMONIE model (Bubnova et al., 1995; De Troch et al., 2013; Gerard et al., 2009), and the COnsortium for Small-scale MOdelling (COSMO) model (Schättler et al., 2014), respectively. They plan to produce a deterministic run and a set of ensembles of reanalyses with up to 20 members each. The data assimilation methods employed by MO are a hybrid 4DVar (Clayton et al., 2013) for the deterministic run and an ensemble of 4D-Vars (Rawlins et al., 2007) for the ensemble runs. SMHI has implemented a 3D-Var (Berre, 2000; Fischer et al., 2005) for the deterministic run and an ensemble of (2) 3D-Vars (using the ALADIN and ALARO-0 physics versions as described in De Troch et al., 2013) for the ensemble runs. HErZ uses nudging and is developing a Local Ensemble Transform Kalman Filter (LETKF) (Hunt et al., 2007; Harnisch and Keil, 2015) with Ensemble Nudging for the deterministic and ensemble runs, respectively. Ensembles are created either by perturbed initial conditions (MO), disturbed model physics (SMHI), disturbed observations (HErZ), or a combination of these. Météo-France performs an additional statistical and dynamical downscaling of the RRAs produced at SMHI followed by high resolution reanalysis for the surface only in order to drive inter alias hydrological physical models. The data sets are planned to span several decades, and a few years for the more experimental set-ups. SMHI produces in addition a cloudiness reanalysis (based on Häggmark et al., 2000) using a consistent satellite data set.

An estimation of uncertainty is required by users on different temporal and spatial scales for enabling and enhancing applications of RRA products. Section 2 outlines the methods, relying on assimilation feedback, comparison against station observations, gridded station observations, gridded satellite remote sensing based data products, and ensembles. We also discuss deriving uncertainty estimation from user applications. The advantages and disadvantages of the various methods are discussed and they are related to various scientific and user questions. The meteorological parameters for which each method is feasible depend mainly on practical considerations. In Sect. 3, the various skill scores are summarized, the suitability of which depends on the meteorological parameters and the question sought to be answered. In Sect. 4, a summary and recommendations for best practises are given.

2 Methods to estimate uncertainties in RRAs

Table 2 summarizes six methods which have been discussed within the UERRA project and correspond to those discussed in the EU FP7 project Coordinating Earth Observation Data Validation for Re-analysis for Climate Services (CORE-CLIMAX), see Kaiser-Weiss et al. (2015). Each of the methods is discussed in this section, benefits and drawbacks pointed out, and the principle aim pursued with each method condensed into a scientific and a user question. Furthermore, the discussion below addresses both random and systematic uncertainties. A systematic uncertainty (which can depend on time as well as location or meteorological situation) is referred to as bias. Random uncertainties are characterized with the root mean square (RMS) against some data set considered as truth, in case of absence of error estimates of the latter.

2.1 Method A: feedback statistics

Here, we distinguish between virtually independent observations (not assimilated yet) and dependent observations (observations passed through the assimilation system prior to the production of the reanalysis field). Strictly speaking, any observation system biased relative to the model will yield the analysis biased, and in this sense, measurements at later points in time are not strictly independent (as they are subject to this bias, too). Strictly independent data are hard to come by, especially for longer time periods, because they would naturally be selected for assimilation. Integrated parameters, such as precipitation, are strictly independent for most reanalysis systems. Further, minimum and maximum temperatures are not assimilated (though would not be strictly independent in case of biases). In order to maintain a reasonable large data base, we have to loosen our requirements on independence such that we use virtually independent data as if they were strictly independent.

We regard the feedback as the output of the assimilation system in observation space. The feedback comprises the assimilated observations (bias corrected where applicable) (o), the background or "free forecast" (the short range forecast used in the data assimilation) ($\mathbf{H}_b$), and the analysis ($\mathbf{H}_a$). The background and the analysis are brought into observation space with the matrix of the linearized observation operator $\mathbf{H}$, so a direct comparison between observations and model parameters can be performed in an optimal way. We check the feedback for trends and seasonal dependency (which are not desirable).

Usually, feedback statistics are a standard output of the data assimilation system and are frequently used by the producers for quality control. Note that, when comparing different systems, the bias correction might differ. For one system, an approximately Gaussian distribution is expected for both $o - \mathbf{H}_b$ and $o - \mathbf{H}_a$, where the absolute value of $o - \mathbf{H}_b$ is the bias between observations and modelled observations, i.e. even if the difference is zero, both terms could be biased versus the (unknown) truth.

Comparison between the different RRAs has to be done with observations assimilated by all of the UERRA reanaly-

Table 2. Methods and data sources suitable to derive uncertainty estimates for regional reanalyses.

Method	Data source	Parameter	Details	Scientific questions	User questions
A: feedback statistics	Radiosonde soundings	Temperature, wind speed, and relative humidity	Focus on lower troposphere; bias and RMSE of time series; store in ODB format	How stable are the regional reanalyses (RRAs) with respect to multi-annual trends on a spatial scale of roughly 100 km?	How well represented are trends and climatologies of wind speed relevant for wind energy?
B: station observations	B1: (independent) mast station data; B2: (dependent, i.e. assimilated) station data	B1: wind speed B2: T_{min}, T_{max}, and number of days of threshold exceedance of temperature and precipitation	There are many more suitable observations available for B2 than for B1.	At which time scales can we find which correlations between reanalysis fields and station observations?	On which time scales of variability and for which parameters can we use the RRAs similar to the use of station measurements?
C: gridded station observations	Gridded data products for the Nordic region and the UK; E-OBS, APGD	Precipitation; T_{min} and T_{max}	To consider whether a part of underlying station observations was assimilated into the reanalysis.	What differences do we get with different products when determining the effective spatial and temporal scales of the RRAs?	Which scales of the RRAs (temporal, spatial) can be interpreted?
D: gridded satellite data products	Satellite data products of CM-SAF and CCI	Global radiation; cloud liquid water path; total cloud cover; precipitation; snow water equivalent		How well do the RRAs compare to the satellite observations – or exceed their quality?	Does the RRA or the satellite provide the better data product for the user applications?
E: Ensemble based comparison	Data with uncertainty estimates;	Precipitation; T_{min}, T_{max}, T_{mean};	Ensemble based uncertainty estimates will be performed on (1) data with uncertainty estimates.	Does the ensemble provide a useful spatially and temporally resolved uncertainty estimate?	Which uncertainty characteristics can be interpreted from the ensembles, for which user relevant parameters?
	Products as in methods A through D	Parameters as in A through D	(2) the basis of methods A through D		
F: User related models		T_{mean}; T_{max} and T_{min} pseudo analysis; wind speed; precipitation;	SURFEX by Météo France and HYPE by SMHI		Is the result of a user model forced by RRAs significantly better than with the original forcing?

sis systems, which do not differ much in their bias correction. Especially suitable for this purpose are radiosondes (temperature, wind speed and direction, and relative humidity). The core scientific interest in applying feedback statistics for validating the RRAs is how stable the RRAs are with respect to multi-annual trends on different spatial scales for these parameters. For instance, users of wind energy applications want to know how well represented the wind speed at heights relevant to wind energy is, especially with respect to trends and frequency distributions. Feedback statistics can of course be applied to all sorts of assimilated observations in order to perform the afore mentioned task. However, assimilated radiosonde data best fit the purpose of this method due to the

fact that they are used throughout the different RRA systems as anchor-data (i.e. no bias corrections applied). In contrast, aircraft data are handled specifically for each assimilation system (due to data thinning, data selection, and individual pre-processing) which renders inter-comparison of that data source much more difficult.

When, firstly, comparing the mean or RMS of $o - \mathbf{H}_b$ between different reanalysis systems, it is crucial that comparable forecast lengths are selected. If this is valid $o - \mathbf{H}_b$ is a good measure to start with as a means of comparing against virtually independent observations. A smaller $o - \mathbf{H}_b$, i.e. a closer match, can be caused by a number of reasons: (1) the background is closer to the truth – desired from the user perspective, (2) the bias removal prior to the assimilation was successful – desired for the assimilation, (3) the background and the observations have the same bias against the truth – unlikely when a number of observing systems yield similar results, (4) the model is biased due to the assimilation of biased observations in the previous step – also unlikely when a number of observing systems yield similar results. Secondly, in a successful assimilation, the average RMS of $o - \mathbf{H}_a$ should be smaller than the average RMS $o - H_b$ for each observing system. Note that the mean or RMS of $o - \mathbf{H}_a$ is harder to compare between different reanalysis systems.

Desroziers et al. (2005) provide a methodology of the diagnosis in observation space and show that the variance of $o - \mathbf{H}_b$ is the sum of the variance of background error in observation space $(\widetilde{\sigma b})^2$ and the variance of observation error $(\widetilde{\sigma o})^2$, compare their Eq. (1):

$$\frac{1}{N}\left(\sum_{i=1}^{N} \left(o_i - \mathbf{H}_{b_i}\right)\left(o_i - \mathbf{H}_{b_i}\right)\right) = (\widetilde{\sigma b})^2 + (\widetilde{\sigma o})^2.$$

Further, they show that the background error covariance in observation space can be related to the product of $\mathbf{H}_a - \mathbf{H}_b$ and $o - \mathbf{H}_b$, see their Eq. (2), and the observation error covariance can be related to the product of $o - \mathbf{H}_a$ and $o - \mathbf{H}_b$, see their Eq. (3). Finally, if the error covariances are correctly specified in the analyses, the product of $\mathbf{H}_a - \mathbf{H}_b$ and $o - \mathbf{H}_a$ should relate to the analysis-error covariance (see their Eq. 4).

Before Desroziers et al. (2005), Hollingsworth and Lönnberg (1986) and Lönnberg and Hollingsworth (1986) applied a curve fitting of the covariances as a function of distance to separated background errors from observation error based on the assumption that forecasts errors are horizontally homogeneous and observation errors are horizontally uncorrelated, where it is understood that the observation error includes the representativeness error (also known as sampling error) as well as instrumental errors. This means a separation of background bias and observation bias by inspection of the spatial structure of mean background departures mean($o - \mathbf{H}_b$) for, e.g. each season or time of the day. It is well known that the NWP models used in the data assimilation exhibit biases that have both diurnal and seasonal variations. Thus, inspecting horizontal plots of biases at station locations (or possibly from a gridding interpolation procedure) will show if there is a significant background model bias. This is the case if there are spatially consistent mean departures that are sizeable compared with the RMS of the departures. In addition, this can be found by looking at the mean of the analysis increments mean($a - b$) in grid point space.

On the other hand, if there are large variations of the mean background departures from station to station without horizontal consistency one may conclude that the bias is in the observation itself, or in some cases, due to poor representativeness of the NPW model background at the specific loca-

tion. It may be due to occasional large departures between model orography and the one of the station or very different surface properties. The latter is expected to be the case only for a small portion of the stations and manual inspections of the data can reveal if it is the case. Either way, stations with large biases should be excluded for evaluations of random errors and used with special care when studying trends.

Multi-annual trends in the RRAs are determined by the boundary conditions (lateral boundary from the global reanalysis as well as the lower boundary, especially by the soil moisture and sea surface temperature), and the assimilated observations, where any trend in the observation system bias (relative to the model) would influence trends in the reanalysis. The latter might be expected to happen in regions where the observation coverage changed significantly over time.

The advantages of applying method (A) is that it provides a relative measure and can detect discontinuities and breaks as well as slowly increasing or decreasing systematic errors of the analyses (which all would influence trends). These results are dependent on the weight the observations are given in the different data assimilation systems, thus they need to be inter-compared between the different RRAs with care. A drawback of this method is that inter-comparison between the feedback statistics of different reanalysis systems is in principle difficult mainly because the handling of the observations may differ from one centre to another. For some of the parameters of interest (like, e.g. precipitation), the bias and RMSE are not sufficient scores to perform a thorough statistical analysis, thus more than these standard parameters should be evaluated based on the feedback. Note this method is limited in application to parameters which are assimilated.

2.2 Method B: station observations

Method (B) describes the comparison against point measurements from station observations. Interpreting single grid cells from the regional reanalysis and taking it as a proxy for a single point poses several questions from a theoretical point of view, such as that of the representativeness of the NWP model for a given observation location and observing method. However, the benefit of the method is that from a practical point of view this is often the easiest approach. Users of the RRAs are very much interested to answer the question how well the reanalyses compare against (their own) independent observation time series. This boils down to the question at what time scales correlations can be found between RRA fields and station observations.

In contrast to the previously described feedback methodology, this method allows to compare parameters which are not assimilated. For an independent validation and uncertainty estimate the reference data are required not to be assimilated into the RRAs. This leads to the scientific question on which time scales of variability and with which parameters the RRAs can be used to compare with station measurements.

The major drawback of the method is that a point measurement cannot be expected to closely match the nearest reanalysis grid point even if the measurement happens to be located exactly at the centre of the model grid point. This is due to the fact that the point measurement is representative for a limited area around the measurement, whereas the model grid point represents a much larger area, corresponding to the inherent spatial resolution, which can be expected to be larger than the nominal resolution, i.e. span several model grid cells. Not only the difference in the spatial but also the temporal representativeness needs to be considered, because the point measurement takes place more or less instantaneous, whereas the model value represents an average over a longer time period. Hakuba et al. (2014) (and references therein) provide a study on the representativeness of ground-based point measurements of surface solar radiation compared to gridded satellite data products detailing the points mentioned above.

And most importantly, the vertical representation usually is different between the model and the point measurement because the model levels are seldom exactly at the height of the measurement as, e.g. 2 m temperature or 10 m wind speed. This is complicated by the fact that the model topography is smoothed and thus different to the real one and there are limits to which extent small scale variability can be modelled or parameterized.

It is possible to correct for some of the above mentioned deficiencies. For interpolating the model value to the observation height, either the vertical observation operator can be used – as done in the intrinsic handling in method (A) – or some simplified interpolation of the model levels to the height above ground where the observation took place can be applied. Still, the problem remains that the model topography and the real one differ and some local topographic effects are not modelled.

If the assimilated, dependent observations together with the observation operators are used, this method (B) is identical to method (A). Then the difference is more a technical one, in terms of separate data handling of the observations and RRA fields. However, in practice, the full observation operators of the RRA cannot readily be applied off-line. Hence, the advantage of method (B) is that the same observations can be used for all RRA systems, irrespectively how they were used (or not used) in each of the systems.

Most of the representative station observations are assimilated into the RRAs, leaving only few high quality independent measurements for the uncertainty estimation. Therefore, the reference data are divided into two groups, namely (B1) independent observations, mainly wind speed from tall mast stations, and (B2) dependent observations, which were chosen to include T_{min}, T_{max}, and threshold values of temperature as well as precipitation. Refer to Sect. 3 for a discussion on scores and skill scores to use for this method and parameters.

2.3 Method C: gridded station observations

Method (C) describes validation against gridded data fields which are spatially interpolated station observations. This is, similar to method (B), a handy comparison for users of traditional data sources, who consider switching to or also including reanalysis data for their specific applications. Several data products exist which cover the European continent or a sub-region thereof. Mainly two data products, which will be extended and improved within the UERRA project, will be used for the validation performed within this project. The E-OBS data set (Haylock et al., 2008; van den Besselaar et al., 2011) is created by statistical interpolation of European land station observations and consists of daily values of temperature (minimum, mean, and maximum), precipitation, and sea level pressure. The projection of the data is provided either on a regular grid or a rotated pole grid in 0.25° or 0.5° and 0.22° or 0.44° horizontal resolution, respectively. The E-OBS product is continuously updated and new versions of the product are released frequently. The second gridded dataset used for the validation is the Alpine Precipitation Grid Dataset (APGD), see Isotta et al. (2013) for details. The APGD is a high-resolution statistically interpolated precipitation data set, based on roughly 5500 rain gauge measurements, and covers the Alpine region, a sub-European land area.

Advantages with this method of validation are that the gridded data products have already been very carefully prepared, covering desirable parameters such as temperature and precipitation, and are ready to use. In combination with skill scores (see Sect. 3 for details) such as the equitable threat score (ETS) (Gandin and Murphy, 1992) or fractional skill score (FSS) (Roberts and Lean, 2008), uncertainty estimates concerning spatial and temporal scales as well as trend analyses can be performed. The main scientific interest to pursue with this method is to determine the effective spatial and temporal resolution of the RRAs and answer the user question which scales of the RRAs can be interpreted.

Caution needs to be exercised, as with method B, that the underlying station observations are independent of the RRAs. Furthermore, it is hard to answer how representative the gridded data products are in comparison with the reanalyses at the interpolated points. The smoothed topography plays a significant role because the parameters of interest (i.e. precipitation) are temporally and spatially highly variable surface parameters. We have to make sure that we do not automatically arrive at the conclusion that the best reanalysis is the one with the most similar topography, or the most similar correlation length scales, compared to the topography of the gridded data product.

Another principle concern with gridded data products is that the number of station data contributing to each grid cell varies so that the grid cells come with a regional varying quality. One way to reduce this effect is to demand a threshold value of a minimum number of stations contributing to a single grid cell.

The analysis could comprise a simple comparison (bias, RMSE, and frequency distribution) as well as more sophisticated skill scores as the ETS, FSS, and thresholds or the likes as applicable in order to estimate the effective spatial and temporal scale of the product. In addition, the temporal change (trends) of the above mentioned statistical properties needs to be analysed.

Suitable for this method of comparison are fields of precipitation, T_{min}, and T_{max}, because there is large user interest in these variables and products covering whole of Europe exist. Furthermore, the spatial aggregation remedies the high local differences which make the point comparisons in (B) so difficult. Uncertainty information of the gridded data product would be required, to be provided either as a range or an ensemble of grids, to use for evaluation purposes. Refer to Isotta et al. (2015) for a successful example application of this method for precipitation in the Alpine area.

2.4 Method D: gridded satellite data products

Method (D) concerns the validation against satellite based observations. Long-term climate quality satellite data are produced by the Satellite Application Facility on Climate Monitoring (CM SAF) and the European Space Agency's (ESA) climate change initiative (CCI) (Hollmann et al., 2013). The characteristics of satellite data products depend on the observing system, i.e. whether it was produced by geostationary or low Earth orbiting satellites (GEO, LEO) and on the part of the spectrum the observation was taken, i.e. in the optical or microwave. Data products of GEO satellites exhibit a high temporal (up to 15 min) and spatial (about 5 km) resolution but provide only a limited area (non-global) coverage. LEO satellites have the capacity to provide global products but provide data only in swaths and can only provide few samples of a specific location per day (depending on the latitude). Products based on observations taken in the optical and near infra-red spectrum (e.g. global radiation) provide higher spatial resolution (up to a few kilometres) than products based on microwave observations (e.g. precipitation) that have a much coarser resolution of about 25 km. For studies with focus on Europe, such products can be derived from the observations of the SEVIRI-instrument (Spinning Enhanced Visible and InfraRed Imager) which is on-board of EUMETSAT's METEOSAT-satellites (from 2004 onwards only). UERRA will produce a 20 year cloud cover reanalysis (based on Häggmark et al., 2000) from METEOSAT and NOAA AVHRR satellite data.

Potentially useful parameters for RRA uncertainty characterization include global radiation, cloud liquid water path, and snow water equivalent. The CM SAF CLoud property dAtAset using SEVIRI (CLAAS) provides, amongst other, daily means of cloud properties and solar radiation (Stengel et al., 2014) and is freely available through Stengel

et al. (2013). Snow water equivalent (SWE) data (Takala et al., 2011) is freely provided by the European Space Agency's (ESA) GlobSnow and GlobSnow-2 projects (Luojus et al., 2010, http://www.globsnow.info). Precipitation and total cloud cover are available as satellite products but need to be used with care. The precipitation products have a coarser horizontal resolution of $0.25° \times 0.25°$, do not all cover whole of Europe, and underestimate precipitation in all seasons (Kidd et al., 2012). The total cloud cover satellite product is difficult to compare against model output because of different definitions of total cloud cover between model and satellite instrument.

In order to characterize uncertainties, skill scores in addition to the usually applied bias and RMSE are needed to determine the effective temporal and spatial resolution of the RRAs and satellite data products. Climatologies of frequency distributions, numbers of threshold value exceedance, and any variation over time has to be captured. Another possibility is to apply the fractional skill score (refer to Sect. 3) to the RRAs and the satellite data sets by comparing against a high resolved (station based) reference and investigate different temporal and spatial scales. This result would be of value for users of RRAs in order to judge which data product is likely to best suited for their applications. The analysis can be complicated by the fact that the RRAs may feature a finer effective resolution than the satellite data products, thus only scales corresponding to the satellite data (temporal and/or spatial) can be compared with this method.

2.5 Method E: ensemble based comparison

Method (E) describes validation performed on ensembles of regional reanalyses which are developed in the UERRA project. For instance, MO plans a set of about 20 ensemble members for their RRA employing a 4D-Var data assimilation system. The benefit of an ensemble-based reanalysis is that it inherently provides estimates of the analysis error, as noted by Whitaker et al. (2009). Additionally, the spatio-temporal evolution of the background error is estimated from the ensemble-spread. The ensemble spread describes possible variability and uncertainty within the reanalysis. Whitaker and Loughe (1998) have examined the relationship between the ensemble spread and ensemble mean skill and found that the spread can be used as a measure of skill for the ensemble mean if the spread is very large or very small compared to its climatological value. Many more ensemble based verification methods have been developed and are listed in Sect. 3. The drawbacks of an ensemble-based reanalysis is that the Ensemble Prediction System (EPS) costs considerable additional computing time and therefore always results into a trade-off between resolution and the number of ensemble members calculated. Additionally, it cannot be guaranteed that the ensemble members cover the full physical space of possible realizations.

Due to the fact that the ensembles of the RRAs are calculated on a lower resolution (about twice as coarse), one interesting scientific question is to what extent the ensembles provide a better estimate on spatial and temporal uncertainty than the deterministic reanalysis. Specifically, a user of the RRAs would be interested in which uncertainty characteristics can be interpreted from the ensembles for user relevant parameters.

Simple metrics such as the ensemble mean and spread can be used to characterize the inherent variability of each RRA and the possible temporal and spatial variation thereof. If an ensemble has been tested to be reliable (refer Sect. 3), it may be used to produce probability density functions (PDFs) or cumulative distribution functions (CDFs) of a given parameter. This allows users to retrieve, for this parameter, information about the probability of exceedance of a given threshold. In addition, probabilistic scores can be applied against station observations, gridded station data, and satellite data products (methods B, C, and D) where in principle all the above considerations hold true.

When performing verification of deterministic or ensemble based (re)analyses or forecasts, their skill is assessed against observations which are assumed to be exact. There is an inconsistency when accepting that the model may be uncertain while the observations are assumed to be certain which can lead to misguided verification results. Therefore, in recent years, the effect of observation errors on the verification of ensemble prediction systems in particular was analysed and new scores and methods were developed. Saetra et al. (2004) investigated the effects of observation errors on the statistics for ensemble spread and reliability. They show that rank histograms are highly sensitive to the inclusion of observation errors, whereas reliability diagrams are less sensitive. Candille and Talagrand (2008) introduce the "observational probability" method by defining observation uncertainty as a normal distribution which guarantees that the uncertainty is variable in both mean and spread. Santos and Ghelli (2012) extend the previous work and developed the "observed observational probability" method which includes uncertainty in the verification process to variables that are non-Gaussian distributed, in particular precipitation.

2.6 Method F: user models

Users regarding reanalysis as an additional data source simply use the reanalysis data fields and draw conclusions based on whether their application improved compared to their traditional forcing data. Though this is a valid approach for a certain case, it does not give scientifically sound uncertainty estimates. We discuss this method here because it is expected to be a popular way.

As one example for user models driven by RRA fields, the surface and soil model SURFEX (Masson et al., 2013) from Météo-France is considered in the UERRA project. The SURFEX model computes soil variables such as surface and

deep soil temperature, soil moisture, and snow characteristics. The SURFEX drainage and run-off forces the hydrological model TRIP (Oki and Sud, 1998) to compute river discharges.

For validation, the discharge of catchments is compared to RRA precipitation over the same catchments. Though a modelled discharge closer to the observed ones is highly desirable for the application, conclusions from the validation results are not straightforward. Poor validation would not necessarily mean the RRA of the UERRA project was poor. It may be that the user model was tuned to good performance with the traditional input data. Likewise, good validation results may reflect the model tuning rather than the quality of the RRA.

3 Verification skill

For the various methods described above, verification scores and skill scores can be applied in order to provide an estimate of uncertainty for the RRAs. Which score can be applied also depends on the parameter but mainly on the question which is sought to be answered. Here, a short summary is given of suitable scores and skill scores which are intended to be applied within the various methods and to the appropriate parameters.

In Sect. 2.2, the described method is based on point (station) reference measurements. Frequency distributions of climatologically relevant threshold values can be used as verification measures such as how often, e.g. a temperature, precipitation, or wind speed threshold was exceeded (or fell below). In addition, there are a number of different skill scores which are suitable to quantify the uncertainty estimation and which are applicable to this method. To start with, continuous metrics such as the bias, RMSE, and correlation are good measures to get a first impression of the data product in relation to its reference. For instance, for temperature or wind speed these measures can easily be applied. However, for precipitation more advanced but commonly used scores and skill scores need to be applied based on contingency tables. These include but are not limited to the probability of (false) detection POD (POFD), false alarm ratio, critical success index, the equitable threat score, and Heidke skill score, (for a comprehensive discussion of statistical methods refer to Wilks, 2011).

When the reference data source is not a single station observation but a gridded data product, as introduced in Sects. 2.3 and 2.4, then the uncertainty estimation can be performed in two ways. It is possible to either verify pointwise by applying the above mentioned classical metrics or use new spatial-based verification approaches. These new verification approaches include fuzzy verification methods, e.g. the fractional skill score (FSS) (Roberts and Lean, 2008), as summarized by Ebert (2008); an added value index which quantifies the added value of a high resolution product compared to a lower resolution product as introduced by Kanamitsu and

DeHaan (2011); and an object-based approach for specifically verifying precipitation estimates as explained in Li et al. (2015).

As introduced in Sect. 2.5, there will be an ensemble based realization of the regional reanalyses. For the verification of ensemble based products, ensemble based and probabilistic verification procedures have been developed. These procedures cover certain aspects of the uncertainty estimation, so usually many scores and skill scores need to be applied to come to a comprehensive result. One classical way of checking the reliability (or statistical consistency) of the ensemble is to build Talagrand (or rank) histograms (Talagrand et al., 1999; Hamill, 2001). It answers the question how well the ensemble represent the true variability (uncertainty) of the observations. Using a reliability diagram (Hamill, 1997), the reliability, sharpness, and resolution of the ensembles (here: ensembles of regional reanalyses) are evaluated. Another diagram which is widely used is the relative operating characteristics (ROC) plot (Masson, 1982; Candille and Talagrand, 2008). Here, POD is plotted against POFD and it answers the question what the ability of the forecast is to discriminate between events and non-events. And finally, there are different skill scores which test the skill of the probabilistic information in varying ways, such as the Brier (skill) score (Brier, 1950; Candille and Talagrand, 2008), which relates to the user question: "What is the magnitude of the probability forecast errors?" or the (continuous) ranked probability score (Hersbach (2000) and references therein), which relates to the user question: "how well did the probability forecast predict the category that the observation fell into?".

4 Summary

In this paper we summarized methods to characterize a regional reanalysis. Several RRAs are produced in the FP7 UERRA project. Uncertainty characterization is among the main foci of the project. The aim was to arrive at uncertainty estimates which are suitable for end users and allow comparisons between different reanalyses. Following methods have been discussed: (A) feedback statistics, the comparison against (B) station observations, (C) gridded station observations, and (D) satellite data, followed by (E) ensemble based comparisons when the UERRA ensemble become available, as well as (F) evaluating user model output driven by RRA input.

We outlined the benefits and drawbacks of each method. Specifically, the benefit of method (A) feedback statistics is that it can be arranged as output during the reanalysis production, and deals with suitable observation operators to arrive at a scientifically sound comparison between reanalysis output and (independent) observations. The drawback of (A) is that it is not easily comparable between different reanalysis systems. A comparison against station measurements (method B) can be easily performed with different reanalysis systems, the result will depend on model resolution, topographic resolution, and representativeness of the station. Comparisons against gridded observations (method C) have the advantage that they fully cover the area of interest. The drawback of (C) is that the result of the comparison might be influenced by how close the spatial smoothing (or correlation length scales) in the gridding procedure resemble the ones in the reanalyses. The advantage of comparing to (not assimilated) satellite data (method D) is that here again the full area is covered, the disadvantage is that the satellite retrieval itself is a statistical procedure which has to rely on a number of assumptions, thus the uncertainty of the satellite product might be larger than that of the reanalyses in many circumstances. Finally, the advantage of method (E) ensembles is that the random part of reanalysis uncertainty can be estimated (if the ensemble is generated with one model) and additionally assess the model uncertainty by combining several models. We pointed out that method (F) investigates advantages in applications, is user friendly, but does not allow to generalize conclusions about uncertainties. The methods are related to scientific and user questions which might be answered with the respective method.

We discussed for which meteorological parameters which methods are most suitable by focusing especially on: air temperature, humidity, and wind speed at the near-ground levels as these are of primary user interest targeted by the regional reanalysis efforts. For the satellite data, suitable parameters include global radiation, cloud liquid water path, and snow water equivalent. Precipitation and total cloud cover from satellite are harder to interpret. Depending on the meteorological parameter of interest, several skill scores are recommended.

Acknowledgements. This study was supported through the UERRA project (grant agreement no. 607193 within the European Union Seventh Framework Programme). We acknowledge all UERRA WP3 partners for their discussion and scientific input, with special thanks for valuable contributions to all participants of the user's workshop within the UERRA project meeting in June 2014 at DWD, Offenbach, Germany.

References

Berre, L.: Estimation of Synoptic and Mesoscale Forecast Error Covariances in a Limited-Area Model, Mon. Weather Rev., 128, 644–667, doi:10.1175/1520-0493(2000)128<0644:EOSAMF>2.0.CO;2, 2000.

Bollmeyer, C., Keller, J. D., Ohlwein, C., Wahl, S., Crewell, S., Friederichs, P., Hense, A., Keune, J., Kneifel, S., Pscheidt, I., Redl, S., and Steinke, S.: Towards a high-resolution regional reanalysis for the European CORDEX domain, Q. J. Roy. Meteorol. Soc., 141, 1–15, doi:10.1002/qj.2486, 2015.

Brier, G. W.: Verification of forecasts expressed in terms of probability, Mon. Weather Rev., 78, 1–3 , 1950.

Brunet, M., Jones, P. D., Jourdain, S., Efthymiadis, D., Kerrouche, M., and Boroneant, C.: Data sources for rescuing the rich heritage of Mediterranean historical surface climate data, Geosci. Data J., 1, 61–73, doi:10.1002/gdj3.4, 2013.

Bubnova, R., Hello, G., Benard, P., and Geleyn, J.-F.: Integration of the fully elastic equations cast in the hydrostatic pressure terrain-following in the framework of the ARPEGE/ALADIN NWP system, Mon. Weather Rev., 123, 515–535, 1995.

Candille, G. and Talagrand, O.: Impact of observational error on the validation of ensemble prediction systems, Q. J. Roy. Meteorol. Soc., 134, 959–971, doi:10.1002/qj.268, 2008.

Clayton, A. M., Lorenc, A. C., and Barker, D. M.: Operational implementation of a hybrid ensemble/4D-Var global data assimilation system at the Met Office, Q. J. Roy. Meteorol. Soc., 139, 1455–1461, doi:10.1002/qj.2054, 2013.

Dahlgren, P., Kållberg, P., Landelius, T., and Undén, P.: Comparison of the regional reanalyses products with newly developed and existing state-of-the art systems, EURO4M project report D2.9, http://www.euro4m.eu/ (last access: 23 October 2015), 2014.

Davies, T., Cullen, M. J. P., Malcolm, A. J., Mawson, M. H., Staniforth, A., White, A. A., and Wood, N.: A new dynamical core for the Met Office's global and regional modelling of the atmosphere, Q. J. Roy. Meteorol. Soc., 131, 1759–1782, doi:10.1256/qj.04.101, 2006.

Dee, D. P., Uppala, S. M., Simmons, A. J., et al.: The ERA-Interim reanalysis: configuration and performance of the data assimilation system, Q. J. Roy. Meteorol. Soc., 137, 553–597, doi:10.1002/qj.828, 2011.

Dee, D. P., Balmaseda, M., Balsamo, G., Engelen, R., Simmons, A. J., and Thepaut, J.-N.: Towards a consistent reanalysis of the climate system, Bull. Amer. Meteor. Soc., 95, 1235–1248, doi:10.1175/BAMS-D-13-00043.1, 2014.

Desroziers, G., Berre, L., Chapnik, B., and Poli, P.: Diagnosis of observation, background and analysis-error statistics in observation space, Q. J. Roy. Meteorol. Soc., 131, 3385–3396, doi:10.1256/qj.05.108, 2005.

De Troch, R., Hamdi, R., Vyver, H., Geleyn, J.-F., and Termonia, P.: Multiscale Performance of the ALARO-0 Model for Simulating Extreme Summer Precipitation Climatology in Belgium, J. Climate, 26, 8895–8915, doi:10.1175/JCLI-D-12-00844.1, 2013.

Ebert, E. E.: Fuzzy verification of high-resolution gridded forecasts: a review and proposed framework, Meteorol. Appl., 15, 51–64, doi:10.1002/met.25, 2008.

Fischer, C., Montmerle, T., Berre, L., Auger, L., and Stefanescu, S. E.: An overview of the variational assimilation in the ALADIN/France numerical weather-prediction system, Q. J. Roy. Meteorol. Soc., 131, 3477–3492, doi:10.1256/qj.05.115, 2005.

Gandin, L. and Murphy, A.: Equitable Skill Scores for Categorical Forecasts, Mon. Weather Rev., 120, 361–370, doi:10.1175/1520-0493(1992)120<0361:ESSFCF>2.0.CO;2, 1992.

Gerard, L., Piriou, J.-M., Brožková, R., Geleyn, J.-F., and Banciu, D.: Cloud and Precipitation Parameterization in a Meso-Gamma-Scale Operational Weather Prediction Model, Mon. Weather Rev., 137, 3960–3977, doi:10.1175/2009MWR2750.1, 2009.

Häggmark, L., Ivarsson, I., Gollvik, S., and Olofsson, O.: Mesan, an operational mesoscale analysis system, Tellus A, 52, 2–20, doi:10.3402/tellusa.v52i1.12250, 2000.

Hakuba, M.Z., Folini, D., Sanchez-Lorenzo, A., and Wild, M.: Spatial representativeness of ground-based solar radiation measurements – Extension to the full Meteosat disk, J. Geophys. Res.-Atmos., 119, 11760–11771, doi:10.1002/2014JD021946, 2014.

Hamill, T. M.: Reliability diagrams for multicategory probabilistic forecasts, Weather Forecast., 12, 736–741, 1997.

Hamill, T. M.: Interpretation of rank histograms for verifying ensemble forecasts, Mon. Weather Rev., 129, 550–560, doi:10.1175/1520-0493(2001)129<0550:IORHFV>2.0.CO;2, 2001.

Harnisch, F. and Keil, C.: Initial conditions for convective-scale ensemble forecasting provided by ensemble data assimilation, Mon. Weather Rev., 143, 1583–1600, doi:10.1175/MWR-D-14-00209.1, 2015.

Haylock, M. R., Hofstra, N., Klein Tank, A. M. G., Klok, E. J., Jones, P. D., and New, M.: A European daily high-resolution gridded data set of surface temperature and precipitation for 1950–2006, J. Geophys. Res., 113, D20119, doi:10.1029/2008JD010201, 2008.

Hersbach, H.: Decomposition of the Continuous Ranked Probability Score for Ensemble Prediction Systems, Weather Forecast., 15, 559–570, doi:10.1175/1520-0434(2000)015<0559:DOTCRP>2.0.CO;2, 2000.

Hollingsworth, A. and Lönnberg, P.: The statistical structure of short-range forecast errors as determined from radiosonde data, Part I: The wind field, Tellus A, 38, 111–136, doi:10.1111/j.1600-0870.1986.tb00460.x, 1986.

Hollmann, R., Merchant, C. J., Saunders, R., Downy, C., Buchwitz, M., Cazenave, A., Chuvieco, E., Defourny, P., de Leeuw, G., Forsberg, R., Holzer-Popp, T., Paul, F., Sandven, S., Sathyendranath, S., van Roozendael, M., and Wagner, W.: The ESA Climate Change Initiative: Satellite Data Records for Essential Climate Variables, B. Am. Meteorol. Soc., 94, 1541–1552, doi:10.1175/BAMS-D-11-00254.1, 2013.

Horányi, A., Ihász, I., and Radnóti, G.: ARPEGE/ALADIN: a numerical weather prediction model for Central-Europe with the participation of the Hungarian Meteorological Service, Időjárás, 100, 277–301, 1996.

Hunt, B. R., Kostelich, E. J., and Szunyogh, I.: Efficient data assimilation for spatiotemporal chaos: A local ensemble transformation Kalman filter, Physics D, 230, 112–126, doi:10.1016/j.physd.2006.11.008, 2007.

Isotta, F. A., Frei, C., Weilguni, V., Perčec Tadič, M., Lassègues, P., Rudolf, B., Pavan, V., Cacciamani, C., Antolini, G., Ratto, S.M., Munari, M., Micheletti, S., Bonati, V., Lussana, C., Ronchi, C., Panettieri, E., Marigo, G., and Vertačnik, G.: The climate of daily precipitation in the Alps: development and analysis of a high-resolution grid dataset from pan-Alpine rain-gauge data, Int. J. Climatol., 34, 1657–1657, doi:10.1002/joc.3794, 2013.

Isotta, F. A., Vogel, R., and Frei, C.: Evaluation of European regional reanalyses and downscalings for precipitation in the Alpine region, Meteorol. Z., 24, 15–37, doi:10.1127/metz/2014/0584, 2015.

Jansson, A., Persson, C., and Strandberg, G.: 2D meso-scale reanalysis of precipitation, temperature and wind over Europe-ERAMESAN: Time period 1980–2004, SMHI Rep. Meteorol. Clim., 112, 44, 2007.

Kaiser-Weiss, A. K., Kaspar, F., Heene, V., Borsche, M., Tan, D. G. H., Poli, P., Obregon, A., and Gregow, H.: Comparison of regional and global reanalysis near-surface winds with station observations over Germany, Adv. Sci. Res., 12, 187–198, doi:10.5194/asr-12-187-2015, 2015.

Kanamitsu, M. and DeHaan, L.: The Added Value Index: A new metric to quantify the added value of regional models, J. Geophys. Res., 116, D11106, doi:10.1029/2011JD015597, 2011.

Kidd, C., Bauer, P., Turk, J., Huffman, G. J., Joyce, R., Hsu, K.-L., and Braithwaite, D.: Intercomparison of High-Resolution Precipitation Products over Northwest Europe, J. Hydrometeorol., 13, 67–83, doi:10.1175/JHM-D-11-042.1, 2012.

Li, J., Hsu, K., AghaKouchak, A., and Sorooshian, S.: An object-based approach for verification of precipitation estimation, Int. J. Remote Sens., 36, 513–529, doi:10.1080/01431161.2014.999170, 2015.

Lönnberg, P. and Hollingsworth, A.: The statistical structure of short-range forecast errors as determined from radiosonde data Part II: The covariance of height and wind errors, Tellus A, 38, 137–161, doi:10.1111/j.1600-0870.1986.tb00461.x, 1986.

Luojus, K., Pulliainen, J., Takala, M., Derksen, C., Rott, H., Nagler, T., Solberg, R., Wiesmann, A., Metsämäki, S., Malnes, E., and Bojkov, B.: Investigating the feasibility of the Glob-Snow snow water equivalent data for climate research purposes, IEEE International Geoscience and Remote Sensing Symposium (IGARSS), Honolulu, HI, USA, 4851–4853, doi:10.1109/IGARSS.2010.5741987, 2010.

Masson, I.: A model for assessment of weather forecasts, Aust. Meteorol. Mag., 30, 291–303, 1982.

Masson, V., Le Moigne, P., Martin, E., Faroux, S., Alias, A., Alkama, R., Belamari, S., Barbu, A., Boone, A., Bouyssel, F., Brousseau, P., Brun, E., Calvet, J.-C., Carrer, D., Decharme, B., Delire, C., Donier, S., Essaouini, K., Gibelin, A.-L., Giordani, H., Habets, F., Jidane, M., Kerdraon, G., Kourzeneva, E., Lafaysse, M., Lafont, S., Lebeaupin Brossier, C., Lemonsu, A., Mahfouf, J.-F., Marguinaud, P., Mokhtari, M., Morin, S., Pigeon, G., Salgado, R., Seity, Y., Taillefer, F., Tanguy, G., Tulet, P., Vincendon, B., Vionnet, V., and Voldoire, A.: The SURFEXv7.2 land and ocean surface platform for coupled or offline simulation of earth surface variables and fluxes, Geosci. Model Dev., 6, 929–960, doi:10.5194/gmd-6-929-2013, 2013.

Oki, T. and Sud, Y. C.: Design of Total Runoff Integration Pathways (TRIP) – A global river channel network, Earth Interact., 2, 1–37, doi:10.1175/1087-3562(1998)002<0001:DOTRIP>2.3.CO;2, 1998.

Poli, P., Hersbach, H., Tan, D., Dee, D., Thépaut, J.-N., Simmons, A., Peubey, C., Laloyaux, P., Komori, T., Berrisford, P., Dragani, R., Trémolet, Y., Hólm, E., Bonavita, M., Isaksen, L., and Fisher, M.: The data assimilation system and initial performance evaluation of the ECMWF pilot reanalysis of the 20th-century assimilating surface observations only (ERA-20C), ERA Rep. Ser., 14, 59, 2013.

Rawlins, F., Ballard, S. P., Bovis, K. J., Clayton, A. M., Li, D., Inverarity, G. W., Lorenc, A. C., and Payne, T. J.: The Met Office global four-dimensional variational data assimilation scheme, Q. J. Roy. Meteorol. Soc., 133, 347–362, doi:10.1002/qj.32, 2007.

Renshaw, R.: New state-of-the-art NAE-based regional atmospheric data assimilation reanalysis system, EURO4M project report, D2.1, D2.2, http://www.euro4m.eu/ (last access: 23 October 2015), 2013.

Roberts, N. M. and Lean, H. W.: Scale-selective verification of rainfall accumulations from high-resolution forecast of convective events, Mon. Weather Rev., 136, 78–97, doi:10.1175/2007MWR2123.1, 2008.

Saetra, Ø., Hersbach, H., Bidlot, J.-R., and Richardson, S.: Effects of Observation Errors on the Statistics for Ensemble Spread and Reliability, Mon. Weather Rev., 132, 1487–1501, doi:10.1175/1520-0493(2004)132<1487:EOOEOT>2.0.CO;2, 2004.

Santos, C. and Ghellie, A.: Observational probability method to assess ensemble precipitation forecasts, Q. J. Roy. Meteorol. Soc., 138, 209–221, doi:10.1002/qj.895, 2012.

Schättler, U., Doms, G., and Schraff, C.: A description of the non-hydrostatic regional COSMO-model – Part VII: User's guide, Technical report, Deutscher Wetterdienst, Offenbach, Germany, http://www.cosmo-model.org/ (last access: 23 October 2015), 2014.

Simmer, C., Adrian, G., Jones, S., Wirth, V., Göber, M., Hohenegger, C., Janjic, T., Keller, J., Ohlwein, C., Seifert, A., Trömel, S., Ulbrich, T., Wapler, K., Weissmann, M., Keller, J., Masbou, M., Meilinger, S., Riß, N., Schomburg, A., Vormann, A., and Weingärtner, C.: HErZ – The German Hans-Ertel Centre for Weather Research, B. Am. Meteorol. Soc., doi:10.1175/BAMS-D-13-00227.1, in press, 2015.

Soci, C., Landelius, T., Bazile, E., Undén, P., Mahfouf, J.-F., Martin, E., and Besson, F.: Comparison of existing ERAMESAN with SAFRAN downscaling, EURO4M project report D2.10, http://www.euro4m.eu/ (last access: 23 October 2015), 2011.

Soci, C., Bazile, E., Besson, F., Landelius, T., Mahfouf, J.-F., Martin, E., and Durand, Y.: Report describing the new MESAN-SAFRAN downscaling system, EURO4M project report D2.6, http://www.euro4m.eu/ (last access: 23 October 2015), 2013.

Stengel, M., Kniffka, A., Meirink, J. F., Riihelä, A. Trentmann, J., Müller, R., Lockhoff, M., and Hollmann, R.: CLAAS: CM SAF CLoud property dAtAset using SEVIRI – Edition 1 – Hourly/Daily Means, Pentad Means, Monthly Means/Monthly Mean Diurnal Cycle/Monthly Histograms, Satellite Application Facility on Climate Monitoring, Offenbach, Germany, doi:10.5676/EUM_SAF_CM/CLAAS/V001, 2013.

Stengel, M., Kniffka, A., Meirink, J. F., Lockhoff, M., Tan, J., and Hollmann, R.: CLAAS: the CM SAF cloud property dataset using SEVIRI, Atmos. Chem. Phys., 14, 4297–4311, doi:10.5194/acp-14-4297-2014, 2014.

Takala, M., Luojus, K., Pulliainen, J., Derksen, C., Lemmetyinen, J., Karna, J. P., Koskinen, J., and Bojkov, B.: Estimating northern snow water equivalent for climate research through assimilation of space-borne radiometer data and ground-based measurements, Remote Sens. Environ., 115, 3517–3529, doi:10.1016/j.rse.2011.08.014, 2011.

Talagrand, O., Vautard, R., and Strauss, B.: Evaluations of probabilistic evaluation systems, Proceedings, ECMWF Workshop on Predictability, October 1997, Reading, UK, 1–25, 1999.

Uppala, S. M., Kållberg, P. W., Simmons, A. J., et al.: The ERA-40 re-analysis, Q. J. Roy. Meteorol. Soc., 131, 2961–3012, doi:10.1256/qj.04.176, 2005.

van den Besselaar, E. J. M., Haylock, M. R., van der Schrier, G., and
    Klein Tank, A. M. G.: A European daily high-resolution obser-
    vational gridded data set of sea level pressure, J. Geophys. Res.,
    116, D11110, doi:10.1029/2010JD015468, 2011.

Whitaker, J. S. and Loughe, A. F.: The Relationship be-
    tween Ensemble Spread and Ensemble mean Skill,
    Mon. Weather Rev., 126, 3292–3302, doi:10.1175/1520-
    0493(1998)126<3292:TRBESA>2.0.CO;2, 1998.

Whitaker, J. S., Compo, G. P., and Thépaut, J.-N.: A Comparison
    of Variatinal and Ensemble-Based Data Assimilation Systems
    for Reanalysis of Sparse Observations, Mon. Weather Rev., 137,
    1991–1999, doi:10.1175/2008MWR2781.1, 2009.

Wilks, D. S.: Statistical Methods in the Atmospheric Sciences,
    3rd Edn., International Geophysics, Academic Press, Oxford,
    UK, 704 pp., 2011.

Permissions

The contributors of this book come from diverse backgrounds, making this book a truly international effort. This book will bring forth new frontiers with its revolutionizing research information and detailed analysis of the nascent developments around the world.

We would like to thank all the contributing authors for lending their expertise to make the book truly unique. They have played a crucial role in the development of this book. Without their invaluable contributions this book wouldn't have been possible. They have made vital efforts to compile up to date information on the varied aspects of this subject to make this book a valuable addition to the collection of many professionals and students.

This book was conceptualized with the vision of imparting up-to-date information and advanced data in this field. To ensure the same, a matchless editorial board was set up. Every individual on the board went through rigorous rounds of assessment to prove their worth. After which they invested a large part of their time researching and compiling the most relevant data for our readers.

The editorial board has been involved in producing this book since its inception. They have spent rigorous hours researching and exploring the diverse topics which have resulted in the successful publishing of this book. They have passed on their knowledge of decades through this book. To expedite this challenging task, the publisher supported the team at every step. A small team of assistant editors was also appointed to further simplify the editing procedure and attain best results for the readers.

Apart from the editorial board, the designing team has also invested a significant amount of their time in understanding the subject and creating the most relevant covers. They scrutinized every image to scout for the most suitable representation of the subject and create an appropriate cover for the book.

The publishing team has been an ardent support to the editorial, designing and production team. Their endless efforts to recruit the best for this project, has resulted in the accomplishment of this book. They are a veteran in the field of academics and their pool of knowledge is as vast as their experience in printing. Their expertise and guidance has proved useful at every step. Their uncompromising quality standards have made this book an exceptional effort. Their encouragement from time to time has been an inspiration for everyone.

The publisher and the editorial board hope that this book will prove to be a valuable piece of knowledge for researchers, students, practitioners and scholars across the globe.

List of Contributors

F. Kaspar, K. Zimmermann and C. Polte-Rudolf
Deutscher Wetterdienst, National Climate Monitoring, Frankfurter Str. 135, 63067 Offenbach, Germany

S. Gaztelumendi, M. Martija, O. Principe and V. Palacio
TECNALIA, Energy and Environment Division, Meteorology Area, Miñano, Basque Country, Spain
Basque Meteorology Agency (EUSKALMET), Miñano, Basque Country, Spain

M. Journée, C. Delvaux and C. Bertrand
Royal Meteorological Institute of Belgium, Brussels, Belgium

N. Pérez-Zanón, J. Sigró, P. Domonkos and L. Ashcroft
Center for Climate Change (C3), Campus Terres de l'Ebre, Universitat Rovira i Virgili, Tortosa, Spain

H. Nakayama and H. Nagai
Japan Atomic Energy Agency, Ibaraki, Japan

T. Takemi
Disaster Prevention Research Institute, Kyoto University, Kyoto, Japan

C. Bertrand, L. González Sotelino and M. Journée
Royal Meteorological Institute of Belgium, Brussels, Belgium

J. Bilbao, R. Román, D. Mateos and A. de Miguel
University of Valladolid, Spain, Atmosphere & Energy Laboratory, Faculty of Sciences, Valladolid, Spain

C. Yousif
University of Malta, Institute for Sustainable Energy, Marsaxlokk, Malta

P. Martano and F. Grasso
CNR-Istituto di Scienze dell'Atmosfera e del Clima – UOS Lecce, Via Monteroni, 73100 Lecce, Italy

C. Elefante
Ripartizione Informatica, Università del Salento, Viale Gallipoli 49, 73100 Lecce, Italy

J. Lindén and J. Esper
Department of Geography, Johannes-Gutenberg University, Mainz, Germany

C.S.B. Grimmond
Department of Meteorology, University of Reading, Reading, UK

P. Jokinen, A. Vajda and H. Gregow
Finnish Meteorological Institute, P.O. Box 503, 00101 Helsinki, Finland

P. Cerralbo
Maritime Engineering Laboratory, Polytechnic University of Catalonia (LIM/UPC), c./Escar, 6,08039 Barcelona, Spain

M. Grifoll and M. Espino
Maritime Engineering Laboratory, Polytechnic University of Catalonia (LIM/UPC), c./Escar, 6,08039 Barcelona, Spain
International Centre of Coastal Resources Research, Polytechnic University of Catalonia (CIIRC/UPC), c./Jordi Girona, 1–3, 08034 Barcelona, Spain

J. Moré, M. Bravo and A. Sairouní Afif
Servei Meteorològic de Catalunya (SMC), c./Berlín, 38–46, 08029 Barcelona, Spain

C. Delvaux, M. Journée and C. Bertrand
Royal Meteorological Institute of Belgium, Brussels, Belgium

O. Hyvärinen, K. Pilli-Sihvola, A. Venäläinen and H. Gregow
Finnish Meteorological Institute, Helsinki, Finland

L. Mtilatila
Department of Climate Change and Meteorological Services, Lilongwe, Malawi

F. Besson, J.-M. Soubeyroux and G. Ouzeau
Meteo-France, DCSC/AVH, Toulouse, France

E. Bazile and C. Soci
Meteo-France, CNRM/GAME, Toulouse, France

M. Perrin
University of Toulouse III – Paul Sabatier, Toulouse, France

L. Tiriolo, R. C. Torcasio, S. Montesanti and C. R. Calidonna
Institute of Atmospheric Sciences and Climate of the Italian National Council of Research, ISAC-CNR, UOS of Lamezia Terme, zona Industriale Comparto 15, 88046 Lamezia Terme, Italy

A. M. Sempreviva
Wind Energy Department, Danish Technical University, Frederiksborgvej 399, 4000-Roskilde, Denmark

C. Transerici and S. Federico
ISAC-CNR, UOS of Rome, via del Fosso del Cavaliere 100, 00133-Rome, Italy

F. Kaspar, R. Posada and J. Riede
Deutscher Wetterdienst, National Climate Monitoring, Frankfurter Str. 135, 63067 Offenbach, Germany

J. Helmschrot, G. Muche, T. Hillmann, K. Josenhans and N. Jürgens
University of Hamburg; Biodiversity, Evolution and Ecology of Plants; Hamburg, Germany

A. Mhanda, M. Butale and M. Castro Matsheka
Department of Meteorological Services (DMS), Gaborone, Botswana

W. de Clercq
Stellenbosch University, Stellenbosch, South Africa

J. K. Kanyanga
Zambia Meteorological Department (ZMD), Lusaka, Zambia

F. O. S. Neto
Instituto Nacional de Meteorologia e Geofisica (INAMET), Luanda, Angola

S. Kruger
National Botanical Research Institute, Windhoek, Namibia

M. Seely
Gobabeb Research and Training Centre, Walvis Bay, Namibia

C. Ribeiro
Instituto Superior Politécnico Tundavala, Lubango, Angola

P. Kenabatho
University of Botswana, Gaborone, Botswana

R. Vogt
University of Basel, Basel, Switzerland

Christian Viel, Anne-Lise Beaulant, Jean-Michel Soubeyroux and Jean-Pierre Céron
DCSC, Météo-France, Toulouse, France

A. Lehmann
GEOMAR Helmholtz Centre for Ocean Research, Kiel, Germany

P. Post
Institute of Physics, University of Tartu, Tartu, Estonia

Cédric Bertrand, Luis González Sotelino and Michel Journée
Royal Meteorological Institute of Belgium, Brussels, Belgium

Alan K. Betts
Atmospheric Research, Pittsford, Vermont, USA

Raymond L. Desjardins and Devon E. Worth
Science and Technology Branch, Agriculture and Agri-Food Canada, Ottawa, Ontario, Canada

A. K. Kaiser-Weiss, F. Kaspar, V. Heene, M. Borsche and A. Obregon
Deutscher Wetterdienst, Frankfurter Straße 135, 63067 Offenbach, Germany

D. G. H. Tan and P. Poli
European Centre for Medium-Range Weather Forecasts, Shinfield Park, Reading, RG2 9AX, UK

H. Gregow
Finnish Meteorological Institute, P.O. Box 503, 00101 Helsinki, Finland

Rachel Honnert
CNRM-Météo-France, CNRM/GMAP, Toulouse, France

Mireille Lefèvre and Lucien Wald
MINES ParisTech – PSL Research University, Sophia Antipolis, Paris, France

Elenio Avolio, Rosa Claudia Torcasio, Teresa Lo Feudo and Claudia Roberta Calidonna
Institute of Atmospheric Sciences and Climate – Italian National Research Council (ISAC-CNR), Lamezia Terme, 88046, Italy

Daniele Contini
Institute of Atmospheric Sciences and Climate – Italian National Research Council (ISAC-CNR), Lecce, 73100, Italy

Stefano Federico
Institute of Atmospheric Sciences and Climate – Italian National Research Council (ISAC-CNR), Rome, 00133, Italy

P. Blanc and L. Wald
MINES ParisTech – PSL Research University, Sophia Antipolis, France

J. Spinoni
European Commission, Joint Research Centre, Institute for the Protection and Security of the Citizen, Ispra, Italy

G. Naumann
CONICET, National Scientific Technical and Research Council, Buenos Aires, Argentina

J. Vogt
European Commission, Joint Research Centre, Institute for the Environment and Sustainability, Ispra, Italy

Sarah Gallagher, Emily Gleeson and Ray McGrath
Research, Environment and Applications Division, Met Éireann, Glasnevin Hill, Glasnevin, Dublin 9, Ireland

Roxana Tiron
OpenHydro Ltd., Greenore, Co. Louth, Ireland

Frédéric Dias
UCD School of Mathematics and Statistics, University College Dublin, Belfield, Dublin 4, Ireland

P. Stocchi and S. Davolio
Institute of Atmospheric Sciences and Climate, Italian National Research Council, CNR-ISAC, Bologna, 40129, Italy

L. Corre and F. Besson
Direction de la Climatologie, Météo-France, Toulouse, France

P. Dandin
Centre national de recherches météorologiques (CNRM), Météo-France, Toulouse, France

D. L'Hôte
Strate Ecole de Design, Sèvres, France

Raymond L. Desjardins and Devon E. Worth
Science and Technology Branch, Agriculture and Agri-Food Canada, Ottawa, Ontario, Canada

J. Ian MacPherson, Matthew Bastian and Ramesh Srinivasan
National Research Council of Canada, Aerospace, Flight Research Laboratory, Ottawa, Ontario, Canada

M. Borsche, A. K. Kaiser-Weiss and F. Kaspar
Deutscher Wetterdienst, National Climate Monitoring, Frankfurter Str. 135, 63067 Offenbach, Germany

P. Undén
Swedish Meteorological and Hydrological Institute, Folkborgsvägen 17, 601 76 Norrköping, Sweden

Index

www.ingramcontent.com/pod-product-compliance
Lightning Source LLC
Chambersburg PA
CBHW080525200326
41458CB00012B/4340